Multilevel Strategic Interaction Game Models for Complex Networks

Eitan Altman · Konstantin Avrachenkov ·
Francesco De Pellegrini ·
Rachid El-Azouzi · Huijuan Wang
Editors

Multilevel Strategic Interaction Game Models for Complex Networks

Springer

Editors
Eitan Altman
INRIA
Sophia Antipolis, France

Konstantin Avrachenkov
INRIA
Sophia Antipolis, France

Francesco De Pellegrini
University of Avignon
Avignon, France

Rachid El-Azouzi
University of Avignon
Avignon, France

Huijuan Wang
Delft University of Technology
Delft, The Netherlands

ISBN 978-3-030-24457-6 ISBN 978-3-030-24455-2 (eBook)
https://doi.org/10.1007/978-3-030-24455-2

This Springer imprint is published by the registered company Springer Nature Switzerland AG
The registered company address is: Gewerbestrasse 11, 6330 Cham, Switzerland

Preface

Many real-world systems possess rich multi-level structures and exhibit complex dynamics. The source of complexity often relates to the presence of a web of interwoven interactions among elements which have autonomous decision-making capabilities.

In this book, we report on several recent mathematical models and tools, often rooted in game theory, for the analysis, prediction and control of dynamical processes in complex systems. In particular, our aim has been to provide a coherent theoretical framework for understanding the emergence of structures and patterns in complex systems. In this context, it is often necessary to account for interactions which may span various scales in time and space. Also, interactions of various elements may act at different structural and aggregation levels. Because the framework we propose is built around game theoretical concepts, we have been giving special attention to evolutionary and multi-resolution games. Also, techniques drawn from graph theory, from statistical mechanics, control and optimization theory have been combined in order to describe different aspects of the dynamics of complex systems. Finally, specific attention is devoted to systems that are prone to intermittency and catastrophic events due to the effect of collective dynamics.

The book is organized in four main parts. Part I is dedicated to the description of evolutionary games where groups of players may interact and individuals may possess a state which evolves over time. Part II of the book describes dynamics which relate to epidemics or information diffusion over networked structures. Part III addresses games on networks, including several variants of congestion games which relate to current debates on network neutrality. Finally, intermittency phenomena are studied for ecology models where abrupt transitions drive repeated reconfigurations of the system state over time.

Sophia Antipolis, France — Eitan Altman
Sophia Antipolis, France — Konstantin Avrachenkov
Avignon, France — Francesco De Pellegrini
Avignon, France — Rachid El-Azouzi
Delft, The Netherlands — Huijuan Wang

Contents

Part I
Multi-Level Evolutionary Games and Social Values. Innovative Methods and Models are Mentioned as Well

Rachid El-Azouzi

Complex system is relatively new and broadly interdisciplinary field that deals with systems composed of many interacting units, often called "agents". Game theory offers several successful stories in understanding agents' interaction due to the fact that it naturally quantifies rationality, and it can predict the outcome of agent interactions by characterizing stable equilibrium operating points from which no agent has incentive to deviate. Game theory is relevant for multi-scale complex systems: one of the features that game theory is meant to model, is the presence of actors that concurrently participate in the formation of interaction patterns that emerge as a consequence of their objectives, or based on the objectives of groups or actors. In the context of complex systems, the potential for contribution of game theory relates indeed to the possibility of introducing various degrees of correlation among actors and to capture coupling effects through both utilities and strategies. There already exist several successful examples where game theory provides deeper understanding of dynamics and leads to better design of efficient, scalable, and robust networks. Still, there remain many interesting open research problems yet to be identified and explored, and many issues to be addressed. Important analysis and applications have been done in the context of static games [22]. However, some works have been focused on the human behavior in social networks using game theory [38, 160, 64]. It was also introduced as graphical models in which agents in a system are supposed to be nodes of a graph, and their interactions and payoff are limited to neighbors [184]. The emergence of cooperation in graph through game theory has been studied in several papers [63, 227].

Motivated by the dynamic behavior of most of the long-term systems and the understanding of prediction, learning and evolution, dynamic game theory becomes an important tool able to describe strategic interactions among large numbers of agents. Traditionally, predictions of behavior in game theory are based on some notion of equilibrium, typically Cournot equilibrium, Bertrand equilibrium, Nash equilibrium, Stackelberg solution, Wardrop equilibrium or some refinement thereof. These notions require the assumption of knowledge, which posits that each user correctly anticipates how other agents will act or react. The knowledge assumption is too strong and is difficult to justify in particular in contexts with large numbers of agents. As an alternative to the equilibrium approach, an explicitly dynamic updating choice is proposed, a model in which agents myopically update their strategies in response to their opponents' current behavior. This dynamic procedure does not assume an automatic coordination of agent's actions and beliefs, and it can derive many specifications of agents' choice procedures. These procedures are specified formally by defining a revision of pure strategies called revision protocol. This revision is flexible enough to incorporate a wide variety of paradigms, including ones based on imitation, adaptation, learning, optimization, etc. The revision protocols describe the procedures agents follow in adapting their behavior in the dynamic evolving environment such as evolving networks. The evolutionary games was first used by Fisher [103]. This formalism identifies and studies two concepts: the Evolutionary Stability, and the Evolutionary Game Dynamics. The unbeatable strategy has been defined by Hamilton [130, 131], which is the analogous of strong equilibrium (resilient against multilateral deviations) in large systems. The biologists Maynard Smith and Price [248, 245] have defined a weaker of notion of locally unbeatable strategy, the Evolutionary Stable State or Strategy (ESS). The ESS is characterized by a property of robustness against invaders (mutations). More specifically, (i) if an ESS is reached, then the profile of the population does not change in time. (ii) at ESS, the populations are immune from being invaded by other small population. This notion is stronger than Nash equilibrium in which it is only requested that a single user would not benefit by a change (mutation) of its behavior. The ESS concept helps to understand mixed strategies in games with symmetric payoffs. A mixed strategy can be interpreted as a composition of the population. An ESS can be also interpreted as a Nash equilibrium of the one-shot game but a (symmetric) Nash equilibrium may not be an ESS. As is shown in [245] , ESS has strong refinement properties of equilibria such as proper equilibrium, perfect equilibrium etc.

In the biological context, replicator dynamics models the change of the profile of population(s). It describes how the fraction of the population that uses different actions evolves over time. These equations are obtained by making assumptions on the way that the probability of using an action increases as a function of the fitness of an agent using it. This dynamic may answer fundamental questions such as the lifetime of a certain emergent strategy or behavior and consequently the span of the influence of such an agent over time even if interactions between components happen with spatial constraints [205].

In the Part I, we present a novel foundations in evolutionary games: we accounts for the structure of networks and embed novel dimensions: cooperation levels, rationality, information availability and the agent's state. Part I presents several now contribution with respect to evolutionary games for complex system. In Chap. 2 we provide a new concept to remedy this shortcoming in the standard evolutionary games in order to cover this kind of behavior. Indeed, in many behaviors, many phenomena where individuals do care about the benefits of others in group or about their intentions, can be observed in the real word. Hence, the assumption of selfishness becomes inconsistent with the real behavior of individual in a population. This framework thus excludes species such as bees or ants in which the individual reproduced (the queen) is not the one that interacts with other individuals. We begin by defining the Group Evolutionary Stable Strategy (GESS), deriving it in several ways and exploring its major characteristics. The main focus of this work is to study how this new concept changes the profile of population and to explore the relationship between GESS and Nash equilibrium or ESS. We characterize through the study of many GESS and we show how the evolution and the equilibrium are influenced by the size of the group, as well as by their immediate payoff. In Chap. 3, we extend the theory developed in Chap. 2 to cover several behaviour ranging from altruist behavior to fully non-cooperative behavior. First, we begin by defining this new concept, driving it in several ways and exploring the implications of the model. The major focus is to study how the level of cooperation impacts the profile of the population as well as the global performance of the system. Chapter 4 we extent the evolutionary games framework by considering a population composed of communities with each having its set of strategies and payoff function. Assuming the interactions among the communities occur with different probabilities, we define new evolutionarily stable strategies (ESS) with different levels of stability against mutations. In particular, through the analysis of two-community two-strategy model, we derive the conditions of existence of ESSs under different levels of stability. We also study the evolutionary game dynamics both in its classic form and with de- lays. The delays may be strategic, i.e. associated with the strategies, spatial, i.e. associated with the communities, or spatial-strategic. In Chap. 5, we continue the study on evolutionary games and we tackle the concept of stability of the Evolutionarily Stable Strategy (ESS) in the continuous-time replicator dynamics subject to random time delays. In fact, in many examples the interactions between individuals take place instantaneously but their impacts are not immediate. Their impact may require a certain amount of time, which is usually random. In this work we study the effect of randomly distributed time delays in the replicator dynamics. We show that, under the exponential delay distribution, the ESS is asymptotically stable for any value of the rate parameter. For the uniform and Erlang distributions, we derive necessary and sufficient conditions for the asymptotic stability of the ESS. We also study random discrete delays and we derive a necessary and sufficient delay independent stability condition. Finally, in Chap. 6, we study the coupled dynamics of the policies and the individual states inside a population of interacting individuals. We first define a general model by coupling replicator dynamics and continuous-time Markov Decision Processes and we then consider a particular case of a two-policies and two-states evolutionary game. We first obtain a system of combined dynamics: the rest-points of this system are equilibria profile of our evolutionary game with individual state dynamics. Second, by assuming two different time scales between states and policies dynamics, we can compute explicitly the equilibria. Then, by transforming our evolutionary game with individual states into a standard evolutionary game, we obtain an equilibrium profile which is equivalent, in terms of average sojourn times and expected fitness, to the previous one.

Chapter 1
Altruism in Groups

Ilaria Brunetti, Rachid El-Azouzi and Eitan Altman

Evolutionary Game Theory has been originally developed and formalized by [245], in order to model the evolution of animal species and it has soon become an important mathematical tool to predict and even design evolution in many fields, others than biology. It mainly focuses on the dynamical evolution of the strategies adopted in a population of interacting individuals, where the notion of equilibrium adopted is that of Evolutionarily Stable Strategy (ESS, [245]), implying robustness against a mutation (i.e. a change in the strategy) of a small fraction of the population. This is a stronger condition than the standard Nash equilibrium concept, which requires robustness against deviation of a single user. On the importance of the ESS for understanding the evolution of species, Dawkins writes in his book "The Selfish Gene" [292]: "we may come to look back on the invention of the ESS concept as one of the most important advances in evolutionary theory since Darwin." He further specifies: Maynard Smith's concept of the ESS will enable us, for the first time, to see clearly how a collection of independent selfish entities can come to resemble a single organized whole.

Evolutionary game theory is nowadays considered as an important enrichment of game theory and it's applied in a wide variety of fields, spanning from social sciences [100] to computer science. Some examples of applications in computer science can be found in multiple access protocols [237], multihoming [236] and resources competition in the Internet [300].

I. Brunetti · R. El-Azouzi (✉)
Computer Science Laboratory (LIA), University of Avignon Computer Science Laboratory (LIA), Avignon, France
e-mail: rachid.elazouzi@univ-avignon.fr

E. Altman
Maestro, INRIA Maestro, Sophia Antipolis, France
e-mail: eitan.altman@sophia.inria.fr

E. Altman et al. (eds.), *Multilevel Strategic Interaction Game Models for Complex Networks*, https://doi.org/10.1007/978-3-030-24455-2_1

This theory is usually adopted in situations where individuals belonging to a very large population are matched in random pairwise interactions. In classical evolutionary games (EG), each individual constitutes a selfish player involved in a non-cooperative game, maximizing its own utility, also said fitness, since in EG it's originally assumed that individuals' utility corresponds to the Darwinian fitness, i.e. the number of offsprings. The fitness is defined as a function of both the behavior (strategy) of the individual as well as of the distribution of behaviors among the whole population, and strategies with higher fitness are supposed to spread within the population. A behavior of an individual with a higher fitness would thus result in a higher rate of its reproduction. We observe that since strategies and fitness are associated to the individual, then classical EG is restricted to describe populations in which the individual is the one that is responsible for the reproduction and where the choice of its own strategies is completely selfish. In biology, in some species like bees or ants, the one who interacts is not the one who reproduces. This implies that the Darwinian fitness is related to the entire swarm and not to a single bee and thus, standard EG models excludes these species in which the single individual which reproduces is not necessarily the one that interacts with other players. Furthermore, in many species, we find altruistic behaviors, which favors the group the playing individual belongs to, but which may hurt the single individual. Altruistic behaviors are typical of parents toward their children: they may incubate them, feed them or protect them from predator's at a high cost for themselves. Another example can be found in flock of birds: if a bird sees a predator, then it gives an alarm call to warn the rest of the flock to protect the group, but attracting the predators attention to itself. Also the stinging behavior of bees is another example of altruism, since it serves to protect the hive but its lethal for the bee which strives. In human behavior, many phenomena where individuals do care about other's benefits in their groups or about their intentions can be observed and thus the assumption of selfishness becomes inconsistent with the many real world behavior of individuals belonging to a population.

Founders of classical EG seem to have been well aware of this problem. Indeed, Vincent writes in [279] "Ants seem to completely subordinate any individual objectives for the good of the group. On the other hand, the social foraging of hyenas demonstrates individual agendas within a tight-knit social group (Hofer and East 2003). As evolutionary games, one would ascribe strategies and payoffs to the ant colony, while ascribing strategies and payoffs to the individual hyenas of a pack."

In the case of ants, the proposed solution is thus to model the ant colony as a player. Within the CEG paradigm, this would mean that we have to consider interactions between ant colonies. The problem of this approach is that it doesn't allow to model behavior at the level of the individual.

In this chapter we present a new model for evolutionary games in which the concept of the agent as a single individual is replaced by that of the agent as a whole group of individuals. We define a new equilibrium concept, named Group Equilibrium Stable Strategy (GESS), allowing to model competition between individuals in a population in which the whole group shares a common utility. Even if we still consider pairwise interactions among individuals, our perspective is substantially different: we assume that individuals are simple actors of the game, maximizing the utility of their group instead of their own one. We first define the GESS, deriving it in several ways and exploring its major characteristics. The main interest of this work is to study how this new concept changes the profile of population and to explore the relationship between GESS and standard Nash equilibrium and ESS. We characterize the GESS and we show how the evolution and the equilibrium are influenced by the groups' size as well as by their immediate payoff. We compute some interesting in a particular example of a multiple access games, in which the payoff related to local interactions also depend on the type of individuals that are competing, and not only the strategy used. In such application, we evaluate the impact of altruism behavior on the performance of the system.

The chapter is structured as follows. We first provide in Sect. 1.1 the needed background on evolutionary games. In the Sect. 1.2 we then study the new natural concept GESS and the relationship between GESS and ESS or Nash equilibrium. The characterization of the GESS is studied in Sect. 1.3. Section 1.4 provides some numerical illustration through some famous examples in evolutionary games. In Sect. 1.5 we study the multiple access control in slotted Aloha under altruism behavior. The paper closes with a summary in Sect. 1.6.

1.1 Classical Evolutionary Games and ESS

We consider an infinite population of players and we assume that each individual disposes of the same set of pure strategies $\mathcal{K} = \{1, 2, .., m\}$. Each individual repeatedly interacts with an other individual randomly selected within the population. Individuals may use a mixed strategy $\mathbf{p} \in \Delta(\mathcal{K})$, where $\Delta(\mathcal{K}) = \{\mathbf{p} \in \mathbb{R}_+^m | \sum_{i\in\mathcal{K}} p_i = 1\}$, where each vector $\mathbf{p}$ corresponds to a probability measure over the set of actions $\mathcal{K}$. This serves to represent those cases where an individual has the capacity to produce a variety in behaviors. A mixed strategy $\mathbf{p}$ can also be interpreted as the vector of densities of individuals adopting a certain pure strategy, such that each component p_i represents the fraction of the population using strategy $i \in \mathcal{K}$. However, in the original formulation of evolutionary game theory, it is not necessary to make the distinction between population-level and individual-level variability for infinite population [245].

Let now focus on the case of monomorphic populations, i.e., on the case in which individuals use mixed strategy. We define by $J(\mathbf{p}, \mathbf{q})$ the expected payoff of a given individual if it uses a mixed action $\mathbf{p}$ when interacting with an individual playing the mixed action $\mathbf{q}$. Actions with higher payoff (or "fitness") are thus expected to spread faster in the population. If we define a payoff matrix A and consider $\mathbf{p}$ and $\mathbf{q}$ to be column vectors, then $J(\mathbf{p}, \mathbf{q}) = \mathbf{p}' A\mathbf{q}$ and the payoff function J is bilinear, i.e. it is linear both in $\mathbf{p}$ and in $\mathbf{q}$. A mixed action $\mathbf{q}$ is said to be a Nash equilibrium if

$$\forall \mathbf{p} \in \Delta(\mathcal{K}), \quad J(\mathbf{q}, \mathbf{q}) \geq J(\mathbf{p}, \mathbf{q}) \tag{1.1}$$

As we mentioned above, in evolutionary games the most important concept of equilibrium is the ESS, introduced by [245] as a strategy that, if adopted by the whole population, can not be invaded by a different ("mutant") strategy. More precisely, if we suppose that the entire population adopt a strategy $\mathbf{q}$ and that a small fraction ϵ of individuals (*mutants*) plays another strategy $\mathbf{p}$, then $\mathbf{q}$ is evolutionarily stable against $\mathbf{p}$ if

$$J(\mathbf{q}, \epsilon\mathbf{p} + (1-\epsilon)\mathbf{q}) > J(\mathbf{p}, \epsilon\mathbf{p} + (1-\epsilon)\mathbf{q}) \tag{1.2}$$

The definition of ESS thus corresponds to a robustness property against deviations by a (small) fraction of the population. This is an important difference that distinguishes the equilibrium in a large population as seen in evolutionary games and the standard Nash equilibrium often used in economic context, where the robustness is defined against the possible deviation of each single agent. Since in evolutionary games context we deal with very large populations, it is more likely to expect that some group of individuals may deviate from the incumbent strategy and thus robustness against deviations by a single user would not be sufficient to guarantee that the mutant strategy will not spread among a growing portion of the population. By defining the ESS through the following equivalent definition, it's possible to establish the relationship between ESS and Nash Equilibrium (NE). The proof of the equivalence between the two definitions can be found in ([288], Proposition 2.1) or ([135], Theorem 6.4.1, page 63).

Strategy $\mathbf{q}$ is an ESS if it satisfies the two conditions:

- Nash equilibrium condition:

$$J(\mathbf{q}, \mathbf{q}) \geq J(\mathbf{p}, \mathbf{q}) \qquad \forall \mathbf{p} \in \mathcal{K}. \tag{1.3}$$

- Stability condition:

$$J(\mathbf{p}, \mathbf{q}) = J(\mathbf{q}, \mathbf{q}) \Rightarrow J(\mathbf{p}, \mathbf{p}) < J(\mathbf{q}, \mathbf{p}). \quad \forall \mathbf{p} \neq \mathbf{q} \tag{1.4}$$

The first condition (1.3) corresponds to the condition for a Nash equilibrium. In fact, if inequality (1.3) is satisfied, then the fraction of mutations in will increase, since it hasn't an higher fitness, and thus a lower growth rate. If the two strategies provide the same fitness but condition (1.4) holds, then a population using q is "weakly" immune against mutants using p. Indeed, if the mutant's population grows, then we shall frequently have individuals with action q competing with mutants. In such cases, the inequality in (1.4), $J(\mathbf{p}, \mathbf{p}) < J(\mathbf{q}, \mathbf{p})$, ensures that the growth rate of the original population exceeds that of the mutants. In this sense an ESS can be seen as a refinement of the Nash equilibrium.

1.2 New Natural Concept on Evolutionary Games

In this section we present our new concept for evolutionary games, where we still consider pairwise interactions among individuals but the actual player of the game is a whole group of these individuals. We suppose that the population is composed of N groups, $G_i, i = 1, 2, .., N$, where the normalized size of each group G_i is denoted by α_i, with $\sum_{j=1}^{N} \alpha_i = 1$.

Each individual can meet a member of its own group or of a different one, in random pairwise interactions, and disposes of a finite set of actions, denoted by $\mathcal{K} = \{a_1, a_2, .., a_M\}$. Let p_{ik} be the probability that an individual in the group G_i chooses an action $a_k \in \mathcal{K}$. Each group i is associated to the vector of probabilities $\mathbf{p_i} = (p_{i1}, p_{i2}, .., p_{iM})$ where $\sum_{l=1}^{M} p_{il} = 1$, giving the distribution of actions within the group. By assuming that each individual can interact with any other individual with equal probability, then the expected utility of a player (the group) i is:

$$U_i(\mathbf{p}_i, \mathbf{p}_{-i}) = \sum_{j=1}^{N} \alpha_j J(\mathbf{p}_i, \mathbf{p}_j), \tag{1.5}$$

where $\mathbf{p}_{-i}$ is the strategies profile of all the other groups (but i) and $J(\mathbf{p}_i, \mathbf{p}_j)$ is the immediate expected utility of an individual player adopting strategy $\mathbf{p}_i$ against an opponent playing $\mathbf{p}_j$.

1.2.1 Group Equilibrium Stable Strategy

The definition of GESS is related to a notion of robustness against deviations within each group. In this context, there are two possible equivalent interpretations of an $\epsilon-$deviation toward $\mathbf{p}_i$. If the group G_i plays according to the incumbent strategy $\mathbf{q}_i$, an $\epsilon-$deviation, can be thought as:

1. A small deviation in the strategy by all members of a group, shifting to the new group's strategy $\bar{\mathbf{p}}_i = \epsilon \mathbf{p}_i + (1 - \epsilon \mathbf{q}_i)$;
2. The second is a (possibly large) deviation of a fraction ϵ of individuals belonging to G_i, playing the mutant strategy $\mathbf{p}_i$.

After an $\epsilon-$deviation under both interpretations the profile of the whole population becomes $\alpha_i \epsilon \mathbf{p}_i + \alpha_i(1 - \epsilon)\mathbf{q}_i + \sum_{j \neq i} \alpha_j \mathbf{q}_j$. Then the average payoff of group G_i after mutation is given by:

$$
\begin{aligned}
U_i(\bar{\mathbf{p}}_i, \mathbf{q}_{-i}) &= \sum_{j=1}^{N} \alpha_j J(\bar{\mathbf{p}}_i, \mathbf{p}_j) \\
&= U_i(\mathbf{q}_i, \mathbf{q}_{-i}) + \epsilon^2 \alpha_i \Omega(\mathbf{p}_i, \mathbf{q}_i) + \epsilon\Big(\alpha_i (J(\mathbf{p}_i, \mathbf{q}_i) \\
&+ J(\mathbf{q}_i, \mathbf{p}_i) - 2J(\mathbf{q}_i, \mathbf{q}_i)) + \sum_{j \neq i} (J(\mathbf{p}_i, \mathbf{q}_j) - J(\mathbf{q}_i, \mathbf{q}_j))\Big)
\end{aligned}
\tag{1.6}
$$

where $\Omega(\mathbf{p}_i, \mathbf{q}_i) := J(\mathbf{p}_i, \mathbf{p}_i) - J(\mathbf{p}_i, \mathbf{q}_i) - J(\mathbf{q}_i, \mathbf{p}_i)) + J(\mathbf{q}_i, \mathbf{q}_i)$.

Definition 1.1 A strategy $\mathbf{q} = (\mathbf{q}_1, \mathbf{q}_2, .., \mathbf{q}_N)$ is a GESS if $\forall i \in \{1, \dots, N\}$, $\forall \mathbf{p}_i \neq \mathbf{q}_i$, there exists some $\epsilon_{\mathbf{p}_i} \in (0, 1)$, which may depend on $\mathbf{p}_i$, such that for all $\epsilon \in (0, \epsilon_{\mathbf{p}_i})$

$$
U_i(\bar{\mathbf{p}}_i, \mathbf{q}_{-i}) < U_i(\mathbf{q}_i, \mathbf{q}_{-i}), \tag{1.7}
$$

where $\bar{\mathbf{p}}_i = \epsilon \mathbf{p}_i + (1 - \epsilon)\mathbf{q}_i$.

From Eq. (1.7), this implies that strategy $\mathbf{q}$ is a GESS if the two following conditions hold:

- $\forall \mathbf{p}_i \in [0, 1]^M$

$$
F_i(\mathbf{p}_i, \mathbf{q}) := \alpha_i \Omega(\mathbf{p}_i, \mathbf{q}_i) - U_i(\mathbf{p}_i, \mathbf{q}_{-i}) + U_i(\mathbf{q}_i, \mathbf{q}_{-i}) \geq 0, \tag{1.8}
$$

- $\exists \mathbf{p}_i \neq \mathbf{q}_i$ such that:

$$
\text{If} \quad F_i(\mathbf{p}_i, \mathbf{q}) = 0 \Rightarrow \Omega(\mathbf{p}_i, \mathbf{q}_i) < 0 \tag{1.9}
$$

Remark 1.1 The second condition (1.9) can be rewritten as

$$
U_i(\mathbf{q}_i, \mathbf{q}_{-i}) > U_i(\mathbf{p}_i, \mathbf{q}_{-i})
$$

which coincide to the definition of the strict Nash equilibrium of the game composed by N groups, each of them maximizing its own utility.

1.2.2 GESS and Standard ESS

Here we analyze the relationship between our new equilibrium concept, the $GESS$ and the standard ESS.

Proposition 1.1 *Consider games such that the immediate expected reward is symmetric, i.e.* $J(\mathbf{p}, \mathbf{q}) = J(\mathbf{q}, \mathbf{p})$. *Then any ESS is a GESS.*

Proof Let $\mathbf{q} = (q, .., q)$ be an ESS. From the symmetry of the payoff function and Eq. (1.8), we get:

$$F_i(\mathbf{p}_i, \mathbf{q}) = -\Big(\alpha_i(J(\mathbf{p}_i, q) + J(q, \mathbf{p}_i) - 2J(q, q)$$
$$+ \sum_{j \neq i} \alpha_j (J(\mathbf{p}_i, q) - J(q, q)\Big)$$
$$= -2\alpha_i(J(\mathbf{p}_i, q) - J(q, q)) - \sum_{j \neq i} \alpha_j (J(\mathbf{p}_i, q) - J(q, q))$$
$$= -(1 + \alpha_i)(J(\mathbf{p}_i, q) - J(q, q)) \geq 0$$

where the second equality follows from the symmetry of J and the last inequality follows form the fact that $\mathbf{q}$ is an ESS and satisfies (1.3). This means that $\mathbf{q}$ satisfies the first condition of GESS (1.8). Assume now that $F_i(\mathbf{p}_i, q) = 0$ for some $\mathbf{p}_i \neq \mathbf{q}$, previous equations imply that $J(\mathbf{p}_i, \mathbf{q}) = J(\mathbf{q}, \mathbf{q})$. Thus the second condition (1.9) becomes $\Omega(\mathbf{p}_i, \mathbf{q}_i) = J(\mathbf{p}_i, \mathbf{q}) - J(\mathbf{q}, \mathbf{q}) < 0$ which coincide with the second condition of ESS (1.4). This completes the proof. □

1.2.3 Nash Equilibrium and GESS

In the classical evolutionary games, the ESS can be seen as a refinement of a Nash equilibrium, since all ESSs are Nash equilibria while the converse is not true. In order to study and characterize this relationship in our context, we define the game between groups. There are N players in which each player has a finite set of pure strategies $\mathcal{K} = \{1, 2, .., m\}$; let $U_i(\mathbf{q}_i, \mathbf{q}_{-i})$ be the utility of player i when using mixed strategy $\mathbf{q}_i$ against a population of players using $\mathbf{q}_{-i} = (\mathbf{q}_1, \ldots, \mathbf{q}_{i-1}, \mathbf{q}_{i+1}, \ldots, \mathbf{q}_N)$.

Definition 1.2 A strategy $\mathbf{q} = (\mathbf{q}_1, \mathbf{q}_2, .., \mathbf{q}_N)$ is a Nash Equilibrium if $\forall i \in \{1, \ldots, N\}$

$$U_i(\mathbf{q}_i, \mathbf{q}_{-i}) \geq U_i(\mathbf{p}_i, \mathbf{q}_{-i}) \tag{1.10}$$

for every mixed strategy $\mathbf{p}_i \neq \mathbf{q}_i$. If the inequality is a stric one, then $\mathbf{q}$ is a strict Nash equilibrium.

From the definition of the strict Nash equilibrium, it is easy to see that any strict Nash equilibrium is a GESS defined in Eq. (1.7). But in our context, we address several questions on the relationship between the GESS, ESS and the Nash equilibrium defined in (1.10). For the sake of simplicity, we restrict to the case of two-strategies games. Before studying them, we introduce here some necessary definitions.

Definition 1.3
- A fully mixed strategy $\mathbf{q}$ is a strategy such that all the actions available for a group have positive probability, i.e., $0 < q_{ij} < 1 \; \forall (i, j) \in \mathcal{I} \times \mathcal{K}$.
- A mixer (resp. pure) group i is the group that uses a mixed (resp. pure) strategy $0 < q_i < 1$ (resp. $q_i \in \{0, 1\}$).
- An equilibrium with mixed and non mixed strategies is an equilibrium in which there is at least one pure group and a mixer group.

1.3 Analysis of N-Groups Games with Two Strategies

In this section we present a simple case, with N-groups games disposing of two strategies. The two available pure strategies are A and B and the payoff matrix of pairwise interctions is given by:

$$P = \begin{matrix} & A & B \\ A & a & b \\ B & c & d \end{matrix}, \quad P = \begin{pmatrix} a & b \\ c & d \end{pmatrix},$$

where $P_{ij}, i, j = A, B$ is the payoff of the first (row) individual if it plays strategy i against the second (column) individual playing strategy j. We assume that the two individuals are symmetric and hence payoffs of the column player are given by P^t, i.e. the transposed of P. According to the definition of GESS, $\mathbf{q}$ is a GESS if it satisfies the conditions (1.8)–(1.9), which can be rewritten here as:

- $\forall \mathbf{p}_i \in [0, 1], i = 1, .., N$:

$$F_i(p_i, \mathbf{q}) = (q_i - p_i)\Big(\alpha_i(J(q_i, 1) - J(q_i, 0)) + \sum_{j=1}^{N} \alpha_j(J(1, q_j) - J(0, q_j))\Big) \geq 0 \tag{1.11}$$

- If $F(p_i, \mathbf{q}) = 0$ for some $p_i \neq q_i$, then:

$$(p_i - q_i)^2 \Delta < 0 \Longrightarrow \Delta < 0 \tag{1.12}$$

where $\Delta = a - b - c + d$.

1.3.1 Characterization of Fully Mixed GESS

In this section we are interested in characterizing the full mixed GESS $\mathbf{q}$. According to (3.4), a full mixed equilibrium $\mathbf{q} = (q_1, \ldots, q_N)$ is a GESS if it satisfies the condition (3.5) where the equality must holds for all $p \in [0, 1]$. This yields to the following equation: $\forall i = 1, \ldots N$,

$$\alpha_i(J(q_i, 1) - J(q_i, 0)) + \sum_{j=1}^{N} \alpha_j(J(1, q_j) - J(0, q_j)) = 0$$

which can be rewritten as

$$\alpha_i \Delta q_i + b - d + \alpha_i(c-d) + \Delta \sum_{j=1}^{N} \alpha_j q_j = 0$$

This leads to the following expression of the mixed GESS:

$$q_i^* = \frac{d - b + \big((1+N)\alpha_i - 1\big)(d-c)}{(N+1)\alpha_i \Delta}; \tag{1.13}$$

Proposition 1.2 *If* $\Delta < 0$ *and* $0 < q_i^* < 1, i = 1, \dots, N$, *then there exists a unique fully mixed GESS equilibrium given by (1.13).*

Note that the fully mixed GESS is a strict Nash equilibrium since the condition (3.5) corresponds to the definition of the strict Nash equilibrium (see Remark 1.1), under the condition $F(p, q_1, \dots, q_N) = 0, \forall p \in [0, 1]$.

1.3.2 Characterization of Strong GESS

We say that an equilibrium is a strong GESS if it satisfies the strict inequality (3.4) for all groups. As we did above for the fully mixed GESS, we explicit here the condition for the existence of a strong GESS. Note that all groups have to use pure strategy in a strong GESS. With no loss of generality, we assume that a pure strong GESS can be represented by n_A, where $n_A \in \{1, ..., N\}$ denotes that the n_A first groups use A pure strategy and remaining $N - n_A$ groups chose strategy B. For example $n_A = N$ (resp. $n_A = 0$) means that all groups choose pure strategy A (resp. B).

Proposition 1.3 *If* $a \neq c$ *or* $b \neq d$, *then every N-player game with two strategies has a GESS. We distinguish the following possibilities for the* ***strong GESS****:*

i. If $a - c > \max_i(\alpha_i) \cdot (b - a)$ *then* $n_A = N$ *is a strong GESS;*
ii. If $b - d < \min_i(\alpha_i)(d - c)$ *then* $n_A = 0$ *is a strong GESS;*
iii. Let $H(n_a) := \sum_{j=1}^{n_A} \alpha_j(a - c) + \sum_{j=n_A+1}^{N} \alpha_j(b - d)$. *If* $\alpha_i(d - c) > H(n_a) > \alpha_i(b - a)$ *then* n_A *is a strong GESS.*

Proof In order to prove that a strategy $n_A = N$ is a GESS, we have to impose the strict inequality, i.e.: $\forall p_i \neq 1$ for $i \in \{1, .., n_A\}$ and $\forall p_i \neq 0$ for $i \in \{n_A + 1, .., N\}$

$$F_i(p_i, 1_{n_A}, 0_{N-n_A}) > 0$$

We provide here the necessary conditions of the existence in the case $n_A = N$; the other cases straightforward follow from the players' symmetry assumption.

We now suppose that $(n_A = N)$ is a strong GESS. The inequality (3.4) becomes: $\forall p_i \neq 1, \ \forall i$

$$(p_i - 1)\Big(\alpha_i(a-b) + \sum_{j=1}^{N} \alpha_j (a-c)\Big) = (p_i - 1)\Big(\alpha_i(a-b) + a - c)\Big) < 0,$$

Since $p_i < 1$, one has $\alpha_i(b-a) < a - c \ \forall i$. This completes the proof of (i). To show conditions of the other strong GESSs, we follow the lines of the proof of (i).□

1.3.3 Characterization of Weak GESS

We define a weak GESS as an equilibrium in which at least one group uses a strategy that satisfies the condition (3.5) with equality. Here we distinguish two different situations: the equilibrium with no mixer group and the equilibrium with mixer and no mixer groups. Conditions for the equilibrium with no mixed strategy are given by Proposition 1.3 with at the least one group satisfying it with equality and $\Delta < 0$. In what follows we thus focus only on the equilibrium with mixer and no mixer groups. Without loss of generality, we assume that an equilibrium with mixed and non mixed strategies, can be represented by $(n_A, n_B, \mathbf{q})$ where n_A denotes that group i for $i = 1.., n_A$ (resp. $i = n_A + 1, .., n_A + n_B$) uses strategy A (resp. B). The remaining groups $N - n_A - n_B$ are mixers in which q_i is the probability to choose the strategy A by group i.

Proposition 1.4 *Let either $a \neq c$ or $b \neq d$ and $\Delta < 0$. $(n_A, n_B, \mathbf{q})$ is a weak GESS if:*

$$\begin{cases} \alpha_i \Delta + d - b + \alpha_i(c-d) + \Delta(\alpha_{n_A} + y) \geq 0, \ i = 1, .., n_A \\ d - b + \alpha_i(c-d) + \Delta(\alpha_{n_A} + y) \leq 0, \ i = n_A + 1, .., n_B \\ q_i = \frac{d-b+\alpha_i(d-c)-y\Delta}{\Delta\alpha_i}, \ i = n_A + n_B + 1, .., N \end{cases} \tag{1.14}$$

where $y = \frac{(N-n_A-n_B)(d-b-\sum_{j=1}^{n_A}\alpha_j)+(d-c)\sum_{j=n_A+n_B+1}^{N}\alpha_j}{\Delta(N-N_A-n_B+1)}$.

Proof See technical report □

1.4 Some Examples

In this section we analyze a number of examples with two players and two strategies.

1.4.1 Hawk and Dove Game

One of the most studied examples in EG theory is the Hawk-Dove game, first introduced by Maynard Smith and Price in "The Logic of Animal Conflict". This game serves to represent a competition between two animals for a resource. Each animal can follow one of two strategies, either Hawk or Dove, where Hawk corresponds to an aggressive behavior, Dove to a non-aggressive one. So, if two Hawks meet, they involve in a fight, where one of them obtain the resource and the other is injured, with equal probability. A Hawk always wins against a Dove, but there is no fight, so the Dove looses the resource but it's not injured, whereas if two Doves meet they equally share the resource. The payoff matrix associated to the game is the following:

$$\begin{array}{c} \\ H \\ D \end{array} \begin{array}{c} \begin{array}{cc} H & D \end{array} \\ \begin{pmatrix} \frac{1}{2}(V-C) & V \\ 0 & V/2 \end{pmatrix} \end{array}$$

where C represent the cost of the fight, and V is the benefit associated to the resource. We assume that $C > V$.

In standard GT, this example belongs to the class of *anti-coordination games*, which always have two strict pure strategy NEs and one non-strict, mixed strategy NE. In this case the two strict pure equilibria are (H, D) and (D, H), and the mixed-one is: $q^* = \frac{V}{C}$. The latter is the only ESS: even if the two pure NE are strict, being asymmetric they can't be ESSs. We now set $V = 2$ and $C = 3$ and we study the Hawk and Dove game in our groups framework, considering two groups of normalized size α and $1 - \alpha$.

We find that the GESSs and the strict NE always coincides. More precisely we obtain that:

- for $0 < \alpha < 0.25$ the game has one strong GESS (H, D) and a weak GESS (H, q_2);
- for $0.25 < \alpha < 0.37$: one weak GESS (H, q_2);
- for $0.37 < \alpha < 0.5$ one weak GESS, (q_1^*, q_2^*);

We observe that the size of groups has a strong impact on the behavior of players: in the first interval of $\alpha-$values we remark that the GESS is not unique; if the size of the first group increases (which implies that the second one decreases), the probability that the second group plays aggressively against the pure aggressive strategy of the first one increases until we get into the third interval, where both players are mixers. The mixed equilibrium q_1^* is decreasing in α: this is because, as we supposed that an individual can interact with members of its own group, when increasing α, the probability of meeting an individual in the same group increases and thus the individual tend to adopt a less aggressive behavior (Fig. 1.1).

1.4.2 Stag Hunt Game

We now consider a well-known example which belonging to the class of *coordination games*, the Stag Hunt game. The story modeled by this game has been described by J-J. Rousseau: two individuals go hunting; if they hunt together cooperating, they

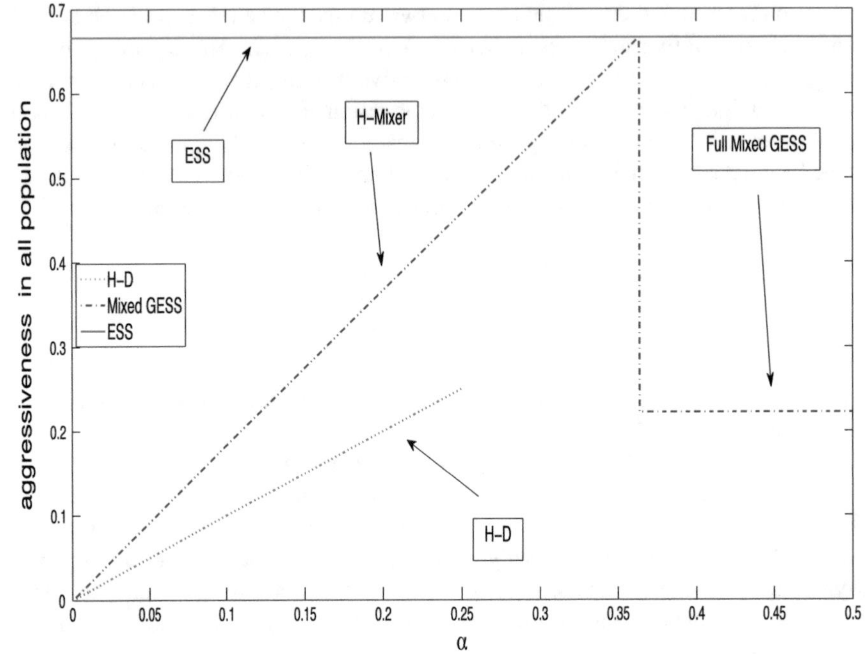

Fig. 1.1 The global level of aggressiveness in the two-groups population for the different GESSs, as a function of α

can hunt a stag; otherwise, if hunting alone, a hunter can only get a hare. In this case, collaboration is thus rewarding for players. It serves to represent a conflict between social and safely cooperation. The payoff matrix is the following:

$$\begin{array}{cc} & S \quad H \\ \begin{array}{c} S \\ H \end{array} & \begin{pmatrix} a & b \\ c & d \end{pmatrix} \end{array}$$

where S and H stand respectively for Stag and Hare and $a > c \geq d > b$. Coordination games are characterized by two strict, pure strategy NEs and one non-strict, mixed strategy NE, which are, respectively, the risk dominant equilibrium (H, H), the payoff dominant one (S, S) and the symmetric mixed equilibrium with $q_1^* = q_2^* = \frac{d-b}{a-b-c+d}$.

We set here $a = 2$, $b = 0$, $c = 1$, $d = 1$ and we determine the equilibria of the two groups game as a function of α. We find that the strict GESSs and the strict NEs don't coincide, since we obtain strict GESSs but not NEs. The two-groups Stag-Hunt Game only have the pure-pure strict NE (S, S), for all values of α, while for the GESSs we obtain that:

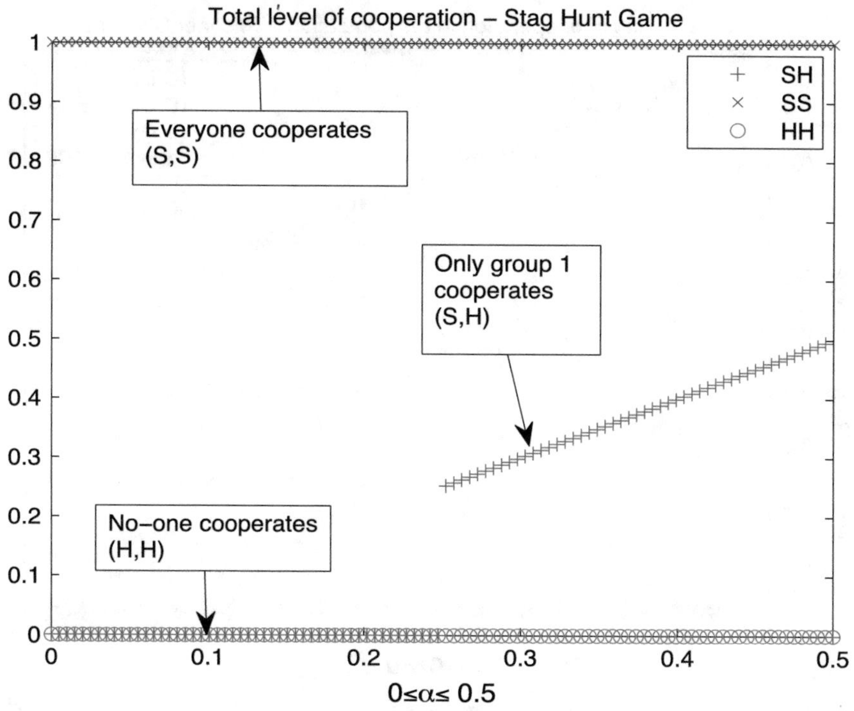

Fig. 1.2 The global level of cooperation in the two-groups population for the different GESSs, as a function of α

- for $0 < \alpha < 0.5$ the game admits two pure-pure strong GESSs: (S, S) and (H, H);
- for $0.25 < \alpha < 0.5$ the game admits three pure-pure strong GESSs: (S, S) and (H, H) and (S, H);
- the game doesn't admit any strict mixed NE (Fig. 1.2).

1.4.3 Prisoner's Dilemma

We consider another classical example in game theory, the Prisoner's Dilemma, which belongs to a third class of games, the *pure dominance class*.

The story to imagine is the following: two criminals are arrested and separately interrogated; they can either accuse the other criminal, either remain silent. If both of them accuse the other (defect—D), then they will be both imprisoned for 2 years. If only one accuse the other, the accused is punished with 3 years of jail while the other is free. If both remain silent (cooperate–C), each of them will serve only one year in jail. The corresponding payoff matrix is the following:

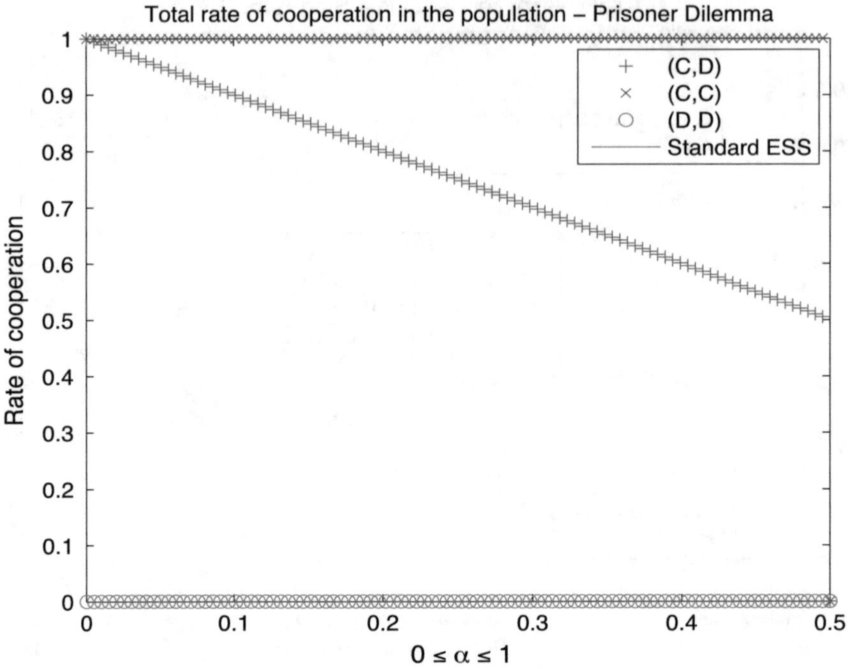

Fig. 1.3 Total rate of collaboration for different GESSs as a function of group 1 size α in the prisoner's Dilemma

$$\begin{array}{c} \\ C \\ D \end{array} \begin{array}{c} \begin{array}{cc} C & D \end{array} \\ \begin{pmatrix} a & b \\ c & d \end{pmatrix} \end{array}$$

where $c > a > d > b$.

In standard GT, pure dominance class games have a unique pure, strict and symmetric NE, which also is the unique ESS; in the Prisoner's Dilemma it's (D, D) (Fig. 1.3).

We set $a = 2$, $b = 0$, $c = 3$, $d = 1$ and we study the two-groups corresponding game. As in the previous example, we find strict GESSs which are not strict NEs. In particular, we have that:

- (C, C) is always a GESS and a strict NE for all values of α;
- (D, D) is a GESS for all values of α but it is never a strict NE;
- (C, D) (symmetrically (D, C)) is always a GESS and a strict NE for $0.5 < \alpha < 1$ (symmetrically $0 < \alpha < 0.5$);
- the game doesn't admit mixed GESSs;

For any value of α, the two groups game thus admits three pure GESSs and two pure strict NEs.

1.5 Multiple Access Control

In this section we shortly introduce a possible refinement of our model, which will be further investigated in future works. The idea is to distinguish between interactions among members of the same group and of two different ones. We study a particular example where we modify the group utility function defined in (1.5), supposing that the immediate payoff matrix depends on the groups the two players belong to (if its the same one or they belong to two different groups).

The utility function of a group i playing q_i against a population profile q_{-i} can be written as follows:

$$U(\mathbf{q_i}, \mathbf{q_{-i}}) = \alpha_i K(\mathbf{q_i}, \mathbf{q_i}) + \sum_{j \neq i} \alpha_i J(\mathbf{q_i}, \mathbf{q_j}), \tag{1.15}$$

where by $K(\mathbf{p}, \mathbf{q})$ we denote the immediate expected payoff of an individual playing p against a member of its own group using q, and by $J(\mathbf{p}, \mathbf{q})$ the immediate expected payoff associated to interactions among individuals of different groups.

We consider a particular application of this model in Aloha system, such that a large population of mobile phones interfere with each other through local pair-wise interactions. The population of mobiles is decomposed into N groups G_i, $i = 1, 2, .., N$ of normalized size α_i with $\sum_{j=1}^{N} \alpha_i = 1$ and each mobile can decide either to transmit (T) or to not transmit (S) a packet to the receiver within transmission range. The interferences occur according to the Aloha protocol, which assume that if more than one neighbor of a receiver transmits a packet at the same time this causes a collision and the failure of transmission. The channel is assumed to be ideal for transmission and the only errors occurring are due to these collisions.

We denote by μ the probability that a mobile k has its receiver $R(k)$ in its range. When the mobile k transmits to the receiver $R(k)$, all the other mobiles within a circle of radius R centered on $R(k)$ cause interference to k and thus the failure of mobile's k packet transmission to $R(k)$.

A mobile belonging to a given group i may use a mixed strategy $\mathbf{p}_i = (p_i, 1 - p_i)$, where p_i (resp. $1 - p_i$) is the probability to choose the action (T) (resp (S)). Let γ be the probability that a mobile is alone in a given local interaction; before transmission, the tagged mobile doesn't know if there is another transmitting mobile within its range of transmission.

Let P_1 (resp. P_2) be the immediate payoff matrix associated to interactions among mobiles belonging to the same group (resp. to two different ones):

$$P_1 \equiv \begin{array}{c} \\ T \\ S \end{array} \begin{array}{c} \begin{array}{cc} T & S \end{array} \\ \begin{pmatrix} -2\delta & 1-\delta \\ 1-\delta & 0 \end{pmatrix} \end{array}, \qquad P_2 \equiv \begin{array}{c} \\ T \\ S \end{array} \begin{array}{c} \begin{array}{cc} T & S \end{array} \\ \begin{pmatrix} \delta & 1-\delta \\ 0 & 0 \end{pmatrix} \end{array}.$$

where $0 < \delta < 1$ is the cost of transmitting the package. The definition of the matrix P_1 implies that, when two mobiles of the same group i interfere, any successful transmission is equally rewarding for the group i. The resulting expected payoffs of a mobile playing $\mathbf{q}_i$ against a member belonging to its own group and to a different one, using respectively the same strategy $\mathbf{q}_i$ and a different one $\mathbf{q}_j$ are, respectively, the following:

$$\begin{aligned} K(\mathbf{q}_i, \mathbf{q}_i) &= \mu\,[q_i[\gamma(1-\delta) + (1-\gamma)((1-\delta)(1-q_i) - 2\delta q_i)] \\ &\quad +(1-\gamma)(1-\delta)(1-q_i)q_i] \\ &= \mu q_i[(1-\delta)(2-\gamma) - 2(1-\gamma)q_i] \end{aligned}$$

$$\begin{aligned} J(\mathbf{q}_i, \mathbf{q}_j) &= \mu q_i[\gamma(1-\delta) + (1-\gamma)((1-\delta)(1-q_j) - \delta q_j] \\ &= \mu q_i[1-\delta-(1-\gamma)q_j] \end{aligned}$$

The expected utility of group i is then given by:

$$U(\mathbf{q}_i, \mathbf{q}_{-i}) = \mu q_i[1-\delta+(1-\gamma)(\alpha_i(1-\delta-q_i) - \sum_{j=1}^{N} \alpha_j q_j)] \tag{1.16}$$

By following the same line of analysis followed in Sect. 1.2, the strategy $\mathbf{q}$ is a GESS if $\forall i = 1, \dots N$ the two following conditions are satisfied:

1. $F_i'(\mathbf{p}_i, \mathbf{q}) \equiv (q_i - p_i)[1-\delta+(1-\gamma)(\alpha_i(1-\delta-2q_i) - \sum_{j=1}^{N} \alpha_j q_j)] \geq 0$ $\forall p_i$,
2. If $F_i'(\mathbf{p}_i, \mathbf{q}) = 0$ for some $p_i \neq q_i$, then $(p_i - q_i)^2(1-\gamma)\alpha_i > 0 \quad \forall p_i \neq q_i$.

We observe that, since the inequality $(p_i - q_i)^2(1-\gamma)\alpha_i > 0$ holds for all values of the parameters, the second condition is always satisfied and thus the first condition is sufficient for the existence of a GESS. We now characterize the GESSs of the presented MAC game. With no loss of generality, we reorder the groups so that $\alpha_1 \leq \alpha_2 \dots \leq \alpha_N$.

Proposition 1.5 *In the presented MAC game, we find that:*

- *The pure symmetric strategy $(S, \dots, S)$ is never a GESS.*
- *If a group G_i adopts pure strategy S, then at the equilibrium, all smaller groups also use S.*
- *If a group G_i adopts pure strategy T, then at the equilibrium, all smaller groups transmit. If the bigger group G_N use strategy T at the equilibrium, then $\gamma > \bar{\gamma}$.*
- *If a group G_i adopts an equilibrium mixed strategy $q_i \in]0, 1[$, then if $q_i > \frac{1-\delta}{2}$, at the equilibrium all smaller groups use pure strategy T, whereas if $q_i < \frac{1-\delta}{2}$, smaller groups play S.*
- *The unique fully mixed GESS of the game is $\mathbf{q}^* = (q_1^*, \dots, q_N^*)$, given by:*

$$q_i^* = \frac{(1-\delta)(1+\gamma+(1-\gamma)(2+N)\alpha_i)}{2(N+2)(1-\gamma)\alpha_i} \tag{1.17}$$

under the condition: $\gamma < \underline{\gamma}$.

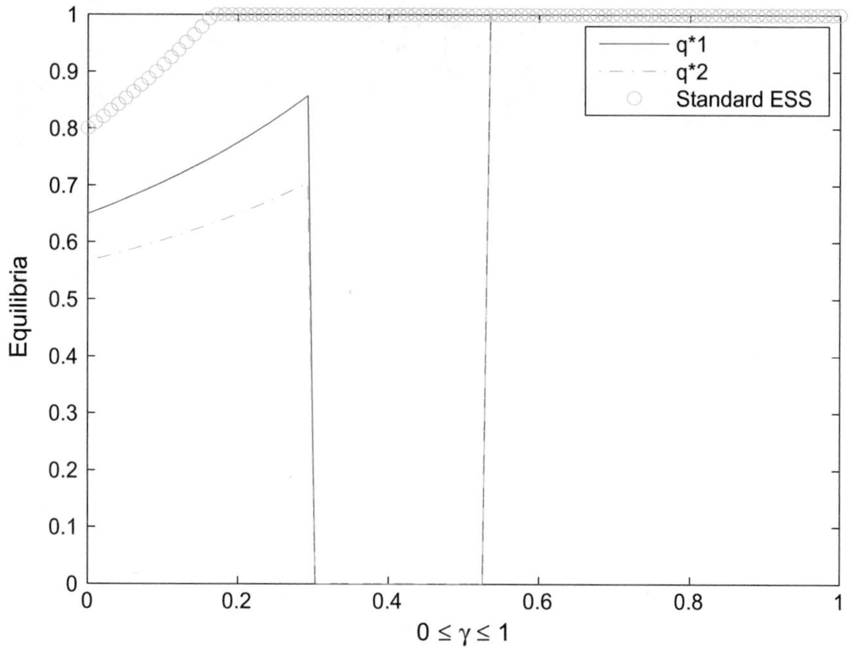

Fig. 1.4 The value of the equilibrium strategy q_1^* and q_2^* in a two groups MAC game as a function of γ for $\alpha = 0.4$ compared to q_{std}^*

The thresholds $\underline{\gamma}$ *and* $\bar{\gamma}$ *are defined as follows:*

$$\underline{\gamma} \equiv \min_{\alpha_i} \frac{\alpha_i(N+2)(1+\delta) - (1-\delta)}{\alpha_i(N+2)(1+\delta) + (1+\delta)},$$

$$\bar{\gamma} \equiv \max_{\alpha_i} \left(1 - \frac{1-\delta}{\alpha_i(\delta+1)+1}\right).$$

Proof The proof is available in [53]. □

In order to provide a better insight, we consider a two groups MAC game, in which we set a low value of the cost of transmission, $\delta = 0.2$, and groups' sizes $\alpha_1 = \alpha = 0.4$, $\alpha_2 = 1 - \alpha = 0.6$, and we let vary the value of the parameter γ. We obtain three different equilibria, depending on the value of γ: a pure GESS $\mathbf{q}_P^*$, a fully mixed GESS $\mathbf{q}_M^*$ and a pure-mixed one $\mathbf{q}_{PM}^*$. In Fig. 1.4 we plot the fully mixed and the pure GESS. For $\gamma < \underline{\gamma} = 0.3$ the game admits a GESS $\mathbf{q}^*\mathbf{q}_M^* = (q_1^*, q_2^*)$, whose components are shown in the plot. Then, for $\gamma = \bar{\gamma} > 0.53$, $\mathbf{q}^*\mathbf{q}_P^* = (T, T)$. We also represent the value of $q_{std}^* := \min(1, \frac{1}{1-\gamma}) - \Delta$. We note that fully mixed equilibrium strategies adopted by the two groups, q_1^*, q_2^*, are both lower then q_{std}^*.

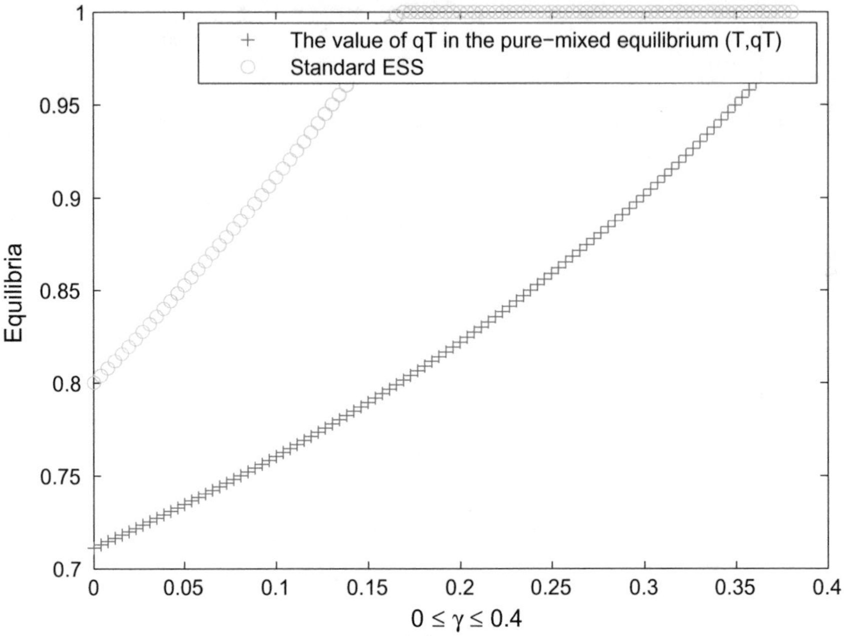

Fig. 1.5 The value of the equilibrium strategy q_2^* of the second group in the pure-mixed equilibrium (T, qT) as a function of γ for $\alpha = 0.85$ compared to q_{std}^*

In Fig. 1.5 we plot the value of the second mixed component of the pure-mixed GESS of the game: (T, q_T), which is an equilibrium in the interval $0 \leq \gamma < 0.4$, and we compare it to q_{std}^*. We observe that, for the second group the probability of transmitting is always lower w.r.t. the standard game.

We denote by $p_S(\mathbf{p})$ the probability of a successful transmission in a population whose profile is $\mathbf{p}$. If $N = 2$, we obtain that:

$$p_S(\mathbf{p}) = \mu[\gamma(\alpha p_1 + (1-\alpha)p_2)] + (1-\gamma)(2\alpha^2 p_1(1-p_1)+$$
$$+\alpha(1-\alpha)((1-p_2)p_1 + (1-p_1)p_2) + 2(1-\alpha)^2 p_2(1-p_2))].$$

In Fig. 1.6 we represent the value of the equilibrium $p_S^* = p_S(\mathbf{q}_M^*)$ as a function of γ for $\alpha = 0.4$. We observe that, $\gamma < \underline{\gamma}$, even if at the equilibrium the probability of the transmission is lower in the groups game.

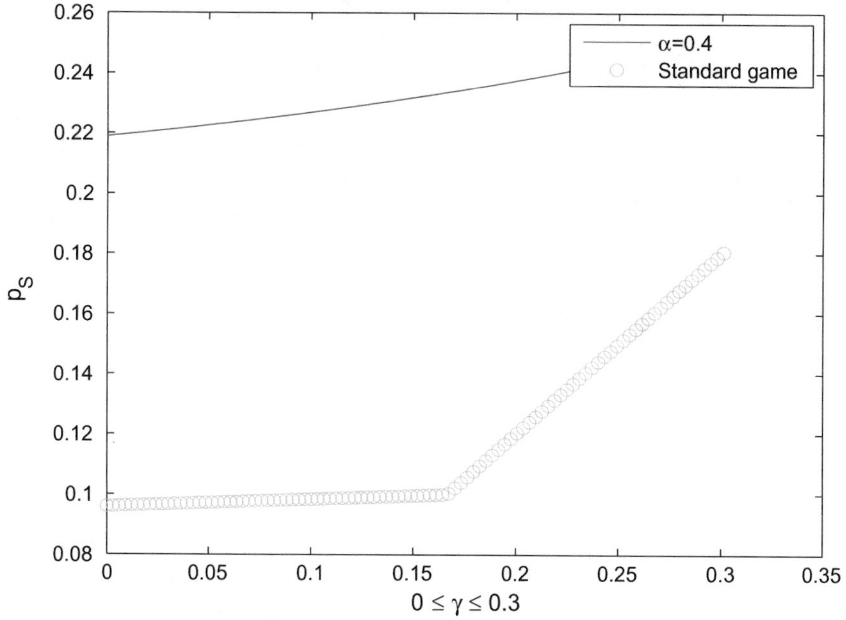

Fig. 1.6 The probability $p_S(\mathbf{q}_M^*)$ of a successful transmission for the fully mixed GESS in a two groups MAC game at the equilibrium as a function of γ

1.6 Conclusions

In this work we defined the new concept of the GESS, an Evolutionarily Stable Strategy in a group-players context and we studied its relationship and the Nash equilibrium and with the standard ESS. We studied some classical examples of two players and two strategies games transposed in our group-players framework and we found that the fact of considering interactions among individuals maximizing their group's utility (instead of their own one) impacts individuals' behavior and changes the structure of the equilibria. We briefly introduce a refinement of our model, where we redefine the utility of a group in order to consider different utilities for interactions among members of the same group or of a different one. Through a particular example in MAC games we showed how the presence of groups can encourage cooperative behaviors. There are many possible issues to be developed in future studies. We are currently studying the replicator dynamics in this group-players games, in order to investigate the existing relationship between the rest point of such dynamics and our GESS, following the lines off he Folk Theorem of EG. At a more theoretical level, the concept of group's utility could be further investigated, and the coexistence of both selfish and altruistic behaviors among individuals could be taken into account.

Chapter 2
From Egoism to Altruism in Groups

Rachid El-Azouzi, Eitan Altman, Ilaria Brunetti, Majed Haddad and Houssem Gaiech

This chapter presents a new formulation which not only covers the fully non- cooperative behavior and the fully cooperative behavior, but also the fully altruistic behaviour. To do so, we make use of the evolutionary game theory which we extend to cover this kind of behavior. The major focus of this work is to study how the level of cooperation impacts the profile of population as well as the global performance of the system. Equilibrium of the system through the notion of Evolutionary Stable Strategies and study the effect of transmission cost and cooperation level. We define and characterize the equilibrium (called Evolutionary Stable Strategy) for these games and establish the optimal level of cooperation that maximizes the probability of successful transmission. Our theoretical results unveil some behaviors. More specifically, we show that when all users increase their level of cooperation, then the performance of the system is not necessary improved. In fact, for some scenarios, the performance of groups may lead to an improvement by adopting selfishness instead of altruism. This happened when the density of agents is high. For low density, the degree of cooperation may indeed improve the performance of all groups. According to the structure of the ESS, we try to evaluate the performance of the global system in order to derive the optimal degree of cooperation.

In this chapter, we explore our finding to study multiple access games with a large population of mobiles decomposed into several groups. Mobiles interfere with each others through many local interactions. We assume that each mobile (or player) cooperates with his group by taking into account the performance of his group.

R. El-Azouzi (✉) · I. Brunetti · M. Haddad · H. Gaiech
Computer Science Laboratory (LIA), University of Avignon Computer Science Laboratory (LIA), Avignon, France
e-mail: rachid.elazouzi@univ-avignon.fr

E. Altman
Maestro, INRIA Maestro, Sophia Antipolis, France
e-mail: eitan.altman@sophia.inria.fr

E. Altman et al. (eds.), *Multilevel Strategic Interaction Game Models for Complex Networks*, https://doi.org/10.1007/978-3-030-24455-2_2

In this chapter, we consider a large population of sensors or relays that are deployed in a large area to detect the occurrence of a specific event. We assume that the population is decomposed into different groups. We consider an Aloha system in which mobiles interfere with each other through many local interactions (e.g., access points, throwboxes) where the collision can happen if more than one mobile transmit a packet in the same time slot. In particular, we consider that each node seeks to maximize some combination between its own performance and the performance of its group. We study the impact of cooperation in the context of multiple access control in many possible behaviors such as altruist behavior and fully non-cooperative behavior. In many problems, assumption about selfishness or rationality has been often questioned by economists and other sciences. Many research works shown that even in a simple game and controlled environment, individuals do not act selfishly. They are rather either altruistic or malicious. Several explanations have been considered for such behavior of players. Fairness reasons are argued by Fehr [94] to consider the joint utility model, while reciprocity among agents are considered in [52]. Cooperation among users, often referred to as altruism, are discussed in [66, 166, 168, 239]. Some of the models in [249] argues that the partial altruism mimics closely users' behavior often observed in practice.

In this chapter, we present a new model for evolutionary games which takes into account both the altruism and selfishness of agents. Firstly, we begin by defining this new concept, driving it in several ways and exploring its major characteristics. The major focus of this chapter is to study how the level of cooperation impacts the profile of population as well as the global performance of the system. Our theoretical results unveil some behaviors. More specifically, we show that when all users increase their level of cooperation, then the performance of the system is not necessary improved. In fact, for some scenarios, the performance of groups may lead to an improvement by adopting selfishness instead of altruism. This happened when the density of nodes is high. For low density, the degree of cooperation may indeed improve the performance of all groups.

The chapter is structured as follows. We first formalize in the next section the system model. We then present, in Sect. 2.2, the evolutionary game model that includes the cooperation aspect. In Sect. 2.3, we compute the expression of the ESS. We study in Sect. 2.4 the replicator dynamics in the classical and delayed forms. In Sect. 2.5, we proceed to some optimization issues through the analysis of the probability of success. Section 2.6 shows some numerical investigations on the equilibrium, the probability of success and the replicator dynamics. Section 2.7 concludes the chapter.

2.1 System Model

Consider an Aloha system composed of a large population of mobiles (or sensors) operating in a low traffic condition. Mobiles are randomly deployed over a plane and each mobile may interact with a mobile in its group or with a mobile in another group. The channel is assumed to be ideal for transmission and all errors are only due to collision. A mobile decides to transmit a packet or not to transmit to a receiver when they are within transmission range of each other. Interference occurs as in the Aloha protocol: if more than one neighbor of a receiver transmit a packet at the same time, then there is a collision.

We assume, in particular, that we can ignore cases of interaction in which more than two sensors or relays transmit simultaneously causing interference to each others. An example where we may expect this to hold is when sensors are deployed in a large area to monitor the presence of some events, e.g., in Delay Tolerant Networks (DTNs) where the network is assumed to be sparse and the relay density is low. Under this setting, communication opportunities arise whenever two nodes are within mutual communication range because of the mobility pattern.

The size of each group of mobiles G_i is denoted by α_i with $\sum_{i=1}^{N} \alpha_i = 1$. Let μ be the probability that a mobile i has its receiver $R(i)$ within its range. When a mobile i transmits to $R(i)$, all mobiles within a circle of radius R centred at $R(i)$ cause interference to the node i for its transmission to $R(i)$. This means that more than one transmission within a distance R of the receiver in the same slot induce a collision and the loss of mobile i packet at $R(i)$. Accordingly, each mobile has two actions: either to transmit T or to stay silent S. A mobile of group G_i may use a mixed strategy $\mathbf{p}_i = (p_i, 1 - p_i)$ where p_i is the probability to choose the action T. If a mobile transmits a packet, it incurs a transmission cost of Δ. The packet transmission is successful if the other users do not transmit (stay silent) in that given time slot. If a mobile transmits successfully a packet, it gets a reward of V. We suppose that the payoff V is greater than the cost of transmission, i.e., $\Delta < V$.

2.2 Utility Functions

As already mentioned, we study a new aspect of evolutionary games for multiple access games where each mobile cooperates with other mobiles of his group in order to improve the performance of his group. Let β be the degree of cooperation. The utility of a tagged mobile choosing action a within group G_i is a convex combination of the utility of his group and his own utility, namely

$$U^i_{\text{user}}(a, \mathbf{p}_{-i}) = \beta U^i_{\text{group}}(a, \mathbf{p}_{-i}) + (1 - \beta) U^i_{\text{self}}(a, \mathbf{p}_{-i}) \tag{2.1}$$

where U_{group} is the utility of the group to which the tagged player belongs and U_{self} is the individual utility of that player.

When the mobile plays T, resp. S, the utility of the group is given resp. by

$$U^i_{\text{group}}(T, p_i, p_{-i}) = \mu\Big[\gamma + (1-\gamma)\sum_{j=1}^{N} \alpha_j (1 - p_j)\Big]$$

$$U^i_{\text{group}}(S, p_i, p_{-i}) = \mu(1-\gamma)\alpha_i p_i$$

with γ being the probability that a mobile is alone in a given local interaction. Analogously, the selfish utility when the mobile chooses strategy T is

$$U^i_{\text{self}}(T, p_i, p_{-i}) = \mu\Big[(1-\Delta) - (1-\gamma)\sum_{j=1}^{N} \alpha_j p_j\Big]$$

while the selfish utility of user i when playing S is zero, namely

$$U^i_{\text{self}}(S, p_i, p_{-i}) = 0$$

Combining the above results, the utility of a mobile of class i using strategy T is given by

$$U^i_{\text{user}}(T, p_i, p_{-i}) = \mu\Big[1 - (1-\beta)\Delta - (1-\gamma)\sum_{j=1}^{N} \alpha_j p_j\Big],$$

while the utility of a mobile of class i when he plays S is

$$U^i_{\text{user}}(S, p_i, p_{-i}) = \mu\beta(1-\gamma)\alpha_i p_i$$

2.3 Computing the ESS

In evolutionary games, the most important concept of equilibrium is the ESS, which was introduced by [244] as a strategy that, if adopted by most members of a population, it is not invadable by mutant strategies in its suitably small neighbourhood. In our context, the definition of ESS is related to the robustness property inside each group. To be evolutionary stable, the strategy $\mathbf{p}^*$ must be resistant against mutations in each group. There are two possible interpretations of $\epsilon-$ deviations in this context:

1. A small deviation in the strategy by all members of a group. If the group G_i plays according to strategy $\mathbf{p}^*_i$, the $\epsilon-$ deviation, where $\epsilon \in (0, 1)$, consist in a shift to the group's strategy $\bar{\mathbf{p}}_i = \epsilon\mathbf{p}_i + (1-\epsilon)\mathbf{p}^*_i$;
2. The second is a deviation (possibly large) of a small number of individuals in a group G_i, that means that a fraction ϵ of individuals in G_i plays a different strategy $\mathbf{p}_i$.

After mutation, the average of a non-mutant will be given $U^i_{\text{user}}(p^*_i, \epsilon p_i + (1-\epsilon)p^*_i, p^*_{-i})$. Analogously, we can construct the average payoff of mutant $U^i_{\text{user}}(p_i, \epsilon p_i + (1-\epsilon)p^*_i, p^*_{-i})$. A strategy $\mathbf{p}^* = (p^*_1, p^*_2, \dots, p^*_N)$ is an ESS if $\forall\, i$ and $p_i \neq p^*_i$, there exists some $\epsilon_i \in (0, 1)$, which may depend on p_i, such that for $\epsilon \in (0, \epsilon_i)$

$$U^i_{\text{user}}(p^*_i, \epsilon p_i + (1-\epsilon)p^*_i, p^*_{-i}) > U^i_{\text{user}}(p_i, \epsilon p_i + (1-\epsilon)p^*_i, p^*_{-i}) \tag{2.2}$$

Equivalently, p^* is an ESS if and only if it meets best reply conditions:

- Nash equilibrium condition:

$$U^i_{\text{user}}(p_i^*, (p_i^*, p_{-i}^*)) > U^i_{\text{user}}(p_i, (p_i^*, p_{-i}^*))$$

- Stability condition:

$$\begin{aligned}\text{If } \mathbf{p_i} \neq \mathbf{p}_i^*, \text{ and } U^i_{\text{user}}(p_i^*, (p_i^*, p_{-i}^*)) = U^i_{\text{user}}(p_i, (p_i^*, p_{-i}^*))\\ \text{then } U^i_{\text{user}}(p_i, (p_i, p_{-i}^*)) < U^i_{\text{user}}(p_i^*, (p_i, p_{-i}^*))\end{aligned}$$

2.3.1 Characterization of the Equilibria

In this section, we provide the exact characterization of the equilibria induced by the game. We distinguish pure ESS equilibria and mixed ESS. Before studying the existence of ESS, we introduce some definitions needed in the sequel.

Definition 2.4

- A fully mixed strategy **p** is the strategy when all actions for each group have to receive a positive probability, i.e., $0 < p_i < 1\ \forall i$.
- A mixer (pure) group i is the group that uses a mixed (pure) strategy $0 < p_i < 1$ (resp. $p_i \in \{0, 1\}$).
- An equilibrium with mixed and non mixed strategies is an equilibrium when there is at least a pure group and a mixer group.

Proposition 1.1 characterizes the condition on the existence of a fully mixed ESS.

Proposition 2.6

1. *For* $\gamma < 1 - \frac{1-(1-\beta)\Delta}{(\beta+N)\min\alpha_i}$ *then there exists a unique fully mixed ESS* $p^* = (p_i^*)_{i=1,\ldots,N}$, *where*

$$p_i^* = \frac{1-(1-\beta)\Delta}{\alpha_i(\beta+N)(1-\gamma)}$$

2. *For* $1 - \frac{1-(1-\beta)\Delta}{(\beta+N)\min\alpha_i} \leq \gamma \leq 1 - \frac{1-(1-\beta)\Delta}{1+\beta\max\alpha_i}$, *then there exists a unique ESS with mixed and non mixed strategies.*
3. *For* $\gamma > 1 - \frac{1-(1-\beta)\Delta}{1+\beta\max\alpha_i}$, *then there exists a fully pure ESS where all groups play pure strategy* T.

Proof 1. From the definition of ESS in (2.2), we have $\forall i \in \{1, \ldots, N\}$,

$$(p_i - p_i^*)\Big[U^i_{\text{user}}(T, \epsilon p_i + (1-\epsilon)p_i^*, p_{-i}^*)$$
$$-U^i_{\text{user}}(S, \epsilon p_i + (1-\epsilon)p_i^*, p_{-i}^*)\Big] < 0$$

Thus

$$(p_i - p_i^*)\Big[1 - (1-\beta)\Delta - (1-\gamma)\sum_{j=1}^{N}\alpha_j p_j^* - (1-\gamma)\beta\alpha_i p_i^*\Big]$$
$$+\epsilon(p_i - p_i^*)^2\Big[-(1-\gamma)(1+\beta)\alpha_i\Big] < 0$$

The mixed Nash equilibrium is obtained when the first term of the previous inequality is strictly negative. While $(p_i - p_i^*)$ can be positive or negative for $p_i^* \notin \{0, 1\}$, the following equation holds

$$1 - (1-\beta)\Delta - (1-\gamma)\sum_{j=1}^{N}\alpha_j p_j^* - (1-\gamma)\beta\alpha_i p_i^* = 0$$

By summing this equation from 1 to N, we get

$$N(1 - (1-\beta)\Delta) - N(1-\gamma)\sum_{j=1}^{N}\alpha_j p_j^*$$
$$-(1-\gamma)\beta\sum_{i=1}^{N}\alpha_i p_i^* = 0$$

and so

$$\sum_{j=1}^{N}\alpha_j p_j^* = \frac{N(1-(1-\beta)\Delta)}{(1-\gamma)(\beta+N)}$$

Finally, we deduce

$$p_i^* = \frac{1-(1-\beta)\Delta}{\alpha_i(1-\gamma)(\beta+N)}$$

Hence p^* is fully mixed Nash equilibrium if $\gamma < 1 - \frac{1-(1-\beta)\Delta}{(\beta+N)\min\alpha_i}$. Furthermore, since $-(1-\gamma)(1+\beta)\alpha_i) < 0$, the stability condition is always satisfied which implies that p^* is an ESS. This complete the proof of (1).

2. Without loss of generality, we assume that the sizes of groups are ordered as follows: $\alpha_1 \leq \alpha_2 \leq \ldots \leq \alpha_N$. We assume that for a given value of γ we have

n_T groups playing strategy T and $N - n_T$ groups play mixed strategy. Hence the profile of population becomes $(T, \ldots, T, p_{n_T+1}, \ldots, p_N)$. For mixed group i ($i \in \{n_T + 1, \ldots, N\}$), we have the following relation

$$1 - (1-\beta)\Delta - (1-\gamma)\sum_{j=1}^{n_T} \alpha_j - (1-\gamma) \cdot$$

$$\sum_{j=n_T+1}^{N} \alpha_j p_j - (1-\gamma)\beta\alpha_i p_i = 0$$

Thus, a strategy of a mixer group is given by

$$p_i^* = \frac{1 - (1-\beta)\Delta - (1-\gamma)\alpha_T}{(1-\gamma)\alpha_i(\beta + N - n_T)}$$

where $\alpha_T = \sum_{j=1}^{n_T} \alpha_j$. For pure groups playing T, the following inequality holds: $\forall i \in \{1, \ldots, n_T\}$

$$1 - (1-\beta)\Delta - (1-\gamma)[\alpha_T + \alpha_i(\beta + N - n_T)] \geq 0$$

which completes the proof of (2).

3. Assume that $\forall i \in \{1, \ldots, N\}$, the group i transmit all the time ($p_i^* = 1$). The Nash equilibrium conditions become: $\forall i \in \{1, \ldots, N\}$,

$$\begin{aligned} 1 - (1-\beta)\Delta - (1-\gamma)(1 + \beta\alpha_i) &> 0 \\ \Rightarrow \gamma > 1 - \frac{1 - (1-\beta)\Delta}{1 + \beta\alpha_i} \end{aligned} \tag{2.3}$$

and the the proof is complete. □

The previous proposition claims that an increased network density results in more transmission which leads to more collision. It also states that, in order to avoid collision between mobiles belonging to the same group, the cooperation degree tends to decrease the probability of transmission within the same group.

Proposition 2.7 *At the Nash equilibrium, there is no group playing pure strategy S.*

Proof Let p^* be the ESS. By contradiction, suppose that there exists a group k playing the strategy S at ESS, i.e., $p_k^* = 0$. The Nash equilibrium condition for group k becomes

$$p_k\Big[1-(1-\beta)\Delta-(1-\gamma)\sum_{j=1,j\neq k}^{N}\alpha_j p_j^*\Big]<0$$

$$\Rightarrow 1-(1-\beta)\Delta-(1-\gamma)\sum_{j=1,j\neq k}^{N}\alpha_j p_j^*<0$$

If all groups use pure strategy S at ESS, the last condition becomes

$$1-(1-\beta)\Delta<0$$

But this contradicts our assumptions on β and Δ. Hence there exists at least a group l playing strategy p_l^* such that $p_l^* \in]0, 1]$. The Nash equilibrium condition is expressed as following:

$$(p_l-p_l^*)\Big[1-(1-\beta)\Delta-(1-\gamma)\cdot \sum_{j=1,j\neq k}^{N}\alpha_j p_j^*-(1-\gamma)\beta\alpha_l p_l^*\Big]<0$$

Thus

$$1-(1-\beta)\Delta-(1-\gamma)\sum_{j=1,j\neq k}^{N}\alpha_j p_j^*\geq(1-\gamma)\beta\alpha_l p_l^* \tag{2.4}$$

Combining conditions (2.3), (2.4) and $p_l^*>0$, we get $(1-\gamma)\beta\alpha_l<0$ which is a contradiction. This completes the proof. □

2.4 Replicator Dynamics

Evolutionary games study not only equilibrium behavior but also the dynamics of competition. We introduce the replicator dynamics which describe the evolution in groups of the various strategies. Replicator dynamic is one of the most studied dynamics in evolutionary game theory. In this dynamic, the frequency of a given strategy in the population grows at a rate equal to the difference between the expected utility of that strategy and the average utility of group i. Hence, successful strategies are more likely to spread over the population.

In this paper, we study the replicator dynamics for the case of two groups. The general case of N groups will be handled in a future work.

2.4.1 Replicator Dynamics Without Delay

The proportion of mobiles in the a group i programmed to play strategy T, denoted p_i, evolves according to the replicator dynamic equation given by:

$$\dot{p}_i(t) = p_i(t)[U^i_{\text{user}}(T, p_i(t), p_{-i}(t)) - \bar{U}^i_{\text{user}}(p_i(t), p_{-i}(t))] \quad (2.5)$$

where

$$\bar{U}^i_{\text{user}}(p_i(t), p_{-i}(t)) = p_i(t)U^i_{\text{user}}(T, p_i(t), p_{-i}(t)) + (1 - p_i(t))U^i_{\text{user}}(S, p_i(t), p_{-i}(t))$$

Thus

$$\dot{p}_i(t) = p_i(t)(1 - p_i(t))\Big[1 - (1-\beta)\Delta - (1-\gamma)\cdot \sum_{j=1}^{2} \alpha_j p_2(t) - (1-\gamma)\beta\alpha_i p_i(t)\Big] \quad (2.6)$$

By expressing Eq. (2.6) for $i = 1, 2$, we obtain a system of two non-linear ordinary differential equations (ODEs) in (2.6). There are several stationary points in which at least one group playing a pure strategy and a unique interior stationary point $p^* = (p_1^*, p_2^*)$ with $0 < p_i^* < 1$. The interior stationary point corresponds to the fully mixed ESS given by Proposition 3.9 and it is the only stationary point at which all mixed strategies coexist. Assuming that the state space is the unit square and that p^* exists, the dynamic properties of this equilibrium point are brought out in the next theorem.

Theorem 2.1 *The interior stationary point p^* is globally asymptotically stable in the replicator dynamics.*

Proof The proof is based on a linearization of the system of non linear ODEs around p^*. We introduce a small perturbation around p^* defined by $x_i(t) = p_i(t) - p_i^*$ for $i = 1, 2$. Keeping only linear terms in x_i, we obtain the following linearized replicator dynamics:

$$\dot{x}_i(t) \approx \rho_i\Big[1 - (1-\beta)\Delta - (1-\gamma)\sum_{j=1}^{2} \alpha_j(x_j(t) + p_j^*) - (1-\gamma)\beta\alpha_i(x_i(t) + p_i^*)\Big] \quad (2.7)$$

with $\rho_i = p_i^*(1 - p_i^*)$ and p^* is the interior stationary point of the ODE system. Equation (2.7) becomes

$$\dot{x}_i(t) \approx -(1-\gamma)\rho_i\Big[\sum_{j=1}^{2}\alpha_j x_j(t) + \beta\alpha_i x_i(t)\Big] \tag{2.8}$$

This linearized system is of the form $\dot{X}(t) = AX(t)$ where

$$A = -(1-\gamma)\begin{pmatrix}\rho_1 & 0\\ 0 & \rho_2\end{pmatrix}\begin{pmatrix}\alpha(1+\beta) & 1-\alpha\\ \alpha & (1-\alpha)(1+\beta)\end{pmatrix}$$

We note that the previous system is asymptotically stable if the eigenvalues of the matrix A has negative real parts. In order to investigate the eigenvalues of the matrix A, we express the following characteristic polynomial of A:

$$\chi_A = det(\lambda I_2 - A) = \lambda^2 - tr(A)\lambda + det(A)$$

Hence, the determinant and the trace of the matrix B are given resp. by

$$\begin{aligned} det(A) &= (1-\gamma)^2\rho_1\rho_2\alpha(1-\alpha)\Big[(1+\beta)^2 - 1\Big]\\ &= (1-\gamma)^2\rho_1\rho_2\alpha(1-\alpha)\beta(\beta+2)\end{aligned}$$

and

$$tr(A) = -(1-\gamma)(1+\beta)\Big(\rho_1\alpha + \rho_2(1-\alpha)\Big)$$

The discriminant of this polynomial is: $D = tr(A)^2 - 4\cdot det(A)$. Let λ_1 and λ_2 be two eigenvalues of A. Thus $\lambda_1+\lambda_2 = tr(A)$ and $\lambda_1\lambda_2 = det(A)$. Since $det(A) \geq 0$ and $tra(A) \leq 0$, it easy to check that the real parts of λ_1 and λ_2 are negative. Hence, the interior fixed point p^* is asymptotically stable in the replicator dynamics. □

2.4.2 Replicator Dynamics with Delay

In the classical replicator dynamics, the fitness of strategy a at time t has an instantaneous impact on the rate of growth of the population size that uses it. A more realistic alternative model for replicator dynamic would be to introduce some delay: a mobile belonging to group i perceives the fitness about his group utility after a given delay τ. Hence, the group utility acquired at time t will impact the rate of growth τ time later. Under this assumption, the replicator dynamics equation for the group i is given by:

$$\dot{p}_i(t) = p_i(t)(1 - p_i(t))\Big[1 - (1-\beta)\Delta - (1-\gamma)\cdot \\ \sum_{j=1}^{N} \alpha_j(\beta p_j(t-\tau) + (1-\beta)p_j(t)) - (1-\gamma)\beta\alpha_i p_i(t-\tau)\Big] \tag{2.9}$$

Similarly to the non-delayed case, we proceed to the linearization of the replicator dynamics equations by introducing a small perturbation around the interior equilibrium p_i^* defined by $x_i(t) = p_i(t) - p_i^*$. We get the following ODEs system:

$$\dot{x}_i(t) \approx -(1-\gamma)\rho_i\Big[\sum_{j=1}^{N} \alpha_j\Big(\beta x_j(t-\tau) + (1-\beta)x_j(t)\Big) \\ + \beta\alpha_i x_i(t-\tau)\Big] \tag{2.10}$$

The Laplace transform of the system (2.10) is given by:

$$[\lambda + (1-\gamma)\rho_i\alpha_i\beta e^{-\tau\lambda}]X_i + (1-\gamma)\rho_i(1-\beta+\beta e^{-\tau\lambda}) \\ \cdot \sum_{j=1}^{N} \alpha_j X_j = 0$$

For the case of two groups, the characteristic equation of the ODEs system is given by:

$$\lambda^2 + \lambda(1-\gamma)(1-\beta+2\beta e^{-2\tau\lambda})(\alpha\rho_1 + (1-\alpha)\rho_2) \\ + (1-\gamma)^2\rho_1\rho_2\alpha(1-\alpha)\Big[2\beta(1-\beta)e^{-\tau\lambda} + 3\beta^2 e^{-2\tau\lambda}\Big] = 0 \tag{2.11}$$

The zero solution of the linearized system above is asymptotically stable if and only if all solutions of the corresponding characteristic equation (2.11) have negative real parts. The form of this equation was studied in [127]. The mixed intermediate ESS is an asymptotically stable state in the time-delayed replicator dynamics if and only if $\tau < \tau_0 = \min(\frac{\pi}{2|\lambda_+|}, \frac{\pi}{2|\lambda_-|})$, with λ_+ and λ_- the roots of the non-delayed characteristic equation ($\tau = 0$). Remember that, according to the proof of Theorem 2.1, the eigenvalues of the differential system have negatives real parts.

2.5 Optimization Issues

According to the structure of the ESS, we try to evaluate the performance of the global system in order to derive the optimal degree of cooperation β. The performance of the system can be presented by the measure of the probability of success in a given local interaction for a mobile randomly selected from all mobiles of all groups. This probability of success is expressed as follows

$$P_{succ} = \gamma \sum_{j=i}^{N} \alpha_i p_i^* + (1-\gamma) \sum_{i=1}^{N} \sum_{j=1}^{N} \alpha_i \alpha_j p_i^* (1 - p_j^*)$$
$$+ p_j^* (1 - p_i^*)$$

Let us now study the expression of the probability of success depending on the structure of the game model considered.

2.5.1 *Fully Mixed ESS*

In the fully mixed ESS, the equilibrium is given $\forall i \in \{1, \ldots, N\}$ by

$$p_i^* = \frac{1 - (1-\beta)\Delta}{\alpha_i (\beta + N)(1 - \gamma)}$$

This gives a probability of success

$$P_{succ} = \frac{N(1 - (1-\beta)\Delta)}{(1-\gamma)(\beta+N)^2} \cdot$$
$$\Big[2(\beta + N(1-\beta)\Delta) - \gamma(\beta + N)\Big]$$

Having this expression, we calculate the level of cooperation β that maximizes the P_{succ}. We find that P_{succ} is maximized for

$$\beta^* = \frac{(4\Delta - \gamma - 2)N}{4N\Delta + \gamma - 2}$$

2.5.2 *Pure-Mixed ESS*

We note that in this structure, there are n_T groups using pure strategy T at the equilibrium and the other $N - n_T$ groups using mixed strategies. Then, the probability of success is expressed by

$$P_{succ} = (\alpha_T + \sum_{j=n_T+1}^{N} \alpha_i p_i^*)$$
$$\cdot \Big(2 - \gamma - 2(1-\gamma)(\alpha_T + \sum_{j=n_T+1}^{N} \alpha_i p_i^*)\Big)$$

with

$$p_i^* = \frac{1-(1-\beta)\Delta-(1-\gamma)\alpha_T}{(1-\gamma)\alpha_i(\beta+N-n_T)}$$

Finally

$$\begin{aligned}P_{succ} &= \frac{\beta\alpha_T(1-\gamma)+(N-n_T)(1-(1-\beta)\Delta)}{(\beta+N-n_T)^2} \\ &\cdot\Big[\beta(1-2\alpha_T)+N-n_T \\ &+\frac{2(N-n_T)(1-\beta)\Delta+\beta-(N-n_T)}{(1-\gamma)}\Big]\end{aligned}$$

We notice here, that $P_{succ}(\beta)$ depends on both n_T and α_T. These two variables are step functions of β. The optimal value β^* which maximizes P_{succ} for each value of γ will be computed through an iterative algorithm.

2.5.3 *Fully Pure ESS*

When all groups play pure strategy T at the equilibrium, i.e. $p_i^* = 1\,\forall i \in \{1, \ldots, N\}$

$$\begin{aligned}P_{succ} &= \sum_{i=1}^{N}\alpha_i p_i^*\Big(2-\gamma-2(1-\gamma)\sum_{i=1}^{N}\alpha_i p_i^*\Big) \\ &= \gamma\end{aligned}$$

2.6 Numerical Results

2.6.1 *Impact of the Transmission Cost on the Equilibrium*

We first investigate the case where two groups compete to access to the medium with $\alpha_1 = \alpha = 0.2$ and $\alpha_2 = 0.8$. In a sparse environment (corresponding to a high value of γ), an anonymous mobile of group G_i is more likely to be alone when transmitting to a destination. This suggests that he will play the strategy T all the time ($p^* = 1$). However, in a dense environment (low values of γ), he is more likely to be in competition with another mobile while transmitting to the destination. In this situation, the strategy played by the mobile i will depend on the cost of transmission Δ. In Fig. 2.1, we consider a low transmission cost ($\Delta = 0.2$). We found that the mobile gives less interest to the effect of collision as the cost of transmitting is very low. In fact, loosing a packet does not affect the mobile's utility compared to what he would

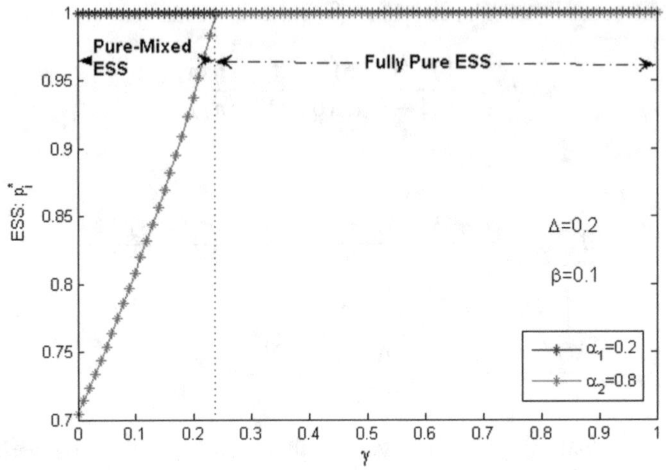

Fig. 2.1 Evolution of the ESS as a function of γ for $\Delta = 0.2$ and $\beta = 0.1$

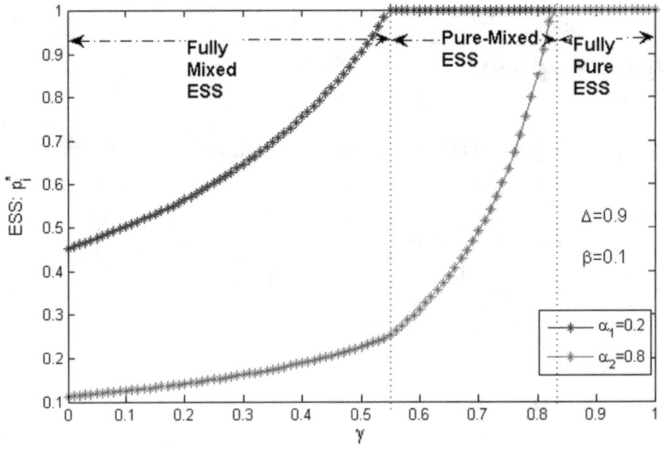

Fig. 2.2 Evolution of ESS as a function of γ for $\Delta = 0.9$ and $\beta = 0.1$

earn if the transmission is successful. This fact justifies the aggressive behavior of the mobile. However, when the transmission cost Δ is high, the equilibrium structure differs. In Fig. 2.2, we consider a higher cost ($\Delta = 0.9$). In this case, the mobile becomes more cautious and take into account the effect of collision since it degrades his utility. Thereby, he lowers his level of transmission.

2.6.2 *Impact of the Cooperation on the Equilibrium*

In Fig. 2.3, we keep a high level of transmission cost ($\Delta = 0.9$) and we change the behaviour of the mobiles. We change the degree of cooperation to pass from a nearly egoistic behavior with $\beta = 0.1$ to a nearly altruistic behaviour with $\beta = 0.9$. We notice that by increasing the degree of cooperation, users have more incentive to use strategy T. This suggests that increasing the degree of cooperation among users induces a coordination pattern in which users have more incentive to use strategy T.

2.6.3 *Impact of the Cooperation and the Transmission Cost on the Probability of Success*

In this section, we investigate the evolution of the P_{succ} according to γ. Intuitively, we can expect that when the mobiles are fully cooperative inside groups, this leads to a better system performance. However, we will show that we obtain a different result. We consider the same system as previously: two interacting groups, the smaller with proportion $\alpha = 0.2$ and the bigger with proportion $1 - \alpha = 0.8$.

We start with low cost of transmission ($\Delta = 0.2$). In this situation, as shown in Fig. 2.4, the more the mobiles cooperate the more the probability of success increases. We found that the full altruistic behaviour is the unique optimal solution up to a value of $\gamma \simeq 0.45$. Beyond this value of γ, $P_{succ} = \gamma$ and becomes, thus, independent of β. Hence, all levels of cooperation give the same performance of the system.

However, when the cost of transmission becomes high ($\Delta = 0.9$), we notice, through Fig. 2.5, that the P_{succ} takes different values according to the level of

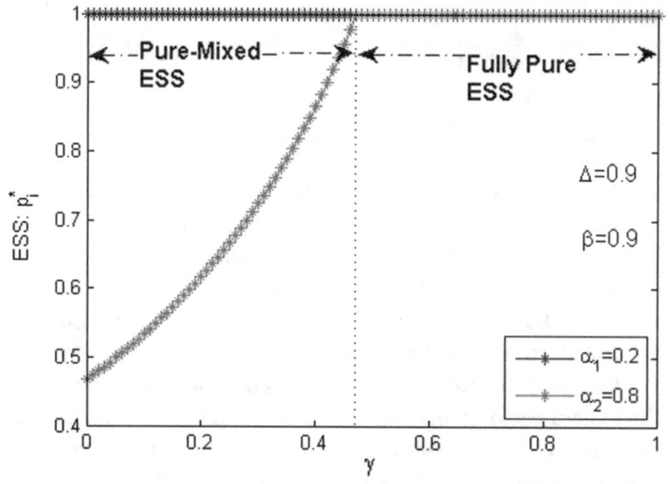

Fig. 2.3 Evolution of ESS as a function of γ for $\Delta = 0.9$ and $\beta = 0.9$

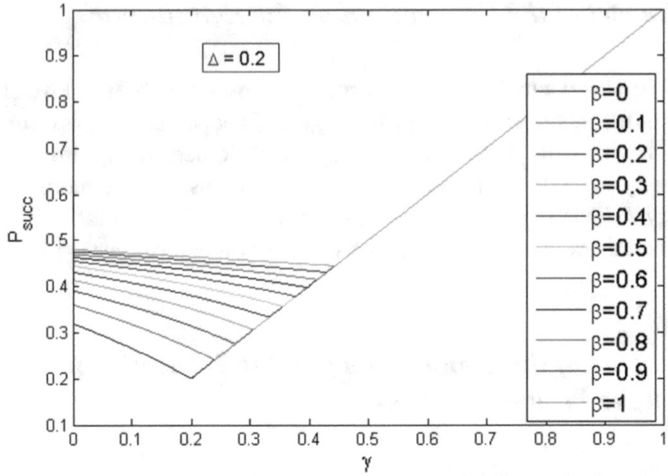

Fig. 2.4 Variation of P_{succ} with different levels of β for $\Delta = 0.2$

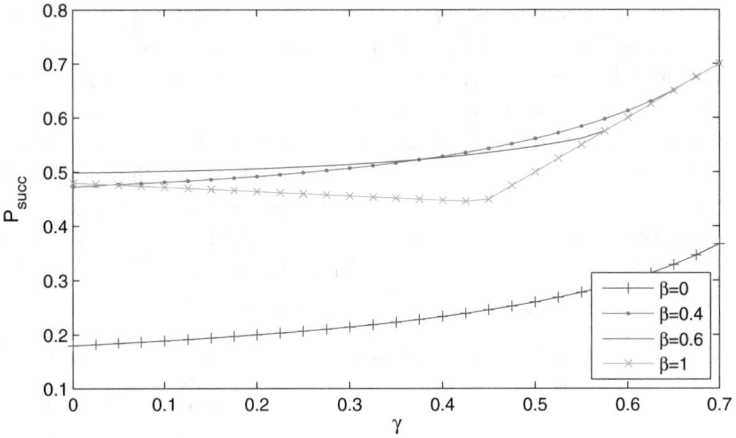

Fig. 2.5 Variation of P_{succ} with different levels of β for $\Delta = 0.9$

cooperation β and we remark that the fully altruistic behavior is no more the optimal solution. In fact, for low values of γ, the level of cooperation that optimizes the performance of the system is unique. The level of cooperation β_{opt} is a decreasing function of γ, which confirms the analytical result. The uniqueness of the level of β_{opt} remains until a value of $\gamma \simeq 0.68$ beyond which several levels of cooperation give the same system performance. Hence, we deduce a counter-intuitive result. We would expect that the fully altruistic behavior is always the best decision that should be adopted to maximize the system performance. However, we found that the mobiles have to be, often, less cooperative. In Fig. 2.6, we represent the margin between the performance of the system when adopting a fully altruistic behavior and this performance when behaving somewhat selfishly but optimally.

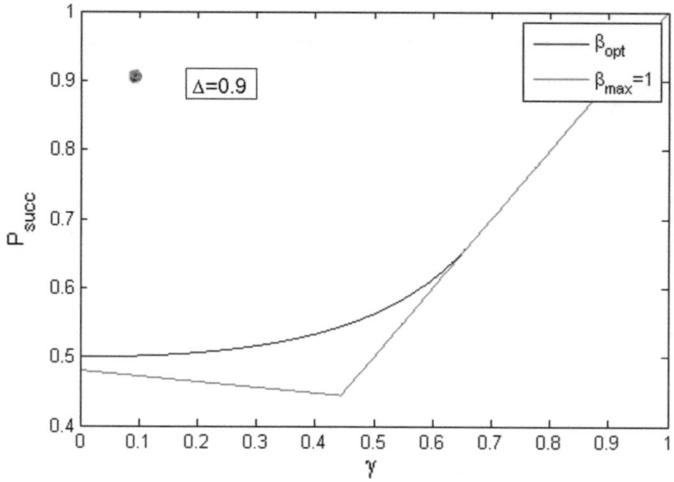

Fig. 2.6 Comparison of P_{succ} with maximal and optimal β for $\Delta = 0.9$

2.6.4 *Impact of the Delay on the Stability of the Replicator Dynamics*

The presence of delay in the replicator dynamic equations does not influence its convergence to the ESS. However, it has an impact on its stability. We investigate this fact through the following numerical example. We consider $N = 2$, $\alpha = 0.4$, $\Delta = 0.7$, $\gamma = 0.2$ and $\beta = 0.75$. This example corresponds to a fully mixed ESS (see Proposition 3.9). In Fig. 2.7, we observe that the replicator dynamics converge

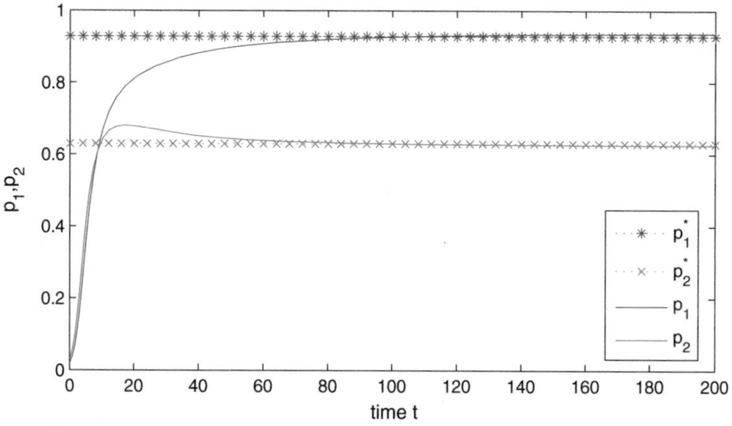

Fig. 2.7 Stability of the replicator dynamics for $\tau = 0$

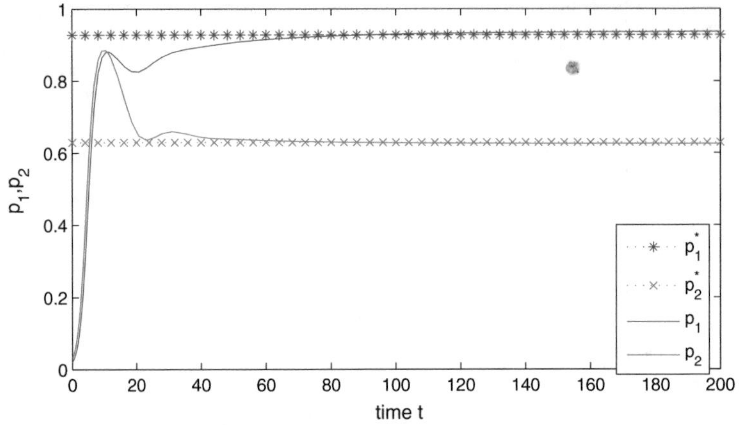

Fig. 2.8 Stability of the replicator dynamics for $\tau = 4$

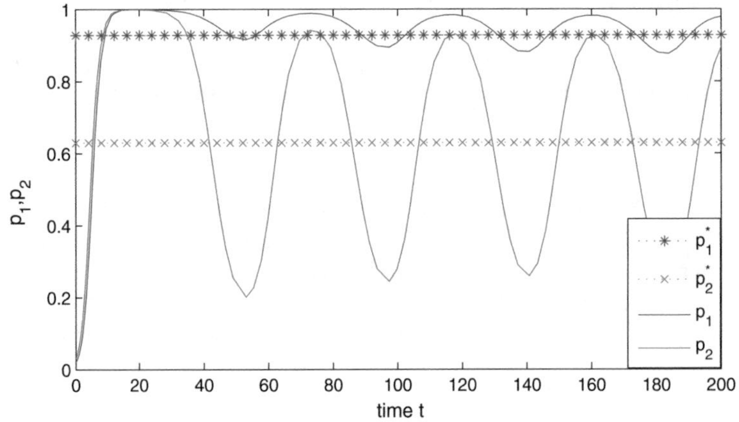

Fig. 2.9 Instability of the replicator dynamics for $\tau = 12$

to the ESS and remain stable, which confirm the Theorem 2.1. However, for $\tau = 4$ in Fig. 2.8, we obtain the stability but the convergence is much slower. The boundary of stability of the replicator dynamics is $\tau_0 \approx 7.5$. In Fig. 2.9, this boundary increases and we observe, that the replicator dynamics oscillate and become no longer stable.

2.7 Conclusion

In this paper, we have presented a new model of Medium Access Control problem through evolutionary game theory which takes in to account pairwise interactions. Our contribution was to include and investigate the aspect of cooperation between agents of the same group. We have studied the equilibrium of the system through the notion of Evolutionary Stable Strategies and study the effect of transmission cost and cooperation level. We have found that the mobiles tend to transmit less when the energy cost is high, whereas they may profit by the cooperation aspect to rise their transmission levels. Thereafter, we have evaluated the performance of the system in terms of the probability of success. We have studied the stability of replicator dynamics in the classical and delayed form. In a future work, we plan to study the general model with specific distributions.

Chapter 3
Evolutionary Games in Interacting Communities

Nesrine Ben Khalifa, Rachid El-Azouzi and Yezekael Hayel

3.1 Introduction

The major question posed in the EGT literature is related to the stability of a steady state which leads to a refinement of the Nash equilibrium. Much of work on evolution has studied the relationship between the steady state of the replicator dynamics and the ESS concept. Taylor and Jonker in [254], established conditions under which one may infer the existence of a stable state under the replicator dynamics given an evolutionarily stable strategy. However, this can fail to be true for multi-populations or more than two strategies [225, 227]. In the last two decades, there have been several attempts to relax the assumption derived by Taylor and Jonker in [254] in order to explore games in which agents only interact with their neighbours on a lattice [83, 131, 196] or on a random graph [198–200, 228, 252]. These modification on the replicators dynamics lead to lose the connection between the stable equilibrium of the replicator and ESS. Indeed, under some payoffs, stable states have no corresponding analogue neither in the replicator dynamics nor in the analysis of ESS. In this paper we aim to connect the analysis of the stability in a static concept and the steady state of the replicator dynamics. Instead of considering well mixed population, we explore games in which the population is composed of several communities or groups. Each group has its own set of strategies, payoff matrix and resulting outcomes. In addition, each group may interact with any other group with different interacting probabilities. Our work focuses on different type of individuals, such that any pairwise interaction does not lead to the same fitness, depending on the type of individuals that are competing, and not only the strategy used. We study the existence of ESSs with different levels of stability defined as follows:

N. B. Khalifa · R. El-Azouzi (✉) · Y. Hayel
Computer Science Laboratory (LIA), University of Avignon Computer Science Laboratory (LIA), Avignon, France
e-mail: rachid.elazouzi@univ-avignon.fr

E. Altman et al. (eds.), *Multilevel Strategic Interaction Game Models for Complex Networks*, https://doi.org/10.1007/978-3-030-24455-2_3

- **Strong ESS**: The stability condition guarantees that no alternative strategy can invade the population. This condition states that all groups or communities have an incentive to remain at the ESS when a rare alternative strategy is used by mutants in all groups that form the population.
- **Weak ESS**: The stability condition guarantees the stability against a fraction of mutants in a single group.
- **Intermediate ESS**: This equilibrium focuses on the global fitness of the whole population instead of a single group. It guarantees that all the population cannot earn a higher total payoff when deviating from the ESS.

We show that any mixed Nash equilibrium is not a strong ESS wherein all communities using this equilibrium cannot be invaded by a small group from all communities with a mutant strategy. But under some assumptions on the payoff and interaction probabilities, this mixed equilibrium is an ESS when we consider the global fitness of all the population instead of the fitness of each community. Furthermore, we analyze one of the most studied examples in evolutionary games, that of the Hawk and the Dove, a model for determining the degree of aggressiveness in the population. Our new model allows to study the evolution of aggressiveness within different species of animals that are interacting in a non-uniform manner.

For the evolutionary dynamics, we introduce the replicator dynamics in structured populations and under non-uniform interactions in communities. We study the relationship between the steady state of the replicator dynamics and the ESSs with different levels of stability. In particular we show that the mixed intermediate ESS is asymptotically stable in the replicator dynamics. In contrast, the condition of weak stability does not ensure the stability condition of the replicator dynamics.

The majority of works in evolutionary dynamics has studied the replicator dynamics without taking into account the delay effect. Many examples in biology and economics show that the impact of actions is not immediate and their effects only appear at some point in the future. A more realistic model of the replicator dynamics would take into consideration some time delay. The authors in [295] studied the effect of symmetric time delays in the replicator dynamics in a population model with two strategies. They assumed that an individual's fitness at the current time is equal to the expected payoff value of the strategy used by this individual at some previous time. The authors proved that the unique interior steady point becomes unstable for sufficiently large values of the delay. A similar result was proved by the authors in [8] in their social-type model. In the Ref. [258], the authors studied the replicator dynamics with asymmetric delays across the strategies and showed that for large delays the instability occurs. On the other hand, the author in [142] examined the effect of heterogeneously distributed delays among the individuals, and he derived conditions under which the mixed ESS is asymptotically stable in the replicator dynamics for any delay distribution.

In this work, we study the effect of time delays on the stability of the replicator dynamics in a population composed of communities. Two types of time delays can be distinguished: strategic delay which is the delay associated with the strategies of players, and spatial delay which is the delay associated with the communities of the competing players. As in one population scenario, we show that for large strategic delays, the mixed intermediate ESS may become an unstable state for the replicator dynamics. In contrast, the replicator dynamics converge to the intermediate ESS for any value of the spatial delay. Spatial delays come from the latency induced by the individuals type when they interact. In fact, we can assume that in some situations, the delay of a pairwise interaction between individuals from the same community can be lower than when individuals from different communities are interacting. For example, in a social network, individuals from the same community will share faster same content/informations as there is some kind of confidence between them. Whereas, a content coming from an individual from another community may yield to a careful behavior and then increases the outcome delay of the interaction.

The chapter is structured as follows. First in Sect. 3.2, we present new ESS definitions with different levels of stability in multi-community settings. Section 3.3 is devoted to the study of non-uniform interactions between two communities. In Sect. 3.4, we study the replicator dynamics, both in its classical form and with strategic and spatial delays. We apply our model to the Hawk-Dove game in Sect. 3.5. In Sect. 3.6, we study the game dynamics on random graphs. Finally, we conclude the paper in Sect. 3.7.

3.2 Evolutionarily Stable Strategies

We consider a large population of players or individuals divided into N communities and each community has its own set of strategies, payoff matrices, and interacting probabilities. Random matching occurs through pairwise interactions and may engage individuals from the same community or from different communities. Let $p = (p_1, .., p_N)$ where $p_i = (p_{i1}, .., p_{iN})$ is the vector describing the interaction probabilities of community i with other communities. Here p_{ij} denotes the probability that an individual in community i involved in an interaction, interacts with an individual in community j and $\sum_j p_{ij} = 1$ (Fig. 3.1). We assume there are n_i pure strategies for each community i and a strategy of an individual is a probability distribution over the pure strategies. We denote by $A^{ij} = (a^{ij}_{kl})_{k=1..n_i, l=1..n_j}$ the payoff matrix. If a player of community i using pure strategy k interacts with a player of community j using pure strategy l, its payoff is a^{ij}_{kl}. Let $\mathbf{s} = (s_1, .., s_N)$ be the population profile where s_i is the column vector describing the distribution of pure strategies in community i (s_{ik} is the frequency of the pure strategy k in community i). We denote by $U_i(k, \mathbf{s}, p)$ the expected payoff of pure strategy k in community i, which depends on the frequency of strategies in community i and in the other communities. The payoff function U_i is given by:

$$U_i(k, \mathbf{s}, p) = \sum_{j=1}^{N} p_{ij} e_k A^{ij} s_j, \tag{3.1}$$

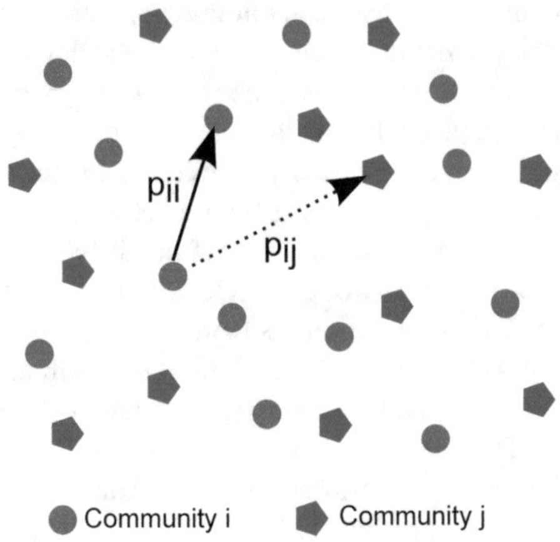

Fig. 3.1 Spatial interactions between two communities

where e_k is a row vector corresponding to the $k-$th element of the canonical basis of $\mathbb{R}^{n_i}$. The expected payoff of a player from community i using a mixed strategy z, when the profile of the population is $\mathbf{s}$, is given by:

$$\bar{U}_i(z, \mathbf{s}, p) = \sum_{k=1}^{n_i} z_k U_i(k, \mathbf{s}, p). \tag{3.2}$$

Our model covers the many situations as:

- The probabilities of interactions may depend on the size of the communities. An individual is more likely to meet and interact with an opponent from the larger community.
- The probabilities of interactions depend on spatial aspects, in which case an individual is more likely to interact with individuals in his neighborhood.

In the following, we present different ESS characterizations that differ in the stability level.

3.2.1 Strong ESS

A strong ESS is a strategy that, when adopted by the entire population, cannot be invaded by a sufficiently small group composed from all communities and using an alternative strategy. The incumbent players following the strong ESS, will get a strictly higher expected payoff when playing against the population composed of incumbents and mutants, than the mutants will get. The following definition can be stated:

Definition 3.5 A strategy $\mathbf{s}^*$ is a strong ESS, if for all $\mathbf{s} \neq \mathbf{s}^*$, there exists an $\epsilon(\mathbf{s}) > 0$ such that for all $i = 1, .., N$ and $\epsilon \leq \epsilon(\mathbf{s})$,

$$\bar{U}_i(s_i, \epsilon\mathbf{s} + (1-\epsilon)\mathbf{s}^*, p) < \bar{U}_i(s_i^*, \epsilon\mathbf{s} + (1-\epsilon)\mathbf{s}^*, p). \tag{3.3}$$

This strong ESS must in fact have a uniform invasion barrier [288] or threshold where any proportion of invaders using an alternative strategy is repelled. An alternative definition can be established as follows:

Definition 3.6 A strategy $\mathbf{s}^*$ is a strong ESS if it meets two conditions for all i and for all $\mathbf{s} \neq \mathbf{s}^*$,

- $\bar{U}_i(s_i, \mathbf{s}^*, p) \leq \bar{U}_i(s_i^*, \mathbf{s}^*, p)$, (3.4)
- if $\bar{U}_i(s_i, \mathbf{s}^*, p) = \bar{U}_i(s_i^*, \mathbf{s}^*, p)$, then $\bar{U}_i(s_i, \mathbf{s}, p) < \bar{U}_i(s_i^*, \mathbf{s}, p)$. (3.5)

Proposition 3.8 *The Definitions 3.5 and 3.6 are equivalent.*

Proof See Sect. 3.8 □

A strong ESS yields a higher expected payoff than any alternative strategy when played against itself (condition (3.4)). If there is a strategy that yields the same payoff as the strong ESS when played against the ESS, then this strategy will yield a strictly lower expected payoff when played against itself than the ESS, and cannot spread in the population (condition (3.5)).

3.2.2 *Weak ESS*

In this subsection, we assume that mutants arise in one community and we introduce an alternative ESS version with a weaker stability condition. A weak ESS is a strategy that, when adopted by the entire population, then each community resists invasion by a sufficiently small group of mutants using an alternative strategy in that community. The definition of the weak ESS is given by:

Definition 3.7 A strategy $\mathbf{s}^*$ is a weak ESS if for all $\mathbf{s} \neq \mathbf{s}^*$ and for all $i = 1, .., N$, there exists $\epsilon_i(\mathbf{s}) > 0$ such that for all $\epsilon_i \leq \epsilon_i(\mathbf{s})$,

$$\bar{U}_i(s_i, (\epsilon_i s_i + (1-\epsilon_i)s_i^*, \mathbf{s}_{-i}^*), p) < \bar{U}_i(s_i^*, (\epsilon_i s_i + (1-\epsilon_i)s_i^*, \mathbf{s}_{-i}^*), p), \tag{3.6}$$

where $(\epsilon_i s_i + (1-\epsilon_i)s_i^*, \mathbf{s}_{-i}^*)$ is the profile of the population where the ith community is composed of the fraction ϵ_i of mutants using an alternative strategy s_i and the fraction $1-\epsilon_i$ of incumbent players using s_i^*, and the remaining of the population follows the ESS $\mathbf{s}_{-i}^*$.

An equivalent definition of the weak ESS can be stated as follows:

Definition 3.8 A strategy $\mathbf{s}^*$ is a weak ESS if, for all i and for all $\mathbf{s} \neq \mathbf{s}^*$,

$$\circ\, \bar{U}_i(s_i, \mathbf{s}^*, p) \leq \bar{U}_i(s_i^*, \mathbf{s}^*, p), \tag{3.7}$$

$$\circ \text{ if } \bar{U}_i(s_i, \mathbf{s}^*, p) = \bar{U}_i(s_i^*, \mathbf{s}^*, p), \text{ then } \bar{U}_i(s_i, (s_i, \mathbf{s}_{-i}^*), p) < \bar{U}_i(s_i^*, (s_i, \mathbf{s}_{-i}^*), p), \tag{3.8}$$

where $\bar{U}_i(s_i, \mathbf{s}^*, p)$ is the expected payoff of a mutant in community i using s_i, and $\bar{U}_i(s_i^*, \mathbf{s}^*, p)$ is the expected payoff of an incumbent player in community i using s_i^* when the profile of the population is $\mathbf{s}^*$.

Proposition 3.9 *The Definitions 3.7 and 3.8 are equivalent.*

Proof See Sect. 3.8. □

This ESS definition is different from that of Cressman, referred to as Cressman ESS in the literature [75, 226], which considers invasion of the communities by a fraction of mutants from all communities. For a state to be a Cressman ESS, it is enough that one community resist invasion from a mutant strategy. In our definition, we consider invasion of a single community by a small **local** group of mutants. In Sect. 3.3, we introduce a particular example with two communities and we show that a weak ESS cannot be a Cressman ESS in this case.

3.2.3 Intermediate ESS

In the intermediate ESS version [253, 288], the focus is the total payoff of the whole population instead of the fitness of each community. An intermediate ESS is a strategy that, when adopted by the entire population, then for any small group using a mutant strategy, the total expected payoff of the incumbent strategies in all the communities is strictly higher than that of the mutant strategy. The formal definition of an intermediate ESS is given by:

Definition 3.9 A strategy $\mathbf{s}^*$ is an intermediate ESS if for all $\mathbf{s} \neq \mathbf{s}^*$, there exists an $\epsilon(\mathbf{s}) > 0$ such that for all $\epsilon \leq \epsilon(\mathbf{s})$,

$$\sum_{i=1}^{N} \bar{U}_i(s_i, \epsilon\mathbf{s} + (1-\epsilon)\mathbf{s}^*, p) < \sum_{i=1}^{N} \bar{U}_i(s_i^*, \epsilon\mathbf{s} + (1-\epsilon)\mathbf{s}^*, p). \tag{3.9}$$

Equivalently, we have the following definition:

Definition 3.10 A strategy $\mathbf{s}^*$ is an intermediate ESS if for all $\mathbf{s} \neq \mathbf{s}^*$,

$$\circ \sum_{i=1}^{N} \bar{U}_i(s_i, \mathbf{s}^*, p) \leq \sum_{i=1}^{N} \bar{U}_i(s_i^*, \mathbf{s}^*, p), \tag{3.10}$$

$$\circ \text{ if} \sum_{i=1}^{N} \bar{U}_i(s_i, \mathbf{s}^*, p) = \sum_{i=1}^{N} \bar{U}_i(s_i^*, \mathbf{s}^*, p), \text{ then} \sum_{i=1}^{N} \bar{U}_i(s_i, \mathbf{s}, p) < \sum_{i=1}^{N} \bar{U}_i(s_i^*, \mathbf{s}, p). \tag{3.11}$$

The condition (3.10) defines the best-reply requirement according to which a mutant strategy cannot yield a better total payoff than the ESS. When the comparison in this condition is an equality, i.e. in case of an alternative best-reply, the condition (3.11) guarantees that the population profile do not shift away from the ESS. It means that all the population have a positive incentive to remain at the ESS when there is a mutant strategy.

Proposition 3.10 *The Definitions 3.9 and 3.10 are equivalent.*

Proof The proof follows by carrying out exactly the same procedure as done in Proposition 3.9 □

3.2.4 Relationship Between the ESSs

This section is a discussion about the relationships between the different concepts of ESS introduced earlier. We explain how these ESS concepts are overlapped one into another.

First, a simple remark shows that a strong ESS is an intermediate ESS. Indeed, if we suppose there is a small fraction of mutants from all the communities using an alternative strategy, the strong ESS, when adopted by all the population, would resist this invasion because incumbent players would get a strictly higher expected payoff than mutants. The total expected payoff of the strong ESS in all the communities would also be strictly higher than that of the mutant strategy, and therefore the strong ESS is an intermediate ESS.

Second, a similar argument explains why an intermediate ESS is also a weak ESS. In fact, if we suppose there is a small fraction of mutants in a single community, an intermediate ESS would resist this invasion by definition and also a weak ESS; therefore an intermediate ESS is a weak ESS. In the next section, we will show through the study of two communities, that (i) an intermediate ESS is not always a strong ESS, and (ii) a weak ESS is not always an intermediate ESS. We then have the following relationships between the different concepts of ESS considering interacting communities:

$$\text{Strong ESS} \quad \Rightarrow \quad \text{Intermediate ESS} \quad \Rightarrow \quad \text{Weak ESS}$$

Note that all these definitions are obviously identical when there is a single community.

Table 3.1 Parameters of the model

Parameter	Value
L_1	$a_1 - b_1 - c_1 + d_1$
L_2	$a_2 - b_2 - c_2 + d_2$
L_{12}	$a_{12} - b_{12} - c_{12} + d_{12}$
L_{21}	$a_{21} - b_{21} - c_{21} + d_{21}$
K_1	$p_1(b_1 - d_1) + (1 - p_1)(b_{12} - d_{12})$
K_2	$p_2(b_2 - d_2) + (1 - p_2)(b_{21} - d_{21})$
Δ	$p_1 p_2 L_1 L_2 - (1 - p_1)(1 - p_2) L_{12} L_{21}$
Δ_1	$4 p_1 p_2 L_1 L_2 - \left((1 - p_1) L_{12} + (1 - p_2) L_{21}\right)^2$

3.3 Two-Community Two-Strategy Model

For the sake of clarity we focus only on the case where there are two communities that interact in a non-uniform manner. An individual from community $i = 1, 2$ involved in an interaction, may interact with an individual from the same community with probability p_i or with an individual from the other community with probability $1 - p_i$. In addition, we assume there are only two strategies $\{G_i, H_i\}$ in each community i. Let $\mathbf{s} = (s_1, s_2)$ be the population profile, with s_i denotes the frequency of strategy G_i in community i (so $1 - s_i$ is the frequency of strategy H_i). The pairwise interactions inside the communities are described by the matrices A and D:

$$A = \begin{matrix} & \begin{matrix} G_1 & H_1 \end{matrix} \\ \begin{matrix} G_1 \\ H_1 \end{matrix} & \begin{pmatrix} a_1 & b_1 \\ c_1 & d_1 \end{pmatrix} \end{matrix}, \quad D = \begin{matrix} & \begin{matrix} G_2 & H_2 \end{matrix} \\ \begin{matrix} G_2 \\ H_2 \end{matrix} & \begin{pmatrix} a_2 & b_2 \\ c_2 & d_2 \end{pmatrix} \end{matrix}.$$

The interactions between individuals from different communities are described by the following matrices:

$$B = \begin{matrix} & \begin{matrix} G_2 & H_2 \end{matrix} \\ \begin{matrix} G_1 \\ H_1 \end{matrix} & \begin{pmatrix} a_{12} & b_{12} \\ c_{12} & d_{12} \end{pmatrix} \end{matrix}, \quad C = \begin{matrix} & \begin{matrix} G_1 & H_1 \end{matrix} \\ \begin{matrix} G_2 \\ H_2 \end{matrix} & \begin{pmatrix} a_{21} & b_{21} \\ c_{21} & d_{21} \end{pmatrix} \end{matrix}.$$

Using Eqs. (3.1) and (3.2), we can derive the expected payoffs of players from either community. In addition, we define in Table 3.1 the parameters which will be used to analyze the model. The parameters L_1, L_2, L_{12}, and L_{21} depend on the payoffs. The parameters K_1, K_2, Δ, and Δ_1 depend on the payoffs and the interaction probabilities.

3.3.1 Dominant Strategies

The players repeatedly get involved in pairwise interactions and at each interaction, they have the opportunity to revise their strategies. A dominant strategy will eventually thrive and displace all dominated strategies. From the model above, the strategy G_1 dominates the strategy H_1 in community 1 if and only if:

$$p_1 s_1 L_1 + (1 - p_1) s_2 L_{12} + K_1 \geq 0, \ \forall (s_1, s_2) \in [0, 1]^2, \tag{3.12}$$

and the strategy G_2 dominates the strategy H_2 in community 2 if and only if:

$$p_2 s_2 L_2 + (1 - p_2) s_1 L_{21} + K_2 \geq 0, \ \forall (s_1, s_2) \in [0, 1]^2. \tag{3.13}$$

We have the same kind of inequality for determining if H_i dominates G_i, $i = 1, 2$. Therefore, we can establish the following theorem.

Theorem 3.2 • *The strategy G_1 dominates the strategy H_1 if and only if:*

$$0 \leq K_1, \ 0 \leq p_1 L_1 + K_1, \ 0 \leq (1 - p_1) L_{12} + K_1, \text{ and } 0 \leq p_1 L_1 + (1 - p_1) L_{12} + K_1.$$

• *The strategy H_1 dominates the strategy G_1 if and only if:*

$$K_1 \leq 0, \ p_1 L_1 + K_1 \leq 0, \ (1 - p_1) L_{12} + K_1 \leq 0, \text{ and } p_1 L_1 + (1 - p_1) L_{12} + K_1 \leq 0.$$

• *The strategy G_2 dominates the strategy H_2 if and only if:*

$$0 \leq K_2, \ 0 \leq p_2 L_2 + K_2, \ 0 \leq (1 - p_2) L_{21} + K_2 \text{ and } 0 \leq p_2 L_2 + (1 - p_2) L_{21} + K_2.$$

• *The strategy H_2 dominates the strategy G_2 if and only if:*

$$K_2 \leq 0, \ p_2 L_{21} + K_2 \leq 0, \ (1 - p_2) L_{21} + K_2 \leq 0 \text{ and } p_2 L_2 + (1 - p_2) L_{21} + K_2 \leq 0.$$

Proof See Sect. 3.8. □

In addition, we have the following result on pure ESSs.

Proposition 3.11 *If strategies $X_1 \in \{G_1, H_1\}$ and $X_2 \in \{G_2, H_2\}$ are dominant in community 1 and 2, respectively, then (X_1, X_2) is a strong ESS.*

Proof See Sect. 3.8. □

3.3.2 Mixed Nash Equilibrium and ESS

In this subsection, we characterize the existence of mixed ESSs under different stability conditions. We study the case of completely mixed ESSs and the case of partially mixed ESSs. At the completely mixed ESS, all strategies in both communities coexist. The following theorem summarizes results on the existence of completely mixed ESSs.

Theorem 3.3 *Let* $\mathbf{s}^* = (s_1^*, s_2^*)$ *with*

$$s_1^* = \frac{(1-p_1)L_{12}K_2 - p_2L_2K_1}{\Delta},$$

and

$$s_2^* = \frac{(1-p_2)L_{21}K_1 - p_1L_1K_2}{\Delta}.$$

We have the following results on $\mathbf{s}^*$*:*

- $\mathbf{s}^*$ *is a unique completely mixed Nash equilibrium, i.e.* $0 < s_1^* < 1$ *and* $0 < s_2^* < 1$, *if*
  - $0 < \Delta$, $0 < (1-p_1)L_{12}K_2 - p_2L_2K_1$, $(1-p_1)L_{12}K_2 - p_2L_2K_1 < \Delta$, $0 < (1-p_2)L_{21}K_1 - p_1L_1K_2$, *and* $(1-p_2)L_{21}K_1 - p_1L_1K_2 < \Delta$, *or*
  - $\Delta < 0$, $(1-p_1)L_{12}K_2 - p_2L_2K_1 < 0$, $\Delta < (1-p_1)L_{12}K_2 - p_2L_2K_1$, $0 < (1-p_2)L_{21}K_1 - p_1L_1K_2 < 0$, *and* $\Delta < (1-p_2)L_{21}K_1 - p_1L_1K_2$.
- $\mathbf{s}^*$ *cannot be a strong ESS.*
- $\mathbf{s}^*$ *is a weak ESS if* $L_1 < 0$ *and* $L_2 < 0$.
- $\mathbf{s}^*$ *is an intermediate ESS if* $L_1 < 0$, $L_2 < 0$ *and* $\Delta_1 > 0$.

Proof See Sect. 3.8. □

From Theorem 3.3, we observe that, if $\Delta_1 > 0$ cannot be satisfied whenever $L_1 < 0$ and $L_2 < 0$ then a weak ESS is not necessarily an intermediate ESS. In fact, for a weak ESS to be an intermediate ESS, it is required that the condition $\Delta_1 > 0$ be satisfied. This condition cannot be always satisfied. As an example, we consider the following payoffs:

$$A = \begin{pmatrix} -1.5 & 1 \\ 0 & 0.5 \end{pmatrix}, D = \begin{pmatrix} -1.5 & 1 \\ 0 & 0.5 \end{pmatrix}, B = \begin{pmatrix} -2 & 1 \\ 0 & 0.5 \end{pmatrix}, C = \begin{pmatrix} -0.5 & 1 \\ 0 & 0.5 \end{pmatrix}.$$

When $p_1 = 0.75$ and $p_2 = 0.4$, there exists a unique purely mixed equilibrium given by $s_1^* = 0.11$ and $s_2^* = 0.53$. The conditions $L_1 < 0$ and $L_2 < 0$ are satisfied, therefore $\mathbf{s}^* = (s_1^*, s_2^*)$ is a weak ESS. However, Δ_1 is strictly negative and consequently,

$\mathbf{s}^*$ is not an intermediate ESS. When $p_1 = 0.4$ and $p_2 = 0.6$, then for the same values of payoffs, we have $\mathbf{s}^* = (0.09, 0.38)$ and $\Delta_1 > 0$. Therefore, $\mathbf{s}^*$ is a weak and intermediate ESS. In addition, $\mathbf{s}^*$ cannot be a Cressman ESS because neither community would resist invasion from a small fraction of mutants composed of all the communities.

When $p_1 = p_2 = 1$, the two communities are completely independent, and $s_1^* = -\frac{b_1-d_1}{L_1}$, $s_2^* = -\frac{b_2-d_2}{L_2}$, and $\Delta_1 = 4L_1L_2$. We find the classical case of a single community: if $L_1 < 0$ and $L_2 < 0$ then $\mathbf{s}^* = (s_1^*, s_2^*)$ is an evolutionarily stable strategy. When $p_1 = p_2 = 0$, the evolutionary game is completely asymmetric and $s_1^* = -\frac{b_{21}-d_{21}}{L_{21}}$, $s_2^* = -\frac{b_{12}-d_{12}}{L_{12}}$ and $\Delta_1 = -(L_{12} + L_{21})^2$ which is strictly negative. Therefore, $\mathbf{s}^* = (s_1^*, s_2^*)$ is neither an intermediate nor a weak ESS. In [136, 235], the authors show that no mixed evolutionarily stable strategy can exist in asymmetric games. In [136], the authors introduced the notion of Nash-Pareto pairs in asymmetric games, which is an equilibrium characterized by the concept of Pareto optimality: it is not possible for players from both groups to simultaneously profit from a deviation from the equilibrium. The mixed equilibrium $\mathbf{s}^*$ is a Nash-Pareto pair if $L_{12}L_{21} < 0$.

Furthermore, we study the existence of partially mixed ESS, that is an ESS which is pure in one community and mixed in the other. In the next proposition, we derive the conditions of existence of partially mixed ESSs.

Proposition 3.12 • $\mathbf{s}^* = (1, s_2^*)$ *with* $s_2^* = -\frac{(1-p_2)L_{21}+K_2}{p_2L_2}$ *is a (partially mixed) strong ESS if* $p_1L_1 + (1-p_1)L_{12}s_2^* + K_1 > 0$ *and* $L_2 < 0$*;*

- $\mathbf{s}^* = (0, s_2^*)$ *with* $s_2^* = -\frac{K_2}{p_2L_2}$ *is a strong ESS if* $(1-p_1)s_2^*L_{12} + K_1 < 0$ *and* $L_2 < 0$*;*
- $\mathbf{s}^* = (s_1^*, 1)$ *with* $s_1^* = -\frac{(1-p_1)L_{12}+K_1}{p_1L_1}$ *is a strong ESS if* $p_2L_2 + (1-p_2)L_{21}s_1^* + K_2 > 0$ *and* $L_1 < 0$*;*
- $\mathbf{s}^* = (s_1^*, 0)$ *with* $s_1^* = -\frac{K_1}{p_1L_1}$ *is a strong ESS if* $(1-p_2)L_{21}s_1^* + K_2 < 0$ *and* $L_1 < 0$.

Proof See Sect. 3.8. □

In the next section, we study the stability of the ESSs in the replicator dynamics.

3.4 Replicator Dynamics

We introduce the replicator dynamics which describe the evolution of the various strategies in the communities. Replicator dynamics is one of the most studied dynamics in evolutionary game theory [136, 254, 288]. In this dynamics, the proportion of a given strategy in the population grows at a rate equal to the difference between the expected payoff of that strategy and the average payoff in the population [138, 288]. Thus, successful strategies increase in frequency in the population.

3.4.1 Replicator Dynamics Without Delay

The replicator dynamics equation writes, for $i = 1, 2$:

$$\begin{aligned} \dot{s}_i(t) &= s_i(t)\big[U_i(G_i, \mathbf{s}(t), p) - \bar{U}_i(s_i(t), \mathbf{s}(t), p)\big], \\ &= s_i(t)(1 - s_i(t))\big[U_i(G_i, \mathbf{s}(t), p) - U_i(H_i, \mathbf{s}(t), p)\big], \end{aligned}$$

with $\mathbf{s}(t) = (s_1(t), s_2(t))$, which yields the following pair of non-linear ordinary differential equations:

$$\begin{aligned} \dot{s}_1(t) &= s_1(t)(1 - s_1(t))\big[p_1 s_1(t) L_1 + (1 - p_1) s_2(t) L_{12} + K_1\big], \\ \dot{s}_2(t) &= s_2(t)(1 - s_2(t))\big[p_2 s_2(t) L_2 + (1 - p_2) s_1(t) L_{21} + K_2\big]. \end{aligned} \tag{3.14}$$

There are nine stationary points at which both $\dot{s}_1 = 0$ and $\dot{s}_2 = 0$, which are: $(0, 0)$, $(1, 1)$, $(0, 1)$, $(1, 0)$, $(0, -\frac{K_2}{p_2 L_2})$, $(-\frac{K_1}{p_1 L_1}, 0)$, $(1, -\frac{(1-p_2)L_{21}+K_2}{p_2 L_2})$, $(-\frac{(1-p_1)L_{12}+K_1}{p_1 L_1}, 1)$, and the interior point $\mathbf{s}^*$ defined in Theorem 3.3.

The dynamic property of $\mathbf{s}^*$ is brought out in the next theorem.

Theorem 3.4 *The interior stationary point* $\mathbf{s}^*$ *is asymptotically stable in the replicator dynamics if* $L_1 < 0$, $L_2 < 0$, *and* $\Delta > 0$.

Proof See Sect. 3.8. □

The next corollary about the asymptotic stability of the mixed intermediate ESS follows.

Corollary 3.1 *The mixed intermediate ESS is asymptotically stable in the replicator dynamics.*

Proof See Sect. 3.8. □

This result confirms that the ESS with the intermediate level of stability is asymptotically stable in the replicator dynamics. In contrast, for the weak ESS to be asymptotically stable, it is required that the condition $\Delta > 0$, which depends on the payoff matrices and the interacting probabilities, be satisfied. In the next theorem we study the asymptotic stability of partially mixed ESSs.

Theorem 3.5 *The partially mixed ESSs are asymptotically stable in the replicator dynamics.*

Proof See Sect. 3.8. □

All partially mixed ESSs, which are strong ESS, are asymptotically stable in the replicator dynamics.

3.4.2 Replicator Dynamics with Strategic Delay

In this section we investigate the impact of time delays of strategies on the dynamics. An action taken today will have some effect after some time [257]. We assume the strategies take a delay τ_{st}. The replicator dynamics for the first community is then given by:

$$\dot{s}_1(t) = s_1(t)(1 - s_1(t))\big[U_1(G_1, \mathbf{s}(t - \tau_{st}), p) - U_1(H_1, \mathbf{s}(t - \tau_{st}), p)\big].$$

Then we get:

$$\dot{s}_1(t) = s_1(t)(1 - s_1(t))\big[p_1 L_1 s_1(t - \tau_{st}) + (1 - p_1)L_{12} s_2(t - \tau_{st}) + K_1\big].$$

Doing the same with the second group, we get:

$$\dot{s}_2(t) = s_2(t)(1 - s_2(t))\big[p_2 L_2 s_2(t - \tau_{st}) + (1 - p_2)L_{21} s_1(t - \tau_{st}) + K_2\big].$$

We introduce a small perturbation around $\mathbf{s}^*$ defined by $x_1(t) = s_1(t) - s_1^*$ and $x_2(t) = s_2(t) - s_2^*$. We then make a linearization of the two above equations around the interior equilibrium point $\mathbf{s}^* = (s_1^*, s_2^*)$ and we study the linearized system. We get the following system:

$$\begin{aligned}\dot{x}_1(t) &= \gamma_1\big(p_1 L_1 x_1(t - \tau_{st}) + (1 - p_1)L_{12} x_2(t - \tau_{st})\big),\\ \dot{x}_2(t) &= \gamma_2\big(p_2 L_2 x_2(t - \tau_{st}) + (1 - p_2)L_{21} x_1(t - \tau_{st})\big),\end{aligned}$$

with $\gamma_1 = s_1^*(1 - s_1^*)$ and $\gamma_2 = s_2^*(1 - s_2^*)$. Taking the Laplace transform of the system above, we obtain the following characteristic equation:

$$\lambda^2 - \lambda[p_1\gamma_1 L_1 + p_2\gamma_2 L_2]e^{-\lambda\tau_{st}} + \gamma_1\gamma_2[p_1 p_2 L_1 L_2 - (1 - p_1)(1 - p_2)L_{12}L_{21}]e^{-2\lambda\tau_{st}} = 0. \quad (3.15)$$

The zero solution is asymptotically stable if all solutions of (3.15) have negative real parts [40]. The Eq. (3.15) is typical for a linear system of two equations of the form $\dot{x}(t) = Ax(t - \tau_{st})$ which was studied by the authors in [55, 126]. Based on their results, we establish the following theorem on the asymptotic stability of the intermediate ESS in presence of symmetric strategic delays.

Theorem 3.6 *The mixed intermediate ESS is asymptotically stable in the delayed replicator dynamics if* $\tau_{st} < \bar{\tau}_{st} = min(\frac{\pi}{2|\lambda_+|}, \frac{\pi}{2|\lambda_-|})$, *with* $\lambda_\pm = \frac{p_1\gamma_1 L_1 + p_2\gamma_2 L_2 \pm \sqrt{D}}{2}$, *and* $D = \big[p_1\gamma_1 L_1 + p_2\gamma_2 L_2\big]^2 - 4\gamma_1\gamma_2\big[p_1 p_2 L_1 L_2 - (1 - p_1)(1 - p_2)L_{12}L_{21}\big]$.

Proof See Sect. 3.8. □

Theorem 3.6 gives an upper bound on strategic delays for which the intermediate ESS remains asymptotically stable in the population. Beyond this delay bound, persistent oscillations around the ESS occur.

3.4.3 Replicator Dynamics with Spatial Delay

In this section we assume the delays are not associated with the strategy used by an individual, but rather with the opponent with which an individual interacts. We consider that inter-community interactions take a delay τ_{sp}. In this case, the replicator dynamics writes:

$$\dot{s}_1(t) = s_1(t)(1 - s_1(t))\big[U_1(G_1, (s_1(t), s_2(t - \tau_{sp})), p) - U_1(H_1, (s_1(t), s_2(t - \tau_{sp})), p)\big],$$
$$\dot{s}_2(t) = s_2(t)(1 - s_2(t))\big[U_2(G_2, (s_1(t - \tau_{sp}), s_2(t)), p) - U_2(H_2, (s_1(t - \tau_{sp}), s_2(t)), p)\big].$$

Following the same procedure as in the previous sections, we get the following characteristic equation:

$$\lambda^2 - (p_1\gamma_1 L_1 + p_2\gamma_2 L_2)\lambda + \gamma_1\gamma_2 p_1 p_2 L_1 L_2 - \gamma_1\gamma_2(1 - p_1)(1 - p_2)L_{12}L_{21}e^{-2\lambda\tau_{sp}} = 0.$$

Or equivalently:

$$\lambda^2 + \alpha\lambda + \beta + \delta e^{-\lambda\tau} = 0, \tag{3.16}$$

with $\tau = 2\tau_{sp}, \alpha = -(p_1\gamma_1 L_1 + p_2\gamma_2 L_2), \beta = \gamma_1\gamma_2 p_1 p_2 L_1 L_2, \delta = -\gamma_1\gamma_2(1 - p_1)$ $(1 - p_2)L_{12}L_{21}$. Now, we summarize the stability property of the mixed ESS for the delayed replicator dynamics in the following theorem which is based on the results of the authors in [99] related to the location of the roots of the characteristic equation (3.16).

Theorem 3.7 *The mixed intermediate ESS is asymptotically stable in the replicator dynamics with spatial delays for any* $\tau_{sp} \geq 0$.

Proof See Sect. 3.8. □

Spatial delays do not affect the stability of the mixed ESS. Indeed, for any value of the delay τ_{sp}, the frequency of strategies in the population converges to the mixed intermediate ESS after some possible damped oscillations.

3.4.4 Replicator Dynamics with Spatial-Strategic Delays

In this section, we study the stability of the replicator dynamics with both strategic and spatial delays. In particular, we aim to study whether the spatial delay has a stabilizing effect on the replicator dynamics with strategic delay. We define the delays as follows:

- τ_{st} is the strategic delay, that is the delay associated with the strategies;
- τ_{sp} is the spatial delay associated with the inter-community interactions.

The expected payoffs of strategies G_1 and H_1 in community 1 then write:

$$U_1(G_1,(s_1(t-\tau_{st}),s_2(t-\tau_{st}-\tau_{sp})),p) = p_1\big(s_1(t-\tau_{st})a_1 + (1-s_1(t-\tau_{st}))\times b_1\big) + (1-p_1)\big(s_2(t-\tau_{st}-\tau_{sp})a_{12} + (1-s_2(t-\tau_{st}-\tau_{sp}))b_{12}\big).$$

And

$$U_1(H_1,(s_1(t-\tau_{st}),s_2(t-\tau_{st}-\tau_{sp})),p) = p_1\big(s_1(t-\tau_{st})c_1 + (1-s_1(t-\tau_{st}))\times d_1\big) + (1-p_1)\big(s_2(t-\tau_{st}-\tau_{sp})c_{12} + (1-s_2(t-\tau_{st}-\tau_{sp}))d_{12}\big).$$

Hence,

$$\dot{s}_1(t) = s_1(t)(1-s_1(t))\big[p_1\alpha_1 s_1(t-\tau_{st}) - p_1\alpha_2 s_1(t-\tau_{st}) + (1-p_1)\alpha_3 s_2(t-\tau_{st} -\tau_{sp}) - (1-p_1)\alpha_4 s_2(t-\tau_{st}-\tau_{sp}) + K_1\big],$$

with $\alpha_1 = a_1 - b_1$, $\alpha_2 = c_1 - d_1$, $\alpha_3 = a_{12} - b_{12}$, $\alpha_4 = c_{12} - d_{12}$. Doing the same with the second community, we obtain:

$$\dot{s}_2(t) = s_2(t)(1-s_2(t))\big[p_2\beta_1 s_2(t-\tau_{st}) - p_2\beta_2 s_2(t-\tau_{st}) + (1-p_2)\beta_3 s_1(t-\tau_{st} -\tau_{sp}) - (1-p_2)\beta_4 s_1(t-\tau_{st}-\tau_{sp}) + K_2\big],$$

with $\beta_1 = a_2 - b_2$, $\beta_2 = c_2 - d_2$, $\beta_3 = a_{21} - b_{21}$, and $\beta_4 = c_{21} - d_{21}$. Following the same procedure in the previous sections, we get the following characteristic equation:

$$\lambda^2 - \lambda[p_1\gamma_1 L_1 + p_2\gamma_2 L_2]e^{-\tau_{st}\lambda} + p_1 p_2 \gamma_1\gamma_2 L_1 L_2 e^{-2\tau_{st}\lambda} - (1-p_1)(1-p_2)\gamma_1\times \gamma_2 L_{12} L_{21} e^{-2(\tau_{st}+\tau_{sp})\lambda} = 0. \quad (3.17)$$

When $\tau_{sp} = 0$, we find the characteristic equation (3.15) obtained when there is only a strategic delay. Equation (3.17) can be solved numerically.

3.5 Application to Hawk-Dove Game

In the classical Hawk-Dove game [21, 248], two individuals compete for a scarce resource. A player may use a Hawk strategy (H) or a Dove strategy (D). The strategy H stands for an aggressive behavior that fights for the resource while the strategy D represents a peaceful behavior which never fights. The matrix that gives the outcome for such competition is given as follows:

$$\begin{array}{cc} & \begin{array}{cc} H & D \end{array} \\ \begin{array}{c} H \\ D \end{array} & \begin{pmatrix} \frac{V-C}{2} & V \\ 0 & \frac{V}{2} \end{pmatrix}. \end{array}$$

with $C > 0$ and $V > 0$. V represents the value of the resource for which the players compete and C represents the cost incurred by a hawk when fighting for the resource against a hawk. The coefficients of the payoff matrix can be interpreted as follows: If two doves meet, they share equally the resource and each one obtains as a payoff $\frac{V}{2}$. If two hawks meet, they fight until one of them gets injured and the other takes the whole resource. When a hawk and a dove meet, the dove withdraws and the hawk takes the whole resource. If $C < V$, then the strategy H is dominant and the entire population will adopt the aggressive behavior. If $C > V$, there exists a mixed ESS given by $(\frac{V}{C}, 1 - \frac{V}{C})$, at which both behaviors coexist.

We apply our model to the Hawk-Dove game played in a population composed of two communities of hawks and doves with different levels of aggressiveness and which interact in a non-uniform fashion. Let:

- p be the probability that an individual (in community 1 or 2), involved in an interaction, competes with an individual from the same community;
- $1 - p$ be the probability that an individual competes with an inter-community opponent.

Furthermore, the interactions inside the communities 1 and 2 are described by the matrices A and D:

$$\begin{array}{c} \\ H_1 \\ D_1 \end{array} \begin{array}{c} \begin{array}{cc} H_1 & D_1 \end{array} \\ \begin{pmatrix} \frac{V-C}{2} & V \\ 0 & \frac{V}{2} \end{pmatrix} \end{array}, \quad \begin{array}{c} \\ H_2 \\ D_2 \end{array} \begin{array}{c} \begin{array}{cc} H_2 & D_2 \end{array} \\ \begin{pmatrix} \frac{V-C}{2} & V \\ 0 & \frac{V}{2} \end{pmatrix} \end{array}.$$

The inter-community interactions are described by the following matrices:

$$\begin{array}{c} \\ H_1 \\ D_1 \end{array} \begin{array}{c} \begin{array}{cc} H_2 & D_2 \end{array} \\ \begin{pmatrix} \frac{V-C_{SW}}{2} & V \\ 0 & \alpha V \end{pmatrix} \end{array}, \quad \begin{array}{c} \\ H_2 \\ D_2 \end{array} \begin{array}{c} \begin{array}{cc} H_1 & D_1 \end{array} \\ \begin{pmatrix} \frac{V-C_{WS}}{2} & V \\ 0 & (1-\alpha)V \end{pmatrix} \end{array}.$$

We introduce the parameters C_{SW}, C_{WS} and α into the payoff matrices to incorporate the disparity in the levels of aggressiveness of the two communities. These parameters can be defined as follows:

- C is the fighting cost incurred by a hawk when fighting against a hawk from the same community;
- C_{SW} is the fighting cost incurred by a hawk from the more aggressive community when fighting against a hawk from the other community;
- C_{WS} is the fighting cost incurred by a hawk from the less aggressive community when fighting against a hawk from the other community;
- α is the resource part that takes a dove from the more aggressive community when competing with a dove from the other community.

The parameters satisfy: $0 < C_{SW} < C < C_{WS}$, and $0.5 \leq \alpha \leq 1$. When two hawks from different communities compete for a resource, they do not incur the same cost as when fighting against an intra-community hawk. Also, when two doves from different communities meet, they share the resource unevenly. In particular, with these parameters, the first community is more aggressive than the second community: hawks in community 1 provoke more injuries than hawks in the other community, and

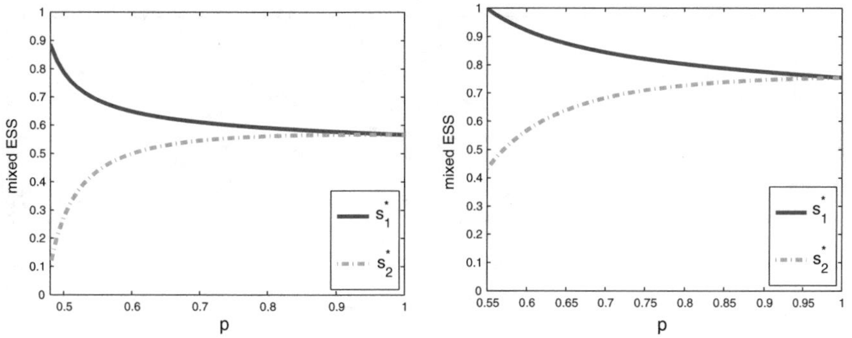

Fig. 3.2 Effect of the resource value on the intermediate ESS. $C = 5.3$, $C_{SW} = 3.5$, $C_{WS} = 6.1$, $\alpha = 0.7$. *Left* $V = 3$. *Right* $V = 4$

doves in community 1 take more resource than those in community 2. When $C_{SW} = C_{WS}$ and $\alpha = 0.5$, the two communities have the same degree of aggressiveness.

First, we aim to study the effect of the resource value on the equilibrium behavior of the two communities. We plot in Fig. 3.2 the mixed intermediate ESSs given by Theorem 3.3 as a function of p for different values of the resource V. We observe that the aggressive behavior in both communities is enhanced by increasing the resource value. Indeed, when the resource value is higher, individuals have a higher propensity to behave aggressively and take the resource. In addition, the range of p over which hawks and doves in both communities coexist as an intermediate (and a weak) ESS becomes smaller as the resource value increases. Beyond this range, the equilibrium exceeds the amount $[0, 1]^2$. For example, for $V = 3$, the range of p of co-existence of all strategies is $[0.48, 1]$. For $V = 4$, this range reduces to $[0.55, 1]$. In Fig. 3.3, we plot the mixed intermediate ESS as a function of p for different values of the parameter α. For $\alpha = 0.52$ (and for any $\alpha \leq 0.8$), the non aggressive behavior is enhanced in the first community while the second community becomes more aggressive as the probability of intra-community interaction increases. For the values of $\alpha > 0.8$, we observe the reverse. When $p = 1$, i.e. the two communities are completely independent, in each community the ESS is given by $(\frac{V}{C}, 1 - \frac{V}{C}) = (0.56, 0.44)$. In addition, for the values of $\alpha \leq 0.8$, community 1 is more aggressive than community 2, i.e., $s_1^* > s_2^*$. For the values of $\alpha > 0.8$, the community 2 becomes more aggressive ($s_2^* > s_1^*$).

3.6 Interacting Communities on Random Graphs

In this section, we study update rules that make the replicator dynamics converge to the ESS in a random graph.

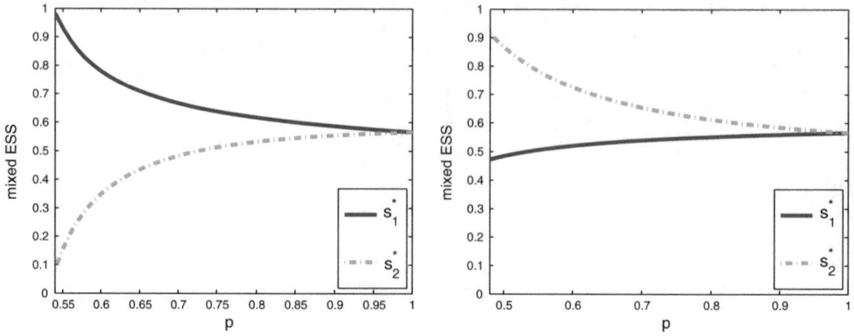

Fig. 3.3 Effect of α on the intermediate ESS. $V = 3$, $C = 5.3$, $C_{SW} = 3.5$, $C_{WS} = 6.1$. *Left* $\alpha = 0.52$. *Right* $\alpha = 1$

3.6.1 Best Response Update Rule

We consider a graph $\mathcal{G}$ composed of uniformly at random vertices that represent the players. Each player can interact, in a non-uniform fashion, with individuals in his spatial neighborhood only (Fig. 3.4). The payoff of a player is the sum of the payoffs obtained in pairwise interactions with the neighbors. At each time step t, a randomly chosen individual from group i revises his strategy. He chooses his current strategy $\hat{A}_i \in \{G_i, H_i\}$ according to:

$$\hat{A}_i(t) = \underset{A \in \{G_i, H_i\}}{\text{argmax}}\; U_i(A, \mathbf{s}_{loc}(t-\tau)),$$

with $\mathbf{s}_{loc}(t-\tau)$ is the local distribution of the strategies in the neighborhood of the selected player at time $t-\tau$, and $U_i(A, \mathbf{s}_{loc}(t-\tau))$ is the sum of payoffs obtained

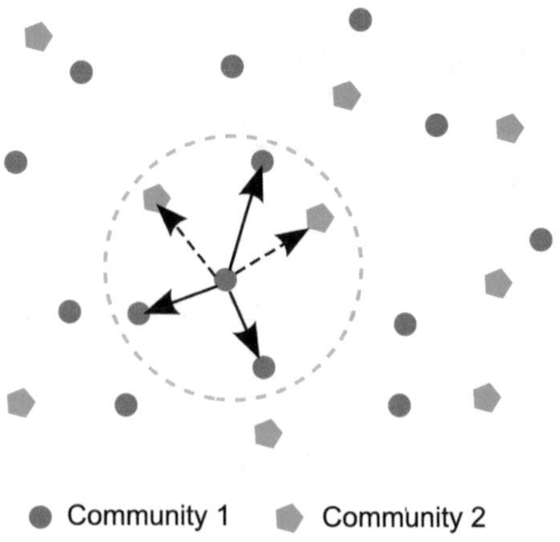

Fig. 3.4 Local spatial interactions on random graphs

by the player in his pairwise interactions with all neighboring players. An individual chooses a strategy that is a best response to the (delayed) profile in his neighborhood. We consider the cases of strategic delay, spatial delay, and strategic-spatial delay. We aim to observe the effect of this update rule and time delays on the convergence to the equilibrium with numerical examples.

3.6.2 Numerical Illustration

We consider a numerical example on a random graph with $N_1 = 1800$ individuals from group 1, and $N_2 = 1200$ individuals from group 2. Then, assuming the interactions depend on the relative sizes of the communities, the probabilities of intra-community interactions and inter-community interactions in the first (resp. second) community are given by $p_1 = 0.6$ and $1 - p_1 = 0.4$ (resp. $p_2 = 0.4$ and $1 - p_2 = 0.6$). We consider the cases of a complete graph, i.e. every individual can interact with any other individual in the graph, and incomplete graph, i.e. an individual may only interact with individuals in his local spatial neighborhood. We aim to study the convergence to the mixed intermediate ESS. When the graph is complete we observe a perfect convergence to the intermediate ESS (Fig. 3.5 *left*). When this is not the case, we observe in Fig. 3.5 *right* a shift between the intermediate ESS and the steady state. In addition, we investigate the effect of time delays on the convergence to the intermediate ESS. We observe that, using the best-response update rule in a complete graph, the population profile converges to the equilibrium in presence of a spatial delay $\tau_{sp} = 4.4$ time units (Fig. 3.6 *top-left*). When a strategic delay of a value $\tau_{st} = 11.2$ time units is introduced, we observe persistent oscillations around the equilibrium (Fig. 3.6 *top-right*). Also, when both types of the delay coexist, oscillations are persistent (Fig. 3.6 *bottom*).

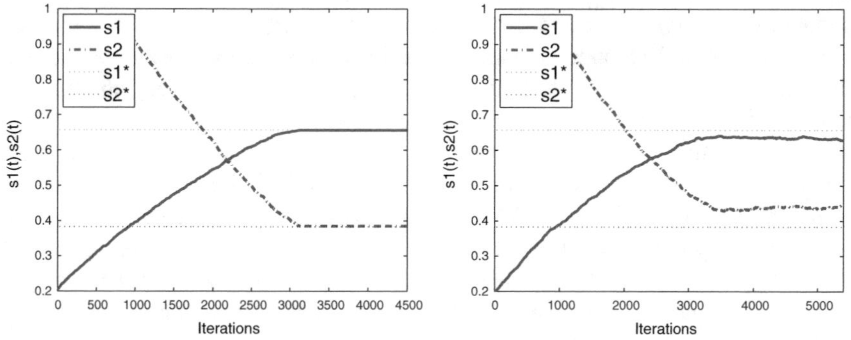

Fig. 3.5 The convergence to the intermediate ESS with the best-response update rule. *Left* Complete graph. *Right* Incomplete graph

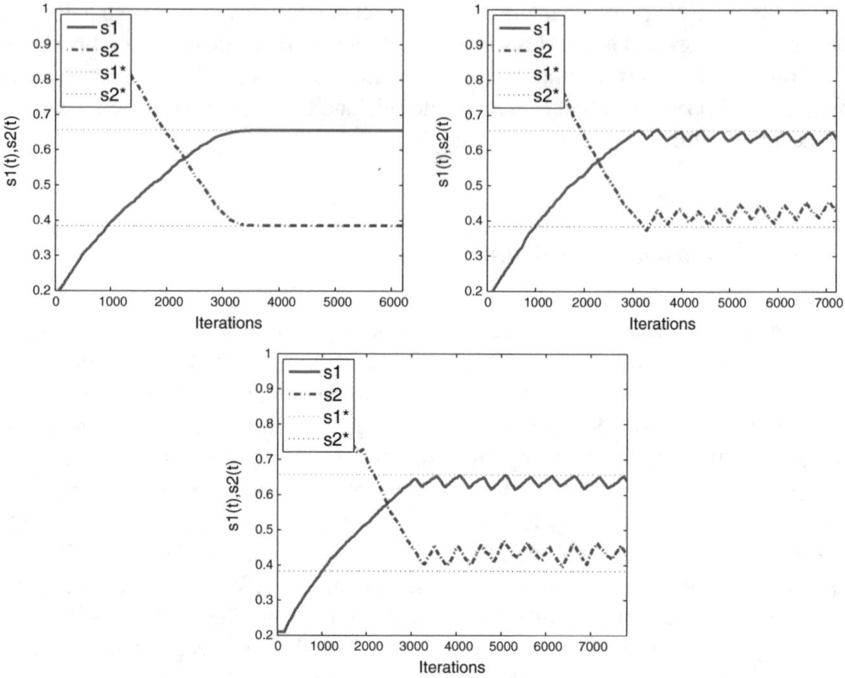

Fig. 3.6 Stability of equilibria in a complete graph. *Top Left*: Purely spatial delay, $\tau_{st} = 0, \tau_{sp} = 4.4$ time units. *Top Right*: Purely strategic delay, $\tau_{st} = 11.2, \tau_{sp} = 0$. *Bottom*: Both strategic and spatial delays, $\tau_{st} = 11.2, \tau_{sp} = 4.4$ time units

3.7 Conclusion

In this chapter we developed a novel mathematical tool to model the evolution of a population composed of several communities that interact in a non-uniform manner. We also extended the replicator dynamics for this context to describe the evolution of strategies in the communities and the impact of the interaction between communities. In particular we identified how the frequency of interaction affects the behavior of individuals inside each group or community. In particular we identified how the frequency of interaction affects the behavior of individuals inside each group or community. We defined three ESSs with different levels of stability against mutations: strong, weak, and intermediate ESS and we showed that the intermediate ESS is asymptotically stable in the replicator dynamics. For the Hawk-Dove game played between two communities with an asymmetric level of aggressiveness, we showed that coexistence of hawks and doves in both communities as a mixed intermediate ESS occurs when some conditions on the payoff matrices (intra- and inter-community fighting costs and resource value) are satisfied and for some range of the interaction probabilities. Otherwise, exclusions occur. In addition, the mixed intermediate ESS remains asymptotically stable for any value of the spatial delay, as opposed to strategic and spatial-strategic delays where persistent oscillations around the equilibrium state occur. We illustrated this result with numerical examples on random graphs.

3.8 Proofs

Proof of Proposition 3.8

Let us first prove that Definition 3.5 implies Definition 3.6. Since the condition in Definition 3.5 holds for any sufficiently small ϵ, as $\epsilon \to 0$, $\bar{U}_i(s_i, \epsilon\mathbf{s} + (1-\epsilon)\mathbf{s}^*, p) < \bar{U}_i(s_i^*, \epsilon\mathbf{s} + (1-\epsilon)\mathbf{s}^*, p)$ for all $i = 1, .., N$, implies $\bar{U}_i(s_i, \mathbf{s}^*, p) \le \bar{U}_i(s_i^*, \mathbf{s}^*, p)$ for all $i = 1, .., N$. Therefore, the first condition in Definition 3.6 is established. Now we suppose there exists i such that $\bar{U}_i(s_i, \mathbf{s}^*, p) = \bar{U}_i(s_i^*, \mathbf{s}^*, p)$. Since the expected utility is linear in $\mathbf{s}$, the condition $\bar{U}_i(s_i, \epsilon\mathbf{s} + (1-\epsilon)\mathbf{s}^*, p) < \bar{U}_i(s_i^*, \epsilon\mathbf{s} + (1-\epsilon)\mathbf{s}^*, p)$ can be written as $\epsilon\bar{U}_i(s_i, \mathbf{s}, p) + (1-\epsilon)\bar{U}_i(s_i, \mathbf{s}^*, p) < \epsilon\bar{U}_i(s_i^*, \mathbf{s}, p) + (1-\epsilon)\bar{U}_i(s_i^*, \mathbf{s}^*, p)$. Since $\bar{U}_i(s_i, \mathbf{s}^*, p) = \bar{U}_i(s_i^*, \mathbf{s}^*, p)$, the last inequality can be written $\epsilon\bar{U}_i(s_i, \mathbf{s}, p) < \epsilon\bar{U}_i(s_i^*, \mathbf{s}, p)$; which yields $\bar{U}_i(s_i, \mathbf{s}, p) < \bar{U}_i(s_i^*, \mathbf{s}, p)$ since $\epsilon > 0$. Therefore, the second condition in Definition 3.6 is established.

Let us now prove that the Definition 3.6 implies Definition 3.5. We have for all i and for any $\mathbf{s} \neq \mathbf{s}^*$, $\bar{U}_i(s_i, \mathbf{s}^*, p) \le \bar{U}_i(s_i^*, \mathbf{s}^*, p)$. If for some i, this inequality is strict, then the condition in Definition 3.5 is satisfied for $\epsilon = 0$ and so for sufficiently small ϵ. If for some i, $\bar{U}_i(s_i, \mathbf{s}^*, p) = \bar{U}_i(s_i^*, \mathbf{s}^*, p)$, then the second condition in Definition 3.6 implies $\bar{U}_i(s_i, \mathbf{s}, p) < \bar{U}_i(s_i^*, \mathbf{s}, p)$. If we multiply this relation by ϵ and add $(1-\epsilon)\bar{U}_i(s_i, \mathbf{s}^*, p)$ to the left-hand side, and $(1-\epsilon)\bar{U}_i(s_i^*, \mathbf{s}^*, p)$ to the right-hand side, we get the condition in Definition 3.5.

Proof of Proposition 3.9

First, let us prove that the Definition 3.7 implies the Definition 3.8. If we take $\epsilon_i \to 0$ in Definition 3.7, we get: $\bar{U}_i(s_i, \mathbf{s}^*, p) \le \bar{U}_i(s_i^*, \mathbf{s}^*, p)$ for all i, and so condition (3.7). Now, to establish the condition (3.8) in Definition 3.8, we suppose there exists i such that $\bar{U}_i(s_i^*, \mathbf{s}^*, p) = \bar{U}_i(s_i, \mathbf{s}^*, p)$, we need to prove that $\bar{U}_i(s_i, (s_i, \mathbf{s}_{-i}^*), p) < \bar{U}_i(s_i^*, (s_i, \mathbf{s}_{-i}^*), p)$. We can write condition (3.6) as follows:

$$\begin{aligned}&\bar{U}_i(s_i, (\epsilon_i s_1^* + (1-\epsilon_i)s_1^*, .., \epsilon_i s_i + (1-\epsilon_i)s_i^*, .., \epsilon_i s_N^* + (1-\epsilon_i)s_N^*), p)\\&< \bar{U}_i(s_i^*, (\epsilon_i s_1^* + (1-\epsilon_i)s_1^*, .., \epsilon_i s_i + (1-\epsilon_i)s_i^*, .., \epsilon_i s_N^* + (1-\epsilon_i)s_N^*), p).\end{aligned}$$

By exploring the linearity of $\bar{U}_i$, we get:

$$\epsilon_i\bar{U}_i(s_i, (s_1^*, .., s_i, .., s_N^*), p) + (1-\epsilon_i)\bar{U}_i(s_i, \mathbf{s}^*, p) < \epsilon_i\bar{U}_i(s_i^*, (s_1^*, .., s_i, .., s_N^*), p) + (1-\epsilon_i)\bar{U}_i(s_i^*, \mathbf{s}^*, p).$$

Since we have $\epsilon_i > 0$ and we suppose $\bar{U}_i(s_i^*, \mathbf{s}^*, p) = \bar{U}_i(s_i, \mathbf{s}^*, p)$, the above inequality yields:

$$\bar{U}_i(s_i, (s_i, \mathbf{s}_{-i}^*), p) < \bar{U}_i(s_i^*, (s_i, \mathbf{s}_{-i}^*), p),$$

and so condition (3.8).

Now we prove that the Definition 3.8 implies the Definition 3.7. We have for all i and for all $\mathbf{s} \neq \mathbf{s}^*$

$$\bar{U}_i(s_i, \mathbf{s}^*, p) \leq \bar{U}_i(s_i^*, \mathbf{s}^*, p).$$

If this inequality is strict for all i, then condition (3.6) holds for $\epsilon_i = 0$, and thus for sufficiently small ϵ_i. If there exists i such that the comparison in (3.7) is an equality, then we obtain $\bar{U}_i(s_i, (s_i, \mathbf{s}^*_{-i}), p) < \bar{U}_i(s_i^*, (s_i, \mathbf{s}^*_{-i}), p)$ (condition (3.8)). We multiply both sides by ϵ_i, and by observing that $\bar{U}_i(s_i, \mathbf{s}^*, p) = \bar{U}_i(s_i^*, \mathbf{s}^*, p)$, we add $(1 - \epsilon_i)\bar{U}_i(s_i, \mathbf{s}^*, p)$ to the left side and $(1 - \epsilon_i)\bar{U}_i(s_i^*, \mathbf{s}^*, p)$ to the right side, we get condition (3.6).

Proof of Theorem 3.2

- The strategy G_1 dominates the strategy H_1 if and only if $\forall \mathbf{s} = (s_1, s_2) \in [0, 1]^2$, $U_1(G_1, \mathbf{s}, p) \geq U_1(H_1, \mathbf{s}, p)$ then as stated before, if and only if $p_1 s_1 L_1 + (1 - p_1) s_2 L_{12} + K_1 \geq 0$, which is $K_1 \geq 0$, $p_1 L_1 + K_1 \geq 0$, $(1 - p_1)L_{12} + K_1 \geq 0$ and $p_1 L_1 + (1 - p_1)L_{12} + K_1 \geq 0$. The same procedure for determining if the strategy G_2 dominates the strategy H_2 in the second group.
- In the same way, strategy H_1 dominates the strategy G_1 if and only if $K_1 \leq 0$, $p_1 L_1 + K_1 \leq 0$, $(1 - p_1)L_{12} + K_1 \leq 0$ and $p_1 L_1 + (1 - p_1)L_{12} + K_1 \leq 0$.

Proof of Proposition 3.11

Let us prove that if G_1 and G_2 are dominant in community 1 and 2 respectively, then $\mathbf{s}^* = (1, 1)$ is a strong ESS. In this case we have the following:

$$U_1(G_1, \mathbf{s}^*, p) > U_1(H_1, \mathbf{s}^*, p);$$
$$U_2(G_2, \mathbf{s}^*, p) > U_2(H_2, \mathbf{s}^*, p).$$

The first condition in Definition 3.6 is a strict inequality, therefore $\mathbf{s}^*$ is a strong ESS. Similarly, we prove that all pure dominant strategies are strong ESSs.

Proof of Theorem 3.3

- There exists a mixed Nash equilibrium strategy $\mathbf{s}^* = (s_1^*, s_2^*)$, when users from any group are indifferent from playing strategy G_i or H_i, i.e. all (pure) strategies are equally fit. At the equilibrium, we have the following system of equations:

$$\begin{cases} U_1(G_1, \mathbf{s}^*, p) = U_1(H_1, \mathbf{s}^*, p), \\ U_2(G_2, \mathbf{s}^*, p) = U_2(H_2, \mathbf{s}^*, p). \end{cases}$$

  Thus, we obtain the following system:

$$\begin{cases} p_1 s_1^* L_1 + (1 - p_1) s_2^* L_{12} + K_1 = 0, \text{ (a)} \\ p_2 s_2^* L_2 + (1 - p_2) s_1^* L_{21} + K_2 = 0, \text{ (b)} \end{cases}$$

where $L_1 = a_1 - b_1 - c_1 + d_1$, $L_{12} = a_{12} - b_{12} - c_{12} + d_{12}$, $L_2 = a_2 - b_2 - c_2 + d_2$, $L_{21} = a_{21} - b_{21} - c_{21} + d_{21}$, $K_1 = p_1(b_1 - d_1) + (1 - p_1)(b_{12} - d_{12})$, $K_2 = p_2(b_2 - d_2) + (1 - p_2)(b_{21} - d_{21})$. The solution of this system is given by $\mathbf{s}^* = (s_1^*, s_2^*)$, with $s_1^* = \frac{(1-p_1)L_{12}K_2 - p_2L_2K_1}{\Delta}$ and $s_2^* = \frac{(1-p_2)L_{21}K_1 - p_1L_1K_2}{\Delta}$; where $\Delta = p_1p_2L_1L_2 - (1 - p_1)(1 - p_2)L_{12}L_{21}$. Clearly, $0 < s_i^* < 1, i = 1, 2$, if:

- $0 < \Delta, 0 < (1 - p_1)L_{12}K_2 - p_2L_2K_1, (1 - p_1)L_{12}K_2 - p_2L_2K_1 < \Delta$, $0 < (1 - p_2)L_{21}K_1 - p_1L_1K_2$, and $(1 - p_2)L_{21}K_1 - p_1L_1K_2 < \Delta$, or
- $\Delta < 0, (1 - p_1)L_{12}K_2 - p_2L_2K_1 < 0, \Delta < (1 - p_1)L_{12}K_2 - p_2L_2K_1$, $0 < (1 - p_2)L_{21}K_1 - p_1L_1K_2 < 0$, and $\Delta < (1 - p_2)L_{21}K_1 - p_1L_1K_2$.

- Let us check for which conditions $\mathbf{s}^* = (s_1^*, s_2^*)$, if exists, is a strong ESS. Assume there is a small proportion of "mutants" that uses another strategy $\mathbf{s} = (s_1, s_2)$. Using the definition of the expected utility, we obtain:

$$\bar{U}_1(s_1^*, \mathbf{s}^*, p) - \bar{U}_1(s_1, \mathbf{s}^*, p) = (s_1^* - s_1)(p_1s_1^*L_1 + (1 - p_1)s_2^*L_{12} + K_1) = 0.$$

Following the same procedure for group 2, we obtain

$$\bar{U}_2(s_2^*, \mathbf{s}^*, p) - \bar{U}_2(s_2, \mathbf{s}^*, p) = 0.$$

From (3.5), $\mathbf{s}^*$ is a strong ESS if $\bar{U}_i(s_i^*, \mathbf{s}, p) - \bar{U}_i(s_i, \mathbf{s}, p) > 0$ for $i = 1, 2$. But

$$\bar{U}_1(s_1^*, \mathbf{s}, p) - \bar{U}_1(s_1, \mathbf{s}, p) = (s_1^* - s_1)\big(p_1s_1L_1 + (1 - p_1)s_2L_{12} + K_1\big),$$
$$\bar{U}_2(s_2^*, \mathbf{s}, p) - \bar{U}_2(s_2, \mathbf{s}, p) = (s_2^* - s_2)\big(p_2s_2L_2 + (1 - p_2)s_1L_{21} + K_2\big).$$

We define f_i, i = 1, 2 as follows:

$$\begin{cases} f_1(s_1, s_2) = (s_1^* - s_1)\big(p_1s_1L_1 + (1 - p_1)s_2L_{12} + K_1\big), \\ f_2(s_1, s_2) = (s_2^* - s_2)\big(p_2s_2L_2 + (1 - p_2)s_1L_{21} + K_2\big). \end{cases}$$

We have $\nabla f_1{}^T = \big[2p_1L_1(s_1^* - s_1) + (1 - p_1)L_{12}(s_2^* - s_2),\ (1 - p_1)L_{12}(s_1^* - s_1)\big]$. Hence, $\frac{\partial^2 f_1}{\partial s_1{}^2}\frac{\partial^2 f_1}{\partial s_2{}^2}-\frac{\partial^2 f_1}{\partial s_1 \partial s_2}\frac{\partial^2 f_1}{\partial s_2 \partial s_1}=-(1-p_1)^2L_{12}^2 < 0$ at $\mathbf{s}^*$ (if $p_1 \neq 1$). Consequently, $\mathbf{s}^*$ is a saddle point. Since $f_1(\mathbf{s}^*) = 0$, f_1 changes the sign around $\mathbf{s}^*$. Therefore, the first community cannot resist invasions by mutants. Following the same procedure with f_2, we find that (s_1^*, s_2^*) is a saddle point. Therefore, the condition of stability (3.5) does not hold and consequently $\mathbf{s}^*$ is not a strong ESS.
- Now, let us study for which condition $\mathbf{s}^* = (s_1^*, s_2^*)$ is a weak ESS. $\mathbf{s}^* = (s_1^*, s_2^*)$ is a weak ESS if $\bar{U}_1(s_1^*, (s_1, s_2^*), p) > \bar{U}_1(s_1, (s_1, s_2^*), p)$ and $\bar{U}_2(s_2^*, (s_1^*, s_2), p) > \bar{U}_2(s_2, (s_1^*, s_2), p)$. But

$$\bar{U}_1(s_1^*, (s_1, s_2^*), p) - \bar{U}_1(s_1, (s_1, s_2^*), p) = -p_1L_1(s_1^* - s_1)^2.$$

which is strictly positive if $L_1 < 0$. Following the same procedure with the second population, we get:

$$\bar{U}_2(s_2^*, (s_1^*, s_2), p) - \bar{U}_2(s_2, (s_1^*, s_2), p) = -p_2L_2(s_2^* - s_2)^2.$$

which is strictly positive if $L_2 < 0$. Therefore, if $L_1 < 0$ and $L_2 < 0$, $\mathbf{s}^*$ is a weak ESS.

- Finally, $\mathbf{s}^*$ is an intermediate ESS if $\bar{U}_1(s_1, \mathbf{s}, p) + \bar{U}_2(s_2, \mathbf{s}, p) < \bar{U}_1(s_1^*, \mathbf{s}, p) + \bar{U}_2(s_2^*, \mathbf{s}, p)$.
  Let $g(s_1, s_2) = \bar{U}_1(s_1^*, \mathbf{s}, p) + \bar{U}_2(s_2^*, \mathbf{s}, p) - \bar{U}_1(s_1, \mathbf{s}, p) - \bar{U}_2(s_2, \mathbf{s}, p)$, we have:

$$g(s_1, s_2) = (s_1^* - s_1)(p_1s_1L_1 + (1 - p_1)s_2L_{12} + K_1) + (s_2^* - s_2)(p_2s_2L_2 + (1 - p_2)s_1L_{21} + K_2).$$

The Hessian matrix of g is given by:

$$\mathcal{H}(g) = \begin{pmatrix} -2p_1L_1 & -(1 - p1)L_{12} - (1 - p_2)L_{21} \\ -(1 - p_1)L_{12} - (1 - p_2)L_{21} & -2p_2L_2 \end{pmatrix}.$$

The determinant of $\mathcal{H}(g)$ is $\Delta_1 = 4p_1p_2L_1L_2 - \left((1 - p_1)L_{12} + (1 - p_2)L_{21}\right)^2$. Hence, g is strictly positive for all $s_1 \neq s_1^*, s_2 \neq s_2^*$, if $L_1 < 0, L_2 < 0$ and $\Delta_1 > 0$.

Proof of Proposition 3.12

- $\mathbf{s}^* = (1, -\frac{(1-p_2)L_{21}+K_2}{p_2L_2})$ is a partially mixed ESS if the following system of equations holds:

$$U_1(G_1, \mathbf{s}^*, p) > U_1(H_1, \mathbf{s}^*, p),$$
$$U_2(G_2, \mathbf{s}^*, p) = U_2(H_2, \mathbf{s}^*, p),$$

which yields

$$p_1L_1 + (1 - p_1)s_2^*L_{12} + K_1 > 0,$$
$$s_2^* = -\frac{(1 - p_2)L_{21} + K_2}{p_2L_2}.$$

Let us check for which conditions $\mathbf{s}^*$ if exists, is a strong ESS. We suppose there is a small fraction of mutants using an alternative strategy $\mathbf{s} = (s_1, s_2)$. We have:

$$U_1(G_1, \mathbf{s}^*, p) - \bar{U}_1(s_1, \mathbf{s}^*, p) = (1 - s_1)[U_1(G_1, \mathbf{s}^*, p) - U_1(H_1, \mathbf{s}^*, p)],$$

which is strictly positive. Now we do the same with the second community. We have:

$$\bar{U}_2(s_2^*, \mathbf{s}^*, p) - \bar{U}_2(s_2, \mathbf{s}^*, p) = (s_2^* - s_2)[U_2(G_2, \mathbf{s}^*, p) - U_2(H_2, \mathbf{s}^*, p)] = 0.$$

Therefore, $\mathbf{s} = (1, s_2)$ is an alternative best reply to $\mathbf{s}^*$. Let us check whether the stability condition holds. Let $\mathbf{s} = (1, s_2)$, we have:

$$\bar{U}_2(s_2^*, \mathbf{s}, p) - \bar{U}_2(s_2, \mathbf{s}, p) = (s_2^* - s_2)[p_2 s_2 L_2 + (1 - p_2)L_{21} + K_2].$$

Let $g(s_2) = \bar{U}_2(s_2^*, \mathbf{s}, p) - \bar{U}_2(s_2, \mathbf{s}, p)$. We have

$$\frac{\partial^2 g(s_2)}{\partial s_2^2} = -2p_2 L_2.$$

Therefore, g is strictly positive for all $s_2 \neq s_2^*$ if $L_2 < 0$. In the same way, we prove that $\mathbf{s}^*$ is an intermediate and a weak ESS. We can prove the other points in the proposition with the same way.

Proof of Theorem 3.4

In order to examine the stability of the interior stationary point, we make a linearization of the system (3.14) around $\mathbf{s}^*$ and observe how the linearized system behaves. We introduce a small perturbation around $\mathbf{s}^*$ defined by $x_1(t) = s_1(t) - s_1^*$ and $x_2(t) = s_2(t) - s_2^*$. The replicator dynamics then writes:

$$\dot{x}_1(t) = (x_1(t) + s_1^*)(1 - x_1(t) - s_1^*)\big(p_1 x_1(t) L_1 + (1 - p_1)x_2(t)L_{12}\big),$$
$$\dot{x}_2(t) = (x_2(t) + s_2^*)(1 - x_2(t) - s_2^*)\big(p_2 x_2(t) L_2 + (1 - p_2)x_1(t)L_{21}\big).$$

Keeping only linear terms in x_1 and x_2, we obtain a linearized system of the form $\dot{x}(t) = Ax(t)$ where $x^t = (x_1, x_2)$,

$$A = \begin{pmatrix} \gamma_1 p_1 L_1 & \gamma_1(1 - p_1)L_{12} \\ \gamma_2(1 - p_2)L_{21} & \gamma_2 p_2 L_2 \end{pmatrix},$$

$\gamma_1 = s_1^*(1 - s_1^*)$, and $\gamma_2 = s_2^*(1 - s_2^*)$. The linearized system is asymptotically stable if all the eigenvalues of A have negative real parts. The eigenvalues of A are the roots of the characteristic polynomial:

$$\mathcal{X}_A = \lambda^2 - tr(A)\lambda + det(A),$$

with $tr(A) = \gamma_1 p_1 L_1 + \gamma_2 p_2 L_2$ and $det(A) = \gamma_1\gamma_2(p_1 p_2 L_1 L_2 - (1 - p_1)(1 - p_2) L_{12}L_{21})$. We check that if $\Delta = p_1 p_2 L_1 L_2 - (1 - p_1)(1 - p_2)L_{12}L_{21} > 0$, $L_1 < 0$ and $L_2 < 0$, then the two eigenvalues of A have negative real parts and the stability follows.

Proof of Corollary 3.1

We aim to prove that a mixed intermediate ESS is asymptotically stable in the replicator dynamics. From Theorem 3.3, the interior equilibrium s^* is an intermediate ESS if $L_1 < 0$, $L_2 < 0$ and $\Delta_1 > 0$. In addition, from Theorem 3.4, s^* is asymptotically stable if $L_1 < 0$, $L_2 < 0$, and $\Delta > 0$. We can then prove that if $\Delta_1 = 4p_1 p_2 L_1 L_2 - ((1 - p_1)L_{12} + (1 - p_2)L_{21})^2 > 0$, then $\Delta = p_1 p_2 L_1 L_2 - (1 - p_1)(1 - p_2)L_{12}L_{21} > 0$ (or equivalently $4\Delta > 0$). We have:

$$4p_1p_2L_1L_2 - ((1-p_1)L_{12} + (1-p_2)L_{21})^2 - 4\big(p_1p_2L_1L_2 - (1-p_1)(1-p_2)L_{12}L_{21}\big) = -((1-p_1)L_{12} - (1-p_2)L_{21})^2 < 0.$$

The proof follows.

Proof of Theorem 3.5

Let us prove that the partially mixed ESS $\mathbf{s}^* = (1, s_2^*)$ with $s_2^* = -\frac{(1-p_2)L_{21}+K_2}{p_2L_2}$ is asymptotically stable. We make a linearization of the replicator dynamics around $\mathbf{s}^*$ and we study the Jacobian matrix. If all the eigenvalues of the Jacobian matrix have negative real parts then the asymptotic stability follows. The Jacobian matrix is given by:

$$A = \begin{pmatrix} -p_1L_1 - (1-p_1)s_2^*L_{12} - K_1 & 0 \\ \gamma_2(1-p_2)L_{21} & \gamma_2p_2L_2 \end{pmatrix},$$

with $\gamma_2 = s_2^*(1-s_2^*)$. The eigenvalues of the Jacobian matrix are the solutions of the following characteristic polynomial:

$$\mathcal{X}_A = \lambda^2 - tr(A)\lambda + det(A).$$

We check that $\mathbf{s}^*$ is asymptotically stable if $p_1L_1 + (1-p_1)s_2^*L_{12} + K_1 > 0$ and $L_2 < 0$. Therefore, the partially mixed ESS $\mathbf{s}^*$ is asymptotically stable.

Similarly, we can prove this result for all other partially mixed ESSs.

Proof of Theorem 3.6

We showed in Sect. 3.8, that the eigenvalues of A, which are solutions of $\lambda^2 - tr(A)\lambda + det(A) = 0$ have negative real parts when $\Delta = p_1p_2L_1L_2 - (1-p_1)(1-p_2)L_{12}L_{21} > 0$, $L_1 < 0$ and $L_2 < 0$; and the mixed intermediate ESS is asymptotically stable when $\tau_{st} = 0$ (Corollary 3.1). For the remaining of the proof that gives the bound on τ_{st} for which the stability is unaffected, the reader should refer to [126], pp.82, Theorem 3.4.

Proof of Theorem 3.7

The proof of this theorem is based on that given by Freedman and Kuang (Theorem 4.1, p. 202), related to the location of roots of the characteristic equation (3.16), and stated as follows:

- If $\beta^2 < \delta^2$, $\Rightarrow$ if $\mathbf{s}^*$ is unstable for $\tau = 0$ then it is unstable for any $\tau \geq 0$; and if $\mathbf{s}^*$ is stable at $\tau = 0$, then it remains stable for τ inferior than some $\tau_s \geq 0$. But, if $\mathbf{s}^*$ is stable at $\tau = 0$, then $\Delta = p_1p_2L_1L_2 - (1-p_1)(1-p_2)L_{12}L_{21} > 0 \Rightarrow \beta^2 > \delta^2$. Therefore, this case is excluded.
- If $\beta^2 > \delta^2$, $2\beta - \alpha^2 > 0$, and $\left(2\beta - \alpha^2\right)^2 > 4(\beta^2 - \delta^2)$, then the stability of the stationary point can change a finite number of times at most as τ is increased, and eventually it becomes unstable. But

$$2\beta - \alpha^2 = 2\gamma_1\gamma_2p_1p_2L_1L_2 - (p_1\gamma_1L_1 + p_2\gamma_2L_2)^2$$
$$= -p_1^2\gamma_1^2L_1^2 - p_2^2\gamma_2^2L_2^2. < 0$$

Therefore, this case is excluded in our model.

- Otherwise, (this is the only case when $\mathbf{s}^*$ is stable at $\tau = 0$), the stability of the stationary point $\mathbf{s}^*$ does not change for any $\tau \geq 0$. $\Rightarrow$ $\mathbf{s}^*$ is asymptotically stable for any $\tau \geq 0$.

Chapter 4
Random Time Delays in Evolutionary Game Dynamics

Nesrine Ben Khalifa, Rachid El-Azouzi and Yezekael Hayel

4.1 Introduction

The majority of works in evolutionary dynamics have studied the replicator dynamics without taking into account the delay effects, assuming that the interactions have an immediate effect on the fitness of strategies. The expected payoff of a strategy is then considered as a function of the frequency of strategies in the population at the current moment. However, many examples in biology and economics show that the impact of actions is not immediate and their effects only appear after some time interval. Therefore, a more realistic model of the replicator dynamics would take into consideration some time delays. In continuous time setting, the authors in [295] studied a model of replicator dynamics with a fixed symmetric delay where the expected payoff of a strategy at time t depends on the frequency of strategies in the population at time $t - \tau$; and proved that for sufficiently large delays the unique mixed ESS can be unstable. The author in [197] studied the best-response dynamics with a symmetric delay and in [256, 258] the authors examined the replicator dynamics with an asymmetric delay over the strategies. Both of these works showed that the delays make the instability possible. In discrete time setting, the authors in [8] proposed two models of replicator dynamics: a social-type model where they proved that large delays result in the instability of the mixed ESS; and a biological-type model where the stability of the ESS is unaffected by any value of delay. In [41, 42], the authors investigated the effect of time delays in the context of multiple interacting communities or groups. They distinguished two types of delays: strategic delays associated with the strategies and may affect the stability of the mixed ESS, and spatial delay associated with the groups of players and do not affect the stability of the ESS.

N. B. Khalifa · R. El-Azouzi (✉) · Y. Hayel
Computer Science Laboratory (LIA), University of Avignon Computer Science Laboratory (LIA), Avignon, France
e-mail: rachid.elazouzi@univ-avignon.fr

E. Altman et al. (eds.), *Multilevel Strategic Interaction Game Models for Complex Networks*, https://doi.org/10.1007/978-3-030-24455-2_4

In [143], author assumed players react to heterogeneously delayed information and proved that there exists an unique mixed ESS is asymptotically stable for any distribution of delays under some class of discrete time replicator dynamics.

Several studies assumed that the time delay is fixed, but this assumption is no longer valid in many applications. In social networks and engineering systems we often face to different delays. In the scenario of social networks, users react to delayed information, and the delay is not the same for all users [143]. To the best of our knowledge, this work is the first attempt to study the randomly delayed replicator dynamics. We aim to study the effect of the delay distribution on the stability of the mixed ESS.

In this research line, we have analyzed the stability of the Evolutionarily Stable Strategy (ESS) in continuous-time for replicator dynamics subject to random time delays. In particular we study the effect of randomly distributed time delays in the replicator dynamics. We show that, under the exponential delay distribution, the ESS is asymptotically stable for any value of the rate parameter. For the uniform and Erlang distributions, we derive necessary and sufficient conditions for the asymptotic stability of the ESS. When only one strategy is delayed, we derive conditions of stability of the mixed ESS under the exponential delay distribution. We hereby summarise the major contribution of this chapter:

The chapter is structured as follows. In Sect. 4.2, we provide a background on Evolutionary Game Theory. In Sect. 4.3 we focus on the replicator dynamics with continuous random delays and we investigate the cases of uniform, exponential, and Erlang distributions. Finally, we study in Sect. 4.5 the stability of the ESS when one strategy is delayed and we conclude the paper in Sect. 4.6.

4.2 Evolutionary Games

Consider a large population of players s who interact repeatedly through random pairwise interactions. A strategy of an individual is a probability distribution over the pure strategies. An equivalent interpretation of strategies is obtained by assuming that individuals choose pure strategies and then the probability distribution represents the fraction of individuals in the population that choose each strategy. Here we assume that at each local interaction, a user or player chooses a (pure) strategy T and S. The following payoff matrix gives the payoff of given player that interacted with another player from the population; interacted player in a local interaction

$$\mathbf{A} = \begin{pmatrix} a & b \\ c & d \end{pmatrix}.$$

A mixed strategy is a probability distribution over the pure strategies (represented as a row vector). If one player chooses a mixed strategy p and interacts with an individual who plays mixed strategy q, the expected fitness F of the first individual is obtained through:

$$F(p, q) = p\mathbf{A}q'.$$

Let s be the population share of strategy T, that is the fraction of the population using strategy T; so $1 - s$ is the fraction of the population using strategy S. Hence the expected payoff of strategy A and B are given, respectively by

$$F_T = (1 \quad 0)\mathbf{A}(s \quad 1 - s)' = as + b(1 - s),$$

and

$$F_S = (0 \quad 1)\mathbf{A}(s \quad 1 - s)' = cs + d(1 - s).$$

The average payoff in the population, denoted by $\bar{F}_s$, is given by:

$$\bar{F}_s = sF_T + (1 - s)F_S.$$

4.2.1 Evolutionarily Stable Strategy

The one important concept in Evolutionary games are the notion of Evolutionary Stable Strategy [248, 279], which includes robustness against a deviation of a whole (possibly small) fraction of the population who may wish to deviate (This is in contrast with the standard Nash equilibrium that only incorporates robustness against deviation of a single user). An ESS has in fact a property of persistence through time since it resists invasion from a rare alternative strategy.

We assume that the whole population adopts a mixed strategy s^*, that a fraction ϵ of *mutants* deviate to mixed strategy x. Strategy x^* is an ESS, if for all $\forall p \neq x$, there exists some $\epsilon_x > 0$ such that $\forall \epsilon \in (0, \epsilon_x)$:

$$J(x^*, \epsilon x + (1 - \epsilon)x^*) > J(x, \epsilon x + (1 - \epsilon)x^*). \tag{4.1}$$

That is, x^* is ESS if, after mutation, mutants are more successful than mutants, in which case mutants cannot invade and will eventually get extinct. n that sense, the equilibrium ESS is said to be more robust than the Nash Equilibrium, because it is robust against the deviation of a fraction of players, and not only one. The following proposition allows to characterize an ESS through its stability properties.

Proposition 4.13 ([137]) *A mixed strategy $q \in \Delta(\mathcal{C})$ is an ESS if and only if it satisfies the following conditions:*

- *Nash Condition:* $J(x, x^*) \leq J(x^*, x^*) \quad \forall x$,
- *Stability Condition:* $J(x, x^*) = J(x^*, x^*) \Rightarrow J(x, x) < J(x^*, x) \quad \forall x \neq x^*$.

The concept of ESS can be viewed from two angles.

An ESS can be viewed from two angles. The first one is through the mixed strategy used by any individual into a global population. A second point of view, is to look at the population profile considering that each individual plays pure strategy. As we are interested in the dynamics of strategies inside a population, we prefer to consider the ESS as a population profile. In this case, it is commonly called an evolutionarily stable state but by abuse of definition concepts, it is also called ESS. In our setting of pairwise interaction where the outcome is defined as a matrix game, the existence and uniqueness of ESS is well known and proved in [137].

4.2.2 Replicator Dynamics

We introduce here the replicator dynamics which describe the evolution in the population of the various strategies. In this dynamics, the growth rate of a strategy is proportional to the difference between the expected payoff of that strategy and the average payoff in the population. The replicator dynamics can be interpreted as an imitation protocol: at each local interaction a player randomly selects another player and switches to his strategy with a probability proportional to the difference between the expected payoff of the two strategies. In the classical replicator dynamics, the payoff of a strategy at time t has an instantaneous impact on the rate of growth of the population size that uses it. As defined in [254], the replicator dynamics is given by:

$$\begin{aligned} \dot{s}(t) &= s(t)[U_A(t) - \bar{U}_s(t)], \\ &= s(t)[1 - s(t)][U_A(t) - U_B(t)]. \end{aligned}$$

The mixed Nash equilibrium is related to the fixed points of the replicator dynamics by the *folk theorem of evolutionary game theory* [137]. It has been shown that any stable fixed point of the replicator dynamics is a Nash equilibrium.

4.3 Replicator Dynamics with Continuous Random Delays

In this section, we introduce in the replicator dynamics when the payoff acquired at time t will impact the rate of growth τ time later. Indeed, when a player uses a strategy at time t, it would receive its payoff after a random delay τ, it means at time $t + \tau$. Then the fitness of that player is determined only at that instant, i.e. $U(t + \tau)$. Equivalently, the expected payoff of a strategy at the current time is a function of the state of the population at some previous random time. Let consider now that the delay is equal to τ, then the expected payoff of strategy S at time t is determined by:

$$U_S(t, \tau) = as(t - \tau) + b(1 - s(t - \tau)),$$

if $t \geq \tau$, it is 0 otherwise. Let $p_d(\tau)$ be the probability distribution of delays whose support is $[0, \infty[$. As we consider a large population, every player can experience a different positive delay. Thus the expected payoff a player choosing strategy S is given by

$$U_S(t) = \int_0^{\infty} p(\tau)U_S(t,\tau)d\tau.$$

The expected payoff of strategy S is then given by:

$$U_S(t) = a\int_0^{\infty} p_d(\tau)s(t-\tau)d\tau + b[1 - \int_0^{\infty} p(\tau)s(t-\tau)d\tau].$$

Similarly, the expected payoff of strategy T is:

$$U_B(t) = c\int_0^{\infty} p(\tau)s(t-\tau)d\tau + d[1 - \int_0^{\infty} p(\tau)s(t-\tau)d\tau].$$

Then the replicator dynamic equation with delay can be written as follows:

Then, considering expected delay, we can write the replicator dynamics as:

$$\dot{s}(t) = s(t)[1-s(t)]\Big[-\delta\int_0^{\infty} p(\tau)s(t-\tau)d\tau + \delta_1\Big]. \tag{4.2}$$

Now we study the stability at the interior equilibrium s^*. To do that, we make a linearization of the replicator dynamics around this interior equilibrium and observe the stability of the linearized equation. We suppose there is a small perturbation around s^* defined by $y(t) = s(t) - s^*$. The replicator dynamics becomes:

$$\dot{y}(t) = -\delta(y(t)+s^*)(1-y(t)-s^*)\int_0^{\infty} p(\tau)y(t-\tau)d\tau.$$

Keeping only linear terms in y in the previous equation, we get the linearized equation:

$$\dot{y}(t) = -\delta s^*(1-s^*)\int_0^{\infty} p(\tau)x(t-\tau)d\tau. \tag{4.3}$$

The characteristic equation corresponding to the delay-differential equation (4.3) is given by:

$$\lambda + \delta s^*(1-s^*)\int_0^{\infty} p(\tau)e^{-\lambda\tau}d\tau = 0. \tag{4.4}$$

It is know from the theory of delay-differential equations, that the steady state s^* is asymptotically stable if and only if all roots of the characteristic equation (4.4) have negative real parts [40, 109]. In the next sections, we shall study the stability by considering different distributions of delay. We consider the uniform, exponential, and Erlang distributions.

4.3.1 Uniform Distribution of Delays

In this section we assume that the delay follows uniform distribution, i.e., $p(\tau) = \frac{1}{\tau_{max}}$ for $0 \leq \tau \leq \tau_{max}$ and 0 otherwise. The characteristic equation 4.4 becomes:

$$\lambda + \frac{D}{\tau_{\max}} \int_0^{\tau_{\max}} e^{-\lambda\tau} d\tau = 0, \tag{4.5}$$

where $D = \delta s^*(1 - s^*)$. Note that $\lambda = 0$ is not a root of (4.5).

Proposition 4.14 *The mixed ESS s^* is asymptotically stable in the replicator dynamics with uniform distribution if and only if $\tau_{\max} < \frac{\pi^2}{2D}$.*

Proof See Sect. 4.7. □

This result gives the upper bound of the maximum value of delay in the uniform distribution so that the stability of the ESS remains unaffected ($\bar{\tau}_{max} = \frac{\pi^2}{2D}$). The mean delay under uniform distribution with the upper bound $\bar{\tau}_{max}$ is $\tau_m = \frac{\pi^2}{4D}$. With fixed delay, it know that the stability of the ESS remains unaffected when $\tau < \frac{\pi}{2D}$ (see [295] for more details). It easy to show that $\frac{\pi}{2D} < \frac{\pi^2}{4D}$, which implies that the mixed ESS is less sensitive to the delay when it is random.

We make numerical simulations of the trajectories of solutions of (4.2) to compare the effect of uniformly at random delays with constant delays on the stability of the ESS. In Fig. 4.1-*left*, we considered a constant delay $\tau = 2.4$ time units, and we observe permanent oscillations around the ESS; whereas in Fig. 4.1-*right* we considered uniformly at random delays with a mean equal to 2.4 time units ($\tau_{max} = 4.8$ time units) and we observe convergence to the ESS after some damped oscillations.

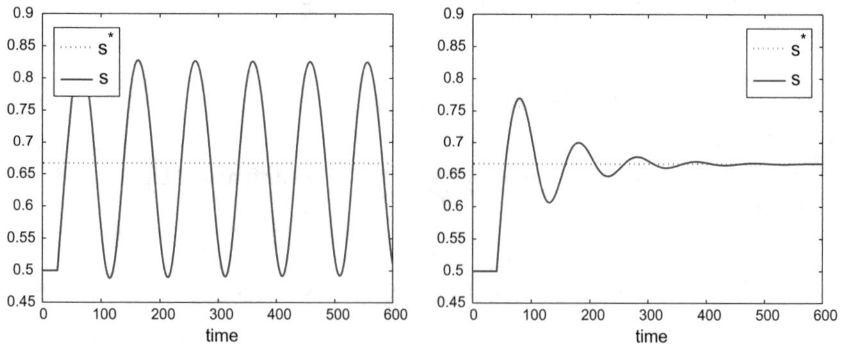

Fig. 4.1 Replicator dynamics with constant delay (left) and uniformly at random delays (right). Left constant delay $\tau = 2.4$. Right $\tau_{\max} = 4.8$. $a = 1, b = 5, c = 2, d = 3, \delta = 3, \bar{\tau}_{\max} = 7.4$

4.3.2 Exponential Distribution of Delays

Now we consider an exponential distribution of delays with parameter $\beta > 0$, i.e, $p(\tau) = \beta e^{-\beta\tau}$ whose support is $[0, \infty[$. Thus, the characteristic equation (4.4) becomes:

$$\lambda + \beta D \int_0^\infty e^{-(\beta+\lambda)\tau} d\tau = 0. \tag{4.6}$$

We have the following result.

Proposition 4.15 *The mixed ESS s^* is asymptotically stable in the replicator dynamics for any value of β of the exponential distribution.*

Proof We shall show that all roots of (4.6) have negative real parts for any $\beta > 0$. Let α be a real solution of (4.6), it is clear that α cannot be positive. Let $\lambda = u + iv$ a complex solution of (4.6) with $u > 0$ and $v > 0$ (without loss of generality, we assume that $v > 0$, since if $u + iv$ a solution of the characteristic equation, then $u - iv$ is also a solution). By contradiction, we assume that there exists a root with positive real part. Separating real and imaginary parts in (4.6), we get:

$$u + \beta D \int_0^\infty e^{-(\beta+u)\tau} \cos(v\tau) d\tau = 0, \tag{4.7}$$

$$v - \beta D \int_0^\infty e^{-(\beta+u)\tau} \sin(\tau v) d\tau = 0. \tag{4.8}$$

Furthermore, we have:

$$\int_0^\infty e^{-(\beta+u)\tau} \sin(\tau v) d\tau = \frac{I}{\beta + u},$$

with $I = \int_0^\infty e^{-(\beta+u)\frac{z}{v}} \cos(z) dz$. Taking into account the above equations, we can write (4.7) and (4.8) as follows:

$$u + \beta D \frac{I}{v} = 0,$$

$$v - \beta D \frac{I}{\beta + u} = 0.$$

The two previous equations yield $u = \frac{-\beta}{2} < 0$. This results proves that the real parts of the roots are always negative, which concludes the proof. □

By virtue of Proposition 4.15, we concludes that the stability of the mixed ESS is unaffected under the exponential distribution for any value of β. This can be explained by the fact that small delays are much more likely to occur than larger delays. Thus, large delays cannot have a destabilising effect.

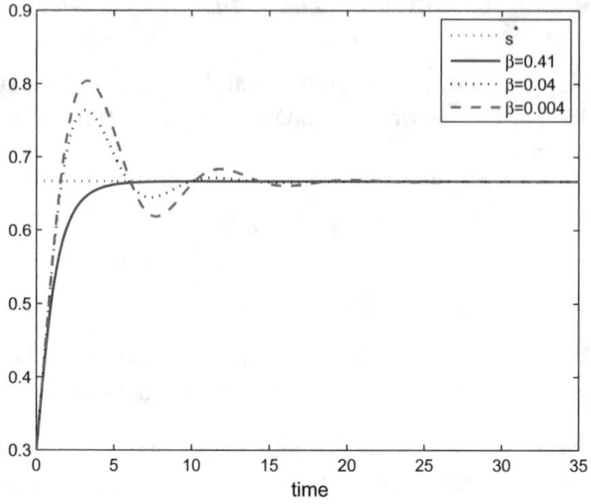

Fig. 4.2 Convergence to the ESS for different values of β

In Fig. 4.2, we plot the evolution of the replicator dynamics for different values of β. When $\beta = 0.41$, then the mean delay of the exponential distribution is 2.4 and we observe convergence to the ESS, unlike the case of a constant delay where we observed permanent oscillations in Fig. 4.1-*Left*. The value of β affects only the convergence rate.

4.3.3 Erlang Distribution

We consider an Erlang distribution (or Gamma distribution) of delays with support $[0, \infty[$ and parameters $k \geq 1$ and $\beta > 0$. The probability distribution is $p(\tau; k, \beta) = \frac{\beta^k \tau^{k-1} e^{-\beta\tau}}{(k-1)!}$ and the mean of this delay distribution is $\frac{k}{\beta}$. See Fig. 4.3.

We propose to find the critical mean delay value for which the stability of the mixed ESS is lost. The characteristic equation associated with this delay distribution is

$$\lambda + D \int_0^\infty \frac{\beta^k}{(k-1)!} \tau^{k-1} e^{-(\beta+\lambda)\tau} d\tau = 0, \tag{4.9}$$

which can be written as:

$$\lambda + D \frac{\beta^k}{(k-1)!(\beta+\lambda)^k} \int_0^\infty z^{k-1} e^{-z} dz = 0.$$

Therefore, the characteristic equation reduces to:

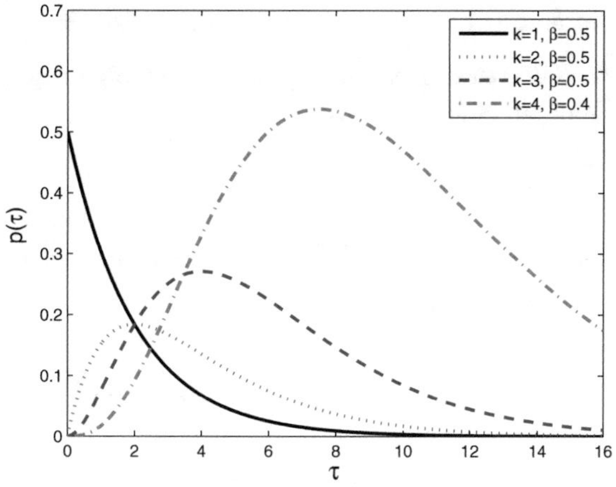

Fig. 4.3 Gamma distribution of delays

$$\lambda + D\frac{\beta^k}{(\beta+\lambda)^k} = 0, \text{ or } \lambda(\beta+\lambda)^k + D\beta^k = 0.$$

The zero solution of the linearized equation is asymptotically stable if $\mathrm{Re}(\lambda) < 0$, and unstable if $\mathrm{Re}(\lambda) > 0$. A stability transition corresponds to the appearance of a purely imaginary solution of the characteristic equation ($\mathrm{Re}(\lambda) = 0$) as β decreases. We aim to find the value of $\bar{\beta}$ at which a purely imaginary root exists. Substituting λ with iw (with $w > 0$) in the above equation we get:

$$iw(\bar{\beta} + iw)^k + D\bar{\beta}^k = 0.$$

Equivalently we have,

$$iw(\bar{\beta}^2 + w^2)^{\frac{k}{2}} e^{ik\theta} + D\bar{\beta}^k = 0,$$

with $\cos(\theta) = \frac{\bar{\beta}}{(\bar{\beta}^2+v^2)^{\frac{1}{2}}}$ and $\sin(\theta) = \frac{w}{(\bar{\beta}^2+w^2)^{\frac{1}{2}}}$. Separating real and imaginary parts in the above equation we get:

$$\cos(k\theta) = 0,$$
$$D\bar{\beta}^k - v(\bar{\beta}^2 + w^2)^{\frac{k}{2}} \sin(k\theta) = 0.$$

which yields:

$$k\theta = \frac{\pi}{2} + 2n\pi, n \in \mathbb{N},$$
$$D\bar{\beta}^k = w(\bar{\beta}^2 + w^2)^{\frac{k}{2}}. \tag{4.10}$$

Since $\cos(\theta) = \frac{\bar{\beta}}{(\bar{\beta}^2+w^2)^{\frac{1}{2}}}$, then $\cos(\frac{\pi}{2k} + \frac{2n\pi}{k}) = \frac{\bar{\beta}}{(\bar{\beta}^2+w^2)^{\frac{1}{2}}}$ and $w^2 = \bar{\beta}^2 \frac{\sin^2(\frac{\pi}{2k}+\frac{2n\pi}{k})}{\cos^2(\frac{\pi}{2k}+\frac{2n\pi}{k})}$. As $\bar{\beta}$ decreases from infinity, the first purely imaginary root appears when $n = 0$. From Eq. (4.10), we have:

$$D^2\bar{\beta}^{2k} = v^2(\bar{\beta}^2 + w^2)^k.$$

Substituting w^2 with $\bar{\beta}^2 \frac{\sin^2(\frac{\pi}{2k})}{\cos^2(\frac{\pi}{2k})}$ in the equation above and taking the square root of both sides, we get:

$$\bar{\beta} = D\frac{\cos^{k+1}(\frac{\pi}{2k})}{\sin(\frac{\pi}{2k})}.$$

At this value of $\bar{\beta}$, a purely imaginary root of the characteristic equation exists. We can establish the following sufficient condition of stability when the delays follow an Erlang distribution with parameter $k \geq 1$.

Proposition 4.16 *The mixed ESS s^* is asymptotically stable in the replicator dynamics if and only if $\beta > \bar{\beta}$ with $\bar{\beta} = D\frac{\cos^{k+1}(\frac{\pi}{2k})}{\sin(\frac{\pi}{2k})}$.*

Proof Let $\beta > \bar{\beta}$. Let $\lambda = u + iv$ be a solution of the characteristic equation with $u > 0$. We suppose without loss of generality that $v > 0$ (since if $u + iv$ is a solution of the characteristic equation than $u - iv$ is also a solution). We aim to find a contradiction and to prove that no root with a positive real part can exist when $\beta > \bar{\beta}$. Substituting λ with $u + iv$ and separating real and imaginary parts in the characteristic equation we get:

$$u + [(\beta + u)^2 + v^2]^{-\frac{k}{2}} D\beta^k \cos(k\theta) = 0,$$
$$v - [(\beta + u)^2 + v^2]^{-\frac{k}{2}} D\beta^k \sin(k\theta) = 0,$$

with $\cos(\theta) = \frac{\beta+u}{[(\beta+u)^2+v^2]^{\frac{1}{2}}}$ and $\sin(\theta) = \frac{v}{[(\beta+u)^2+v^2]^{\frac{1}{2}}}$. The above system can be written:

$$u = \frac{-D\beta^k \cos(k\theta)}{[(\beta + u)^2 + v^2]^{\frac{k}{2}}},$$
$$v = \frac{D\beta^k \sin(k\theta)}{[(\beta + u)^2 + v^2]^{\frac{k}{2}}}.$$

From the equations above, we get:

$$\cos(k\theta) < 0 \Rightarrow \frac{\pi}{2k} < \theta,$$
$$v^2 < D^2 \frac{\beta^{2k}}{[(\beta+u)^2+v^2]^k} < D^2 \frac{\beta^{2k}}{[\beta^2+v^2]^k};$$

which yields $\cos(\theta) < \cos(\frac{\pi}{2k})$. In addition, we check that $\frac{\beta}{[\beta^2+v^2]^{\frac{1}{2}}} < \frac{\beta+u}{[(\beta+u)^2+v^2]^{\frac{1}{2}}}$; and then we have $\frac{\beta}{[\beta^2+v^2]^{\frac{1}{2}}} < \cos(\theta) < \cos(\frac{\pi}{2k})$. From this inequality, we obtain:

$$\beta^2 \frac{\sin^2(\frac{\pi}{2k})}{\cos^2(\frac{\pi}{2k})} < v^2. \tag{4.11}$$

On the other hand we have $v^2 < D^2 \frac{\beta^{2k}}{[\beta^2+v^2]^k}$ and $\frac{\beta^{2k}}{[\beta^2+v^2]^k} < \cos^{2k}(\frac{\pi}{2k})$, which yields:

$$v^2 < D^2 \cos^{2k}(\frac{\pi}{2k}). \tag{4.12}$$

Using (4.27) and (4.28) we get $\beta^2 < D^2 \frac{\cos^{2k+2}(\frac{\pi}{2k})}{\sin^2(\frac{\pi}{2k})}$, which is in contradiction with the initial assumption of $\beta > \bar{\beta}$. Therefore, there is no root with a positive real part when $\beta > \bar{\beta}$, and the local asymptotic stability follows.

Now, let us prove the necessary condition. We differentiate equation 4.9 with respect to β, we get:

$$Re\left(\frac{d\lambda}{d\beta}\right)_{\beta=\bar{\beta}} < 0$$

Therefore, as β decreases, the roots of characteristic equation cross the imaginary axis only from the left to the right and once the stability is lost cannot be regained again. The stability is persistently lost at $\beta = \bar{\beta}$. □

We note that, as the mean of the delay distribution increases and reaches $\frac{k}{\bar{\beta}}$, a purely imaginary root $\lambda^* = iw$ appears. Its imaginary part equals the frequency of oscillations and it is given by:

$$w = \bar{\beta} \frac{\sin(\frac{\pi}{2k})}{\cos(\frac{\pi}{2k})}, \tag{4.13}$$

which can be written as:

$$w = D\cos^k(\frac{\pi}{2k}), \tag{4.14}$$

The maximum possible value of the mean delay is $\frac{k}{\bar{\beta}} = \frac{k\sin(\frac{\pi}{2k})}{D\cos^{k+1}(\frac{\pi}{2k})}$, which is larger than $\frac{\pi}{2D}$, the maximum possible value of a fixed delay. Thus, the ESS is less sensitive to a random delay than a fixed delay. When $k = 2$, $\bar{\beta} = \frac{D}{2}$, and the maximum mean value of the delay is $\frac{4}{D}$. When $k = 1$, $\bar{\beta} = 0$ and the ESS is asymptotically stable for any value of $\beta > 0$. This confirms our result on the exponential delay in the previous section.

4.4 Replicator Dynamics with Several Discrete Random Delays

In this section we consider several random discrete delays. We first study the case of two delays, and then we extend our study to the multiple-delay case.

4.4.1 Two Delays and a Non-delay Term

We suppose that a strategy used by a player would take a delay τ_1 with probability p_1, a delay τ_2 with probability p_2, or no delay ($\tau_0 = 0$) with probability p_0 ($p_0 = 1 - p_1 - p_2$). The expected payoff of strategy A is:

$$U_A(t) = a[p_0 s(t) + p_1 s(t-\tau_1) + p_2 s(t-\tau_2)] + b[1 - p_0 s(t) - p_1 s(t-\tau_1) - p_2 s(t-\tau_2)].$$

Similarly,

$$U_B(t) = c[p_0 s(t) + p_1 s(t-\tau_1) + p_2 s(t-\tau_2)] + d[1 - p_0 s(t) - p_1 s(t-\tau_1) - p_2 s(t-\tau_2)].$$

The replicator dynamics is given by:

$$\dot{s}(t) = s(t)(1-s(t))[-\delta(p_0 s(t) + p_1 s(t-\tau_1) + p_2 s(t-\tau_2)) + \delta_1].$$

Making a linearization of the above equation around s^* we get:

$$\dot{x}(t) \approx -\delta s^*(1-s^*)\big[p_0 x(t) + p_1 x(t-\tau_1) + p_2 x(t-\tau_2)\big]. \tag{4.15}$$

Equation (4.15) was studied in [39, 43, 121, 171]. We use the following proposition in [43] to conclude about the stability of the mixed ESS in the replicator dynamics.

Proposition 4.17 ([43]) *Let* $\dot{x}(t) = -a_0 x(t) - a_1 x(t-h_1) - a_2 x(t-h_2)$, *where* $h_1 > 0$ *and* $h_2 > 0$. *If one of the following conditions holds:*

- $0 < a_0$, $|a_1| + |a_2| < a_0$,
- $0 < a_0 + a_1 + a_2$, $|a_1|h_1 + |a_2|h_2 < \frac{a_0+a_1+a_2}{|a_0|+|a_1|+|a_2|}$,
- $0 < a_0 + a_1$, $|a_1|h_1 < \frac{a_0+a_1-|a_2|}{|a_0|+|a_1|+|a_2|}$,
- $0 < a_0 + a_2$, $|a_2|h_2 < \frac{a_0+a_2-|a_1|}{|a_0|+|a_1|+|a_2|}$.

Then the above equation is asymptotically (and exponentially) stable.

From this proposition, the following corollary results.

Corollary 4.2 *If one of the following conditions holds:*

- $p_1 + p_2 < p_0$,
- $p_1\tau_1 + p_2\tau_2 < \frac{1}{\delta s^*(1-s^*)}$,
- $p_1\tau_1 < \frac{p_0+p_1-p_2}{\delta s^*(1-s^*)}$,
- $p_2\tau_2 < \frac{p_0+p_2-p_1}{\delta s^*(1-s^*)}$,

then s^* is asymptotically stable in the replicator dynamics.

The first sufficient condition of stability is independent of the values of delays, τ_1 and τ_2. When the probability of the non delayed term is sufficiently high, then the asymptotic stability of the ESS cannot be lost for any value of τ_1 and τ_2. The second condition gives an upper bound of the mean delay. The last two conditions give an upper bound of the values of one delay for which the stability of the interior equilibrium is unaffected for any value of the second delay.

4.4.2 Several Delays

We consider that a strategy used by a player would take a delay τ_k with probability p_k, where $k = 0, .., q$, $\tau_0 = 0$ and $\sum_{k=0}^{q} p_k = 1$. We are interested in finding a delay-independent stability condition. Doing the same as in the previous sections, we can write the linearized replicator dynamics as follows:

$$\dot{x}(t) = -\delta s^*(1 - s^*)\big[p_0 x(t) + \sum_{k=1}^{q} p_k x(t - \tau_k)\big].$$

Following [68], we derive the following necessary and sufficient delay-independent stability condition for the asymptotic stability of the interior equilibrium (the proof in [68], Theorem 3.2).

Proposition 4.18 ([68]) *Let $\dot{x}(t) = a_0 x(t) + \sum_{k=1}^{q} b_k x(t - h_k)$ where $a_0 < 0$. The zero solution is asymptotically stable independently of delays if and only if $-a_0 > \sum_{k=1}^{q} |b_k|$ or if $-a_0 = \sum_{k=1}^{q} |b_k|$ but $a_0 + \sum_{k=1}^{q} b_k \neq 0$.*

The next corollary immediately follows.

Corollary 4.3 *The mixed ESS is asymptotically stable in the replicator dynamics for any τ_k, if and only if $p_0 \geq \frac{1}{2}$.*

In Fig. 4.4, we plot the trajectories of solutions of the replicator equation for $\tau_1 = 4.4$ and $\tau_2 = 5.2$. In the left subfigure, we fixed p_0 and p_1 to 0.3 and 0.2 respectively. In the right subfigure, we fixed p_0 and p_1 to 0.6 and 0.2 respectively. The

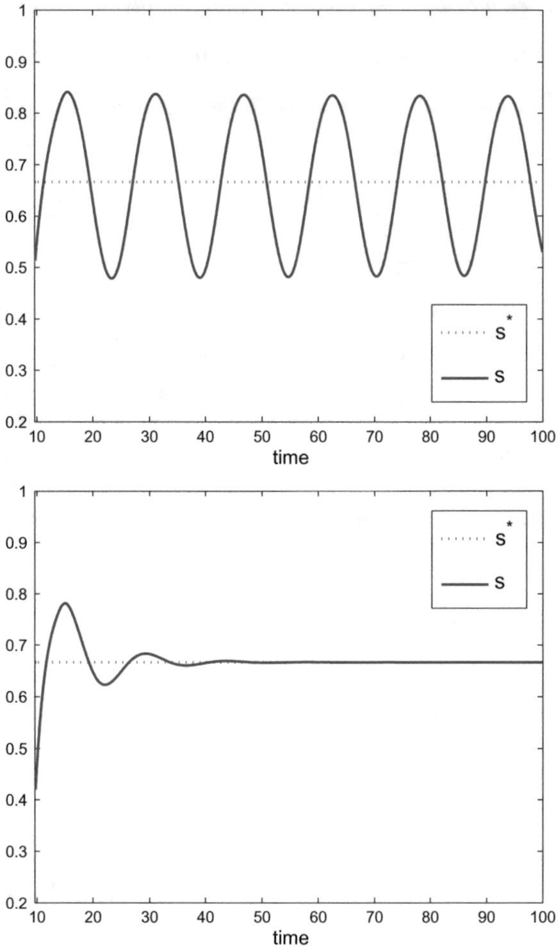

Fig. 4.4 Replicator dynamics with discrete random delays, in both figures: $a = 1, b = 5, c = 2, d = 3, \tau_0 = 0, \tau_1 = 4.4, \tau_2 = 5.2$. *Top* $p_0 = 0.3$, $p_1 = 0.2$ and $p_2 = 0.5$. *Bottom* $p_0 = 0.6$, $p_1 = 0.2$ and $p_2 = 0.2$

results observed validate our previous results. The persistent oscillations observed in the left subfigure disappear in the right subfigure because, in this case, the probability of $\tau = 0$ is sufficiently high, and the delay-independent stability condition is satisfied.

We are also interested in the case when there are several delays which are all positive (i.e. $p_0 = 0$). We aim to look whether the asymptotic stability can be preserved and for which conditions. We use the following result [44]:

Proposition 4.19 *Let* $\dot{x}(t) = -\sum_{k=1}^{q} a_k x(t - \tau_k)$, *with* $\tau_k > 0$ *for all k and* $\sum_{k=1}^{q} a_k > 0$. *If*

$$\sum_{k=1}^{q} (\sum_{i=k}^{q} a_i)|\tau_k - \tau_{k-1}| < \frac{\sum_{k=1}^{q} a_k}{\sum_{k=1}^{q} |a_k|},$$

where $\tau_0 = \frac{1}{e\sum_{k=1}^{q} a_k}$. *Then, the equation above is asymptotically (and exponentially) stable.*

We conclude the following condition:

Corollary 4.4 *If*

$$\sum_{k=1}^{q}[\sum_{i=k}^{q} p_i]|\tau_k - \tau_{k-1}| < \frac{1}{\delta s^*(1-s^*)},$$

where $\tau_0 = \frac{1}{\delta s^*(1-s^*)e}$. *Then the interior equilibrium is asymptotically (and exponentially) stable.*

4.5 Replicator Dynamics with One Delayed Strategy

In this section we study the replicator dynamics with one delayed strategy. There exist many scenarios in which this situation holds. In networking, in particular for the multiple access game [256], the action of transmitting may be delayed, while keeping silent would have no payoff and no delay. For the Hawk-Dove game, the aggressive behavior may be delayed due to the fighting, whereas the peaceful type would not fight and no delay is incurred. The time delay may represent the time of fighting and it is random since it depends on whether the opponent is aggressive or not.

We suppose that only strategy A has a random delay and strategy B has no delay. In this case, the expected payoff to strategy A is given by:

$$U_A(t) = a\int_0^\infty p(\tau)s(t-\tau)d\tau + b[1 - \int_0^\infty p(\tau)s(t-\tau)d\tau].$$

Similarly, the expected payoff of strategy B is:

$$U_B(t) = cs(t-\tau) + d(1-s(t-\tau))].$$

Following the same procedure in the previous sections, we write the linearized replicator dynamics as follows:

$$\dot{x}(t) = D\big[(a-b)\int_0^\infty p(\tau)x(t-\tau)d\tau + (d-c)x(t)\big],$$

with $D = s^*(1-s^*)$. The characteristic equation corresponding to the above equation is given by:

$$\lambda - D(a-b)\int_0^\infty p(\tau)e^{-\lambda\tau}d\tau - D(d-c) = 0.$$

4.5.1 Continuous Delays

We suppose that $p(\tau) = \beta e^{-\beta\tau}$ with $\beta > 0$. The characteristic equation can be written:

$$\lambda - D\beta(a-b)\frac{1}{\lambda+\beta} - D(d-c) = 0. \tag{4.16}$$

Let $\lambda^* = iv$ with $v > 0$ a purely imaginary solution of the characteristic equation. Substituting λ with λ^* in (4.16) and separating real and imaginary parts, we get:

$$1 + \frac{D\beta(a-b)}{\beta^2+v^2} = 0,$$
$$-\frac{D\beta^2(a-b)}{\beta^2+v^2} + D(d-c) = 0,$$

which yields,

$$-D\beta(a-b) = \beta^2 + v^2,$$
$$\beta = D(d-c).$$

Therefore, we conclude that as β decreases, at the value of $\bar{\beta} = D(d-c)$, a stability switch occurs. We can establish the following result:

Theorem 4.8 • *Let $b > a$, then the interior equilibrium s^* is asymptotically stable if and only if $\beta > \bar{\beta} = D(d-c)$.*
• *Let $a > b$, then the interior equilibrium is asymptotically stable for any $\beta > 0$.*

Proof • Let $a - b < 0$. We shall prove that if $\beta > \bar{\beta}$, then all roots of (4.16) have negative real parts, and if $\beta < \bar{\beta}$, there exist a root with a positive real part. Let $\lambda = u + iv$ be a root of (4.16) with $v \geq 0$. Substituting λ with $u + iv$ in (4.16), we get:

$$u + iv - \frac{D\beta(a-b)}{u+\beta+iv} - D(d-c) = 0.$$

Separating real and imaginary parts in the equation above, we get:

$$1 + \frac{D\beta(a-b)}{(u+\beta)^2+v^2} = 0,$$
$$u - \frac{D\beta(a-b)(u+\beta)}{(u+\beta)^2+v^2} - D(d-c) = 0.$$

We obtain

$$(\beta + u)^2 + v^2 = -\beta D(a - b),$$
$$2u = -\beta + D(d - c).$$

Therefore, $u < 0$ if $\beta > \bar{\beta}$. And $u > 0$ if $\beta < \bar{\beta}$.

- Let $a > b$, then all roots of (4.16) are real, and $c - d > 0$ (since $\delta > 0$). Let α be a real root of (4.16), we have:

$$\alpha^2 + \alpha[\beta - D(d - c)] + \beta D\delta = 0.$$

The discriminant of the above equation is given by:

$$\mathcal{X} = [\beta + D(d - c)]^2 + 4\beta D(a - b) > 0.$$

Since $\delta > 0$, and $\beta - D(d - c) > 0$, then the roots of the characteristic equation are negative. □

The exponential delay may affect the stability of the mixed ESS. In fact, in this case, only one strategy is delayed; the asynchronism between the two strategies may increase the possibility of the instability.

4.5.2 *Discrete Delays*

The strategy A would have no delay with probability p_1 or a delay τ_1 with probability $1 - p_1$. The expected payoff to strategy A is given by:

$$\begin{aligned} U_A(s) &= a(p_1 s(t) + (1 - p_1)s(t - \tau)) + b\Big(p_1(1 - s(t)) + (1 - p_1)(1 - s(t - \tau))\Big), \\ &= p_1(a - b)s(t) + (1 - p_1)(a - b)s(t - \tau) + b. \end{aligned}$$

The expected payoff to strategy B remains unchanged and is given by:

$$U_B(t) = cs(t) + d(1 - s(t)).$$

The replicator dynamics is then given by:

$$\dot{s}(t) = s(t)(1 - s(t))\Big(s(t)(p_1(a - b) - c + d) + (1 - p_1)(a - b)s(t - \tau) + b - d\Big). \quad (4.17)$$

The linearized replicator dynamics around the interior stationary point is given by:

$$\dot{x}(t) = s^*(1 - s^*)\Big((p_1(a - b) - c + d)x(t) + (1 - p_1)(a - b)x(t - \tau)\Big). \quad (4.18)$$

Let $\gamma = s^*(1 - s^*)$. We recall that $s^* = \frac{b-d}{b-d+c-a}$, $\gamma = \frac{(b-d)(c-a)}{(b-d+c-a)^2}$, and $b > d$, and $c > a$. (we consider anticoordination games).

Following the same procedure in the previous section, we derive the characteristic equation associated with Eq. 4.18:

$$\lambda - \gamma(p_1(a-b) - c + d) - \gamma(1-p_1)(a-b)e^{-\lambda\tau} = 0 \tag{4.19}$$

Let $\tau = \tau_c$ and $\lambda = iw$. Separating real and imaginary parts in the characteristic equation we get the following system:

$$\cos(w\tau_c) = -\frac{p_1(a-b) - c + d}{(1-p_1)(a-b)}, \tag{4.20}$$

$$\sin(w\tau_c) = -\frac{w}{\gamma(1-p_1)(a-b)}. \tag{4.21}$$

Squaring and summing up the two above equations, we get:

$$w^2 = \gamma^2(b-d+c-a)\Big((2p_1-1)(a-b) - (c-d)\Big).$$

The critical frequency is real if and only if $\big((2p_1-1)(a-b) - (c-d)\big) > 0$. In this case, it is given by:

$$w = |(b-d)(c-a)|\sqrt{\frac{(2p_1-1)(a-b) - c + d}{(b-d+c-a)^3}}. \tag{4.22}$$

From Eq. (4.20), the critical delay is given by:

$$\tau_c = \frac{1}{\gamma\sqrt{(b-d+c-a)\Big((2p_1-1)(a-b) - (c-d)\Big)}}\cos^{-1}\Big(-\frac{p_1(a-b) - c + d}{(1-p_1)(a-b)}\Big)$$

Or equivalently,

$$\tau_c = \frac{1}{|(b-d)(c-a)|}\sqrt{\frac{(b-d+c-a)^3}{(2p_1-1)(a-b) - c + d}}\cos^{-1}\Big(-\frac{p_1(a-b) - c + d}{(1-p_1)(a-b)}\Big).$$

When $p_1 = 0$, we have the deterministic case in which the strategy A has a fixed delay of value τ.

We have the following result on the stability of the mixed ESS.

Theorem 4.9 *The mixed ESS s^* is asymptotically stable in the replicator dynamics with one delayed strategy if $\tau < \tau_c$ where $\tau_c = \frac{1}{(b-d)(c-a)}\sqrt{\frac{(b-d+c-a)^3}{(2p_1-1)(a-b)-c+d}}cos^{-1}$ $\Big(-\frac{p_1(a-b)-c+d}{(1-p_1)(a-b)}\Big)$.*

Proof The characteristic equation is given by:

$$\lambda - \gamma(p_1(a-b) - c + d) - \gamma(1-p_1)(a-b)e^{-\lambda\tau} = 0$$

Let $\lambda = u + iv$ a root of the characteristic equation with $v > 0$ (without loss of generality). We aim to prove that when $u > 0$, then $\tau > \tau_c$. Separating real and imaginary parts in the characteristic equation we get:

$$u - A - B\exp(-u\tau)\cos(v\tau) = 0, \tag{4.23}$$

$$v + B\exp(-u\tau)\sin(v\tau) = 0, \tag{4.24}$$

where $A = \gamma(p_1(a-b) - c + d)$ and $B = \gamma(1-p_1)(a-b)$. We have the relation $A > B$ and $A + B < 0$. First, we study the case when $A < 0$ and $B < 0$. From Eqs. (4.23) and (4.24), we have:

$$\cos(v\tau) \leq -\frac{A}{B} \quad \text{(recall that } u > 0\text{)}, \tag{4.25}$$

$$v^2 \leq B^2 - A^2. \tag{4.26}$$

The system above yields $\tau > \tau_c$. Therefore, we conclude that when $\tau < \tau_c$, the real parts of the roots of the characteristic equation are negative. □

4.6 Conclusions

In this paper, we studied the asymptotic stability of the mixed ESS in presence of random delays. In particular, we investigated the stability of the mixed equilibrium with uniform, exponential and Erlang delay distributions. Furthermore, we studied the case when there is one delayed strategy.

As an extension to this work, we plan to investigate the global stability of the mixed evolutionary stable strategy. We also plan to extend our analysis by considering that an interaction may involve a random number of players.

4.7 Proof of Theorem 4.14

We suppose $\tau_{\max} < \frac{\pi^2}{2D}$. If λ is a real solution of (4.5), then it is clear that λ cannot be positive. Let $\lambda = u + iv$ be a complex root of (4.5) with $u > 0$ and $v > 0$ (without loss of generality we assume that $v > 0$ since if $u + iv$ a solution of (4.5) then $u - iv$ is also a solution). We aim to find a contradiction and hence we prove that no root of (4.5) with a positive real part can exist. We have:

$$v = \frac{D}{v\tau_{\max}} \int_0^{\tau_{\max} v} e^{-\alpha\frac{u}{v}} \sin(\alpha) d\alpha.$$

Let $\tau_{\max} v = 2k\pi + \gamma$ with $k \geq 0$ and $0 \leq \gamma < 2\pi$. We make this change of variable to explore the periodicity of sinus and cosinus functions. Hence we get,

$$\begin{aligned} v^2 &= \frac{D}{\tau_{\max}} \int_0^{2k\pi+\gamma} e^{-\alpha \frac{u}{v}} \sin(\alpha) d\alpha \\ &\leq \frac{D(k+1)}{\tau_{\max}} \int_0^{\pi} \sin(\alpha) d\alpha = \frac{2(k+1)D}{\tau_{\max}}. \end{aligned}$$

Thus we have $v^2 \leq \frac{2(k+1)D}{\tau_{\max}}$; and since $\tau_{\max} < \frac{\pi^2}{2D}$ we get $v\tau_{\max} < (k+1)^{\frac{1}{2}}\pi$. But $v\tau_{\max} = 2k\pi + \gamma$ with $0 \leq \gamma < 2\pi$, which finally yields $k = 0$ and $v\tau_{\max} < \pi$.

On the other hand, we have $u = -\frac{D}{\tau_{\max} v} \int_0^{\tau_{\max} v} e^{-\alpha \frac{u}{v}} \cos(\alpha) d\alpha$. Let us study the sign of the right-hand side of this equation.

If $\tau_{\max} v \leq \frac{\pi}{2}$ then we obtain $u < 0$, this a contradiction.

If $\tau_{\max} v > \frac{\pi}{2}$, then we have

$$\begin{aligned} \int_0^{\tau_{\max} v} e^{-\alpha \frac{u}{v}} \cos(\alpha) d\alpha &= \int_0^{\frac{\pi}{2}} e^{-\alpha \frac{u}{v}} \cos(\alpha) d\alpha + \\ &\quad \int_{\frac{\pi}{2}}^{\tau_{\max} v} e^{-\alpha \frac{u}{v}} \cos(\alpha) d\alpha, \\ &\geq e^{-\frac{\pi u}{2v}} \int_0^{\frac{\pi}{2}} \cos(\alpha) d\alpha + e^{-\frac{\pi u}{2v}} \int_{\frac{\pi}{2}}^{\tau_{\max} v} \cos(\alpha) d\alpha > 0. \end{aligned}$$

We obtain $u < 0$ which is a contradiction with the initial assumption of $u > 0$. This proves the sufficient condition.

Now let us prove the necessary condition. A switch from stability to instability or the inverse corresponds to the existence of a root with a zero real part. We have to prove that when $\tau_{\max} = \frac{\pi^2}{2D}$, the stability is lost and cannot be regained. Let λ a solution of (4.5), and let $\lambda^* = iv$ a purely imaginary solution of (4.5) and let τ^*_{max} the value of τ_{max} when $\lambda = \lambda^*$. We will show that as $\tau_{\max}$ increases, the roots of the characteristic equation can cross the imaginary axis only from the left to the right, and therefore once the stability is lost, it cannot be regained. A stability switch occurs when $\int_0^{vt_{\max}} \cos(\alpha) d\alpha = 0$ and $\int_0^{vt_{\max}} \sin(\alpha) d\alpha \neq 0$ since zero is not a root of (4.5); which yields $v\tau_{\max} = (2k+1)\pi$. When $\tau_{\max} = \frac{\pi^2}{2D}$ we check that the above two conditions are satisfied, and hence this value of $\tau_{\max}$ corresponds to a purely imaginary root. Moreover, after some calculus of the derivative of λ of $\tau_{\max}$ at τ^*_{max}, we get:

$$Re\left(\frac{d\lambda}{d\tau_{\max}}\right)_{\tau_{max}=\tau^*_{max}} = -Dv^2 \cos(v\tau_{\max}) > 0.$$

The derivative of $\lambda(\tau_{max})$ at τ^*_{max} is positive, which means that roots can cross the imaginary axis only from the left to the right. Since the solution of (4.3) is asymptotically stable when $\tau_{\max} < \frac{\pi^2}{2D}$, then the stability is lost at $\tau_{max} = \frac{\pi^2}{2D}$ and cannot be regained. This proves the necessary condition.

4.8 Proof of Theorem 4.16

Let $\beta > \bar{\beta}$. Let $\lambda = u + iv$ be a solution of the characteristic equation with $u > 0$. We suppose without loss of generality that $v > 0$ (since if $u + iv$ is a solution of the characteristic equation than $u - iv$ is also a solution). We aim to find a contradiction and to prove that no root with a positive real part can exist when $\beta > \bar{\beta}$. Substituting λ with $u + iv$ and separating real and imaginary parts in the characteristic equation we get:

$$u + [(\beta + u)^2 + v^2]^{-\frac{k}{2}} D\beta^k \cos(k\theta) = 0,$$
$$v - [(\beta + u)^2 + v^2]^{-\frac{k}{2}} D\beta^k \sin(k\theta) = 0,$$

with $\cos(\theta) = \frac{\beta+u}{[(\beta+u)^2+v^2]^{\frac{1}{2}}}$ and $\sin(\theta) = \frac{v}{[(\beta+u)^2+v^2]^{\frac{1}{2}}}$. The above system can be written:

$$u = \frac{-D\beta^k \cos(k\theta)}{[(\beta + u)^2 + v^2]^{\frac{k}{2}}},$$
$$v = \frac{D\beta^k \sin(k\theta)}{[(\beta + u)^2 + v^2]^{\frac{k}{2}}}.$$

From the equations above, we get:

$$\cos(k\theta) < 0 \Rightarrow \frac{\pi}{2k} < \theta,$$
$$v^2 < D^2 \frac{\beta^{2k}}{[(\beta + u)^2 + v^2]^k} < D^2 \frac{\beta^{2k}}{[\beta^2 + v^2]^k};$$

which yields $\cos(\theta) < \cos(\frac{\pi}{2k})$. In addition, we check that $\frac{\beta}{[\beta^2+v^2]^{\frac{1}{2}}} < \frac{\beta+u}{[(\beta+u)^2+v^2]^{\frac{1}{2}}}$; and then we have $\frac{\beta}{[\beta^2+v^2]^{\frac{1}{2}}} < \cos(\grave{}) < \cos(\frac{\text{B}}{2\text{k}})$. From this inequality, we obtain:

$$\beta^2 \frac{\sin^2(\frac{\pi}{2k})}{\cos^2(\frac{\pi}{2k})} < v^2. \tag{4.27}$$

On the other hand we have $v^2 < D^2 \frac{\beta^{2k}}{[\beta^2+v^2]^k}$ and $\frac{\beta^{2k}}{[\beta^2+v^2]^k} < \cos^{2k}(\frac{\pi}{2k})$, which yields:

$$v^2 < D^2 \cos^{2k}(\frac{\pi}{2k}). \tag{4.28}$$

Using (4.27) and (4.28) we get $\beta^2 < D^2 \frac{\cos^{2k+2}(\frac{\pi}{2k})}{\sin^2(\frac{\pi}{2k})}$, which is in contradiction with the initial assumption of $\beta > \bar{\beta}$. Therefore, there is no root with a positive real part when $\beta > \bar{\beta}$, and the local asymptotic stability follows.

Chapter 5
Coupled State Policy Dynamics in Evolutionary Games

Ilaria Brunetti, Yezekael Hayel and Eitan Altman

In this work we want to extend the EGT models by introducing the concept of individual state. We analyze a particular simple case, in which we associate a state to each player, and we suppose that this state determines the set of available actions. We consider deterministic stationary policies and we suppose that the choice of a policy determines the fitness of the player and it impacts the evolution of the state. We define the interdependent dynamics of states and policies and we introduce the State Policy coupled Dynamics (SPcD) in order to study the evolution of the population profile and we prove the relation between the rest points of our system and the equilibria of the game. We then assume that the processes of states and policies move with different velocities: this assumption allows us to solve the system and then to find the equilibria of our game with two different methods: the singular perturbation method and a matrix approach.

This chapter is organized as follows: in Sect. 5.1 we briefly present standard EGT main definitions and results. In Sect. 5.2 we extend EGT problems introducing individual states. We define the notion of equilibrium profile of the population in such context and the dynamics of states and policies. We show the relation between the rest points of the system and the equilibria of the game. We then provide, in Sect. 5.3, the two different methods to solve this system, which can be applied when assuming a two-timescales behavior. We conclude with some numerical results.

I. Brunetti · Y. Hayel · E. Altman (✉)
University of Cote d'Azur, INRIA, 2004 Route des Lucioles, 06902 Sophia Antipolis, France
e-mail: eitan.altman@inria.fr

E. Altman et al. (eds.), *Multilevel Strategic Interaction Game Models for Complex Networks*, https://doi.org/10.1007/978-3-030-24455-2_5

5.1 Standard Evolutionary Game Theory

5.1.1 Evolutionary Stable Strategy

Consider an infinitely large population of players, where each player repeatedly meets a randomly selected individual within the population. Each individual disposes of a finite pure action space $\mathcal{A}$, $|\mathcal{A}| = K$. Let $\Delta(\mathcal{A}) = \{p \in \mathbb{R}_+^K \sum_{i\in\mathcal{A}} p_i = 1\}$ be the set of mixed strategies, that are probability measures over the action space. We define by $F(p, q)$ the expected payoff of an individual playing p against an opponent using q, where $p, q \in \Delta(\mathcal{A})$. If A is the payoff matrix associated to the pairwise interactions, then $F(p, q) = p^T Aq$.

An ESS is a strategy that, if adopted by the whole population, it is resistant against mutations of a small fraction of individuals in the population. Suppose that the whole population adopts a strategy q, and that a fraction ϵ of *mutants* deviate to strategy p. Strategy q is an ESS if $\forall p \neq q$, there exists some $\epsilon_p > 0$ such that:

$$\forall \epsilon \in (0, \epsilon_p) \qquad F(q, \epsilon p + (1-\epsilon)q) > F(p, \epsilon p + (1-\epsilon)q) \tag{5.1}$$

The following proposition allows to characterize an ESS through its stability properties.

Proposition 5.20 *$q \in \Delta(\mathcal{A})$ is an ESS if and only if it satisfies the following conditions:*

- *Nash Condition: $F(q, q) \geq F(p, q) \quad \forall p$,*
- *Stability Condition: $F(q, q) = F(p, q) \Rightarrow F(q, p) \geq F(p, p) \quad \forall p \neq q$.*

It immediately follows that any strict Nash equilibrium is an ESS,while the converse is not true. When players are programmed to pure actions and mixed ones are not allowed, the notion of ESS is associated to the state of the population instead that to a mixed action, as introduced by Taylor and Jonker in [255]. The population state is given by the vector $q = (q_1, \ldots, q_K)$, where $q_i = \frac{N_i}{N}$ is the proportion of individuals in the population playing pure action i. We observe that $q \in \Delta(\mathcal{A})$, so it is formally equivalent to a mixed action in $\Delta(\mathcal{A})$. Let $F(q, p)$ denotes the immediate expected payoff of a group of individuals in state q playing against a population in state p, and let $F(i, p)$ denotes the immediate expected payoff of an individual playing pure strategy i against a population in state p. We have that: $F(q, p) = \sum_{i=1}^K q_i F(i, p)$.

The definition of the ESS concerning states is then equivalent to that of the ESS concerning strategies: state q is an ESS if $\forall p \neq q$, there exists some $\epsilon_p > 0$ such that $F(q, \epsilon p + (1-\epsilon)q) > F(p, \epsilon p + (1-\epsilon)q), \forall \epsilon < \epsilon_p$.

5.1.2 *Replicator Dynamics*

Let the population be of large but finite size N and let players be programmed to pure actions $\mathcal{A}$. Let N_i be the number of individuals adopting $i \in \mathcal{A}$. The population profile at time t is given by the vector $q(t) = (q_1(t), q_2(t), \ldots, q_K(t)), q(t) \in \Delta(\mathcal{A})$, where $q_i = \frac{N_i}{N}$ is the fraction of individuals playing pure action i. The replicator dynamics describes how the distribution of pure actions evolves in time depending on interactions between individuals. Replicator dynamics of action $i \in \mathcal{A}$ is expressed by the following equation:

$$\dot{q}_i(t) = q_i(t)(F_i(q(t)) - \bar{F}(q(t))), \tag{5.2}$$

where $F_i(q(t))$ is the immediate fitness of an individual playing i and $\bar{F}(q(t))$ is the average immediate fitness of the population. In the two-actions case, with $\mathcal{A} = \{x, y\}$, if $q(t)$ indicates the share of the population playing action x at time t, the latter equation can be rewritten as follows: $\dot{q}(t) = q(t)(1 - q(t))(\bar{F}_x(q(t)) - \bar{F}_y(t))$.

The replicator equation has numerous properties and there is a close relationship between its rest points and the equilibria. The *folk theorem of evolutionary game theory* [139] states that:

1. any Nash equilibrium is a rest point of the replicator equation;
2. if a Nash equilibrium is strict then it's asymptotically stable;
3. if a rest point is the limit of an interior orbit for $t \to \infty$, then it is a Nash equilibrium;
4. any stable rest point of the replicator dynamics is a Nash equilibrium.

Any ESS is an asymptotically stable rest point and an interior ESS is globally stable, but the converse does not hold in general.

5.2 Individual State in EGT Framework

5.2.1 *Individual State and Its Dynamics*

In this section, we introduce the concept of *individual state* in EGT framework and we present the consequent dynamics of our model. We consider a population of N individuals, where each individual can be in one of two possible states, $\mathcal{S} = \{1, 0\}$; every individual goes through a cycle that starts at state 1, moves to states 0 after some random time at a rate that depends on its policy. After some exponentially distributed time it returns to state 1 and so on. At each pairwise interaction, the set of available actions of a player depends on its state: in state 1, $\mathcal{A}_1 = \{x, y\}$, whereas in state 0 an individual can only use y.

We consider the set of deterministic stationary policies, which are functions that associate to each state an action in $\mathcal{A}$ and do not depend on time. Let u_x

(resp. u_y) be the deterministic stationary policy which consists in always playing action x (resp. y) in state 1. In state 0, an individual always plays y. Each player chooses one deterministic stationary policy and we denote by $q_x(t)$ the proportion of individuals in the population that play the deterministic stationary policy u_x at time t. In the same way, $q_y(t)$ denotes the proportion of individuals that play the deterministic stationary policy u_y at time t and we have $q_y(t) = 1 - q_x(t)$.

We suppose that the policy chosen impacts the utility of the player and also the time he spends in state 1. We define by μ_i the rate of decay from state 1 to state 0 when using policy u_i, $i \in \{x, y\}$, where $\mu_x > \mu_y$, and by μ the rate of change from state 0 to state 1. The rates depend on the action induced by the policy chosen by the player. Then, by abuse of description, we say that the rates are controlled by the policies, but in fact, they are controlled by the actions induced by the policies.

We define the proportion of individuals that are in state 1 at time t as $p_1 \equiv \frac{N_1}{N} = \frac{1}{N}\sum_{i=1}^{N} \mathbb{1}_{\{x_i=1\}}$, where x_i denotes the state of player $i \in \{1, \ldots, N\}$ and $\mathbb{1}_{\{x_i=1\}} = 1$ when $x_i = 1$ and it's zero otherwise. We denote by Y_i the random variable $Y_i \equiv \mathbb{1}\{x_i = 1\} \in \{0, 1\}$. Y_i has a Bernoulli distribution, such that $\mathbb{P}(Y_i = 1) = 1 - \mathbb{P}(Y_i = 0) = p_1^i$, and thus $\mathbb{E} = p_1^i$. The individual state dynamics can be described by the following differential equation:

$$\dot{p}_1^i(t) = -\mu_x p_1^i(t) q_x(t) + \mu(1 - p_1^i(t)) - \mu_y p_1^i(t)(1 - q_x(t)),$$

If each individual plays policy u_x with a probability q_x, $(Y_i)_i$ are i.i.d. and thus, for the strong law of large numbers, when $N \to \infty$, $p_1 \to p_1^i$ and consequently:

$$\dot{p}_1(t) = -\mu_x p_1(t) q_x(t) + \mu(1 - p_1(t)) - \mu_y p_1(t)(1 - q_x(t)), \tag{5.3}$$

where $q_x(t)$ is the proportion of individuals in the population that play the deterministic stationary policy u_x at time t.

5.2.2 *Individual Fitness*

At each pairwise interaction, the immediate fitness obtained by an individual depends on his current action and the current action of its opponent. It is given by the following fitness matrix:

$$A := \begin{array}{c} \\ x \\ y \end{array} \begin{array}{c} \quad x \qquad\quad y \\ \begin{pmatrix} (a, a) & (b, c) \\ (c, b) & (d, d) \end{pmatrix} \end{array} \tag{5.4}$$

where x and y are the available actions and the matrix entry $A_{i,j}$ indicates the payoff respectively of the first (row) and the second (column) player. The immediate expected fitness of a player interacting at time t, depends on the population profile at that time. As we added a state component in our framework, the population profile at time t is now expressed by the couple $\xi(t) := (p_1(t), q_x(t))$.

We define $\bar{J}_i^1(\xi(t))$ (resp. $\bar{J}_i^0(\xi(t))$) as the immediate expected fitness of an individual who is in state 1 (resp. 0) and plays deterministic stationary policy u_i against a population whose profile is $\xi(t)$. It follows that the value of $\bar{J}_x^1(\xi(t))$ corresponds to the immediate expected fitness of an individual playing action x, against a population whose profile is $\xi(t)$. We observe that $\bar{J}_y^1(\xi(t)) = \bar{J}_y^0(\xi(t)) = \bar{J}_x^0(\xi(t))$, being all equals to the value of the immediate expected fitness of an individual playing action y; we thus denote it simply as $\bar{J}_y(\xi(t))$. We obtain the following expressions:

$$\bar{J}_x^1(\xi(t)) := p_1(t)(q_x(t)a + (1 - q_x(t))b) + (1 - p_1(t))b,$$

$$\bar{J}_y(\xi(t)) := p_1(t)(q_x(t)c + (1 - q_x(t))d) + (1 - p_1(t))d.$$

The immediate expected fitnesses $\bar{F}_i(\xi(t))$ at time t of an individual playing deterministic stationary policy u_i, $i \in \{x, y\}$, are given by:

$$\bar{F}_x(\xi(t)) = p_1^i(t)\bar{J}_x^1(\xi(t)) + (1 - p_1^i(t)))\bar{J}_y(\xi(t))$$

$$\bar{F}_y(\xi(t)) = \bar{J}_y(\xi(t)).$$

The average expected fitness of the whole population with profile $\xi(t) = (p_1(t), q_x(t))$ is defined as:

$$\bar{F}(\xi(t)) = q_x(t)\bar{F}_x(\xi(t)) + (1 - q_x(t))\bar{F}_y(\xi(t)).$$

5.2.3 Evolutionary Stable Strategies

We want to study the properties of stability of the population profile, supposing that individuals can play only deterministic policies.

We then define the equilibrium profile of the population as follows:

Definition 5.11 (*Equilibrium profile*) A population profile $\xi^* = (p_1^*, q_x^*)$ is an equilibrium profile if $\forall u_i \in supp(q_x^*)$ we have that:

$$\bar{F}_i(\xi^*) \geq \bar{F}_j(\xi^*) \qquad j \neq i,$$

where $supp(q_x^*) = \{u_i,\ i \in \mathcal{A} | q_i^* > 0 \text{ given the state } q_x^*\}$.

Proposition 5.21 *If the population profile $\xi^* = (\tilde{p}_1^*, \tilde{q}_x^*)$ satisfies the indifference principle:*

$$\bar{F}_x(\xi^*) = \bar{F}_y(\xi^*),$$

then it is an equilibrium profile.

We now look for the stability condition of the equilibrium state. This leads to the formal definition of the ESS: in our model it corresponds to an evolutionary stable population profile. Given $\tilde{q}_x \in [0, 1]$ and a population profile $\xi = (p_1, q_x)$, the average fitness of a group of individuals such that a proportion $\tilde{q}_x$ of the group play stationary deterministic policy u_x against a population whose profile is $\xi = (p_1, q_x)$ is given by:

$$\bar{F}_{\tilde{q}_x}(\xi) = \tilde{q}_x \bar{F}_x(\xi) + (1 - \tilde{q}_x)\bar{F}_y(\xi).$$

Definition 5.12 (*ESS*) A population profile $\xi^* = (p_1^*, q_x^*)$ is an ESS if $\forall \xi = (p_1, q_x)$, the two following conditions hold:

1. $\bar{F}_{q_x^*}(\xi^*) \geq \bar{F}_{q_x}(\xi^*)$,
2. $\bar{F}_{q_x^*}(\xi^*) = \bar{F}_{q_x}(\xi^*) \Rightarrow \bar{F}_{q_x^*}(\xi) \geq \bar{F}_{q_x}(\xi)$.

5.2.4 *Policy Based Replicator Dynamics*

As we focus here on policies instead that on strategies, we introduce a policy based replicator dynamics (PbRD), to study the evolution of the share of individuals $q_x(t)$ using pure policy u_x at time t. The PbRD is given by the following equation:

$$\dot{q}_x(t) := q_x(t)(\bar{F}_x(\xi(t)) - \bar{F}(\xi(t))). \tag{5.5}$$

Then, the growth rate of the population share using policy u_x is:

$$\frac{\dot{q}_x(t)}{q_x(t)} = \bar{F}_x(\xi(t)) - \bar{F}(\xi(t)), \tag{5.6}$$

The PbRD can be written as:

$$\begin{aligned} \dot{q}_x(t) &= q_x(t)(1 - q_x(t))(\bar{F}_x(\xi(t)) - \bar{F}_y(\xi(t))), \\ &:= g(p_1(t), q_x(t)). \end{aligned}$$

We now investigate the dynamics of actions, where the fitness is a function of the population profile depending on policies and states. If we pick one random individual in the population at time t, the probability that he plays pure action x, is given by the product $q(t) = q_x(t)p_1(t)$, which leads to: $\dot{q}(t) = \dot{q}_x(t)p_1(t) + q_x(t)\dot{p}_1(t)$. By carrying out the expression of $\dot{q}_x(t)$, after some basic algebra, we get the following equation for the growth rate of the proportion of individuals playing action x in the population at time t:

$$\frac{\dot{q}(t)}{q(t)} = (\bar{F}_x(\xi(t)) - \bar{F}(\xi(t))) + \frac{\dot{p}_1(t)}{p_1(t)} \tag{5.7}$$

Equation (5.7) shows how the evolution of states impacts the dynamics of actions in our context. The growth rate of action x is increasing in the growth rate of state 1. We observe that a sufficiently high growth rate of state 1 can leads to a growing rate of action x even if policy u_x is non-optimal.

5.2.5 State-Policy Coupled Dynamics

The replicator dynamics of our model are defined by the following system of State-Policy Coupled Dynamics (SPcD) which combines the dynamics of the individual state and the dynamics of the policies used in the population:

$$(S)\begin{cases} \dot{p}_1 = h(\xi(t)) \\ \dot{q}_x = g(\xi(t)) \end{cases}$$

where $\xi(t) = (p_1(t), q_x(t))$ corresponds to the population profile. The rest points of the SPcD is the couple $\xi^* = (p_1^*, q_x^*)$ satisfying:

$$\begin{cases} h(\xi^*) = 0 \\ g(\xi^*) = 0. \end{cases} \tag{5.8}$$

Lemma 5.1 *Any interior rest point of the SPcD (S) is a state equilibrium of the state-policy game.*

Remark 5.2 Note that the converse does not necessarily hold. Any equilibrium state is a rest point of the PbRD in (5.5), but it's not necessarily a rest point of the individual state dynamics.

Lemma 5.2 *Any stable rest point of the SPcD (S) is an equilibrium profile of the state-policy game.*

Finally, in order to guarantee that a rest point is an ESS, we need more properties on the rest point of the SPcD. A sufficient conditions to guaratee evolutionarly stability of a rest point is the strong stability [139]. Another method to verify that a rest point ξ^* is an ESS is to construct a suitable local Lyapunov function [140] for the raplicator dynamics in ξ^*.

5.3 Two Time-Scales Behavior

We assume here that the state and the policy dynamics move with different velocities. The individual state dynamics, given by Eq. (5.3), are supposed to move very fast compared to the slow updating strategy process modeled through the SPcD (5.3).

This assumption allows us to find the equilibrium profile of the population with two different approaches. As we showed the relation between the equilibria of the game and the rest points of the SPcD, we can solve the system (S) to find the equilibria. Alternatively, we can consider the stationary distribution of states and rewrite our model as a matrix game.

5.3.1 Singular Perturbations

If we consider the two-time-scales behavior of the system (S), we can approximate its solution using the standard Singular Perturbation Model [157] to find the rest points of the SPcD. We introduce the parameter $\epsilon > 0$, such that:

$$\epsilon \dot{p}_1 := h(p_1, q_x).$$

We rewrite the system of the two coupled differential equations:

$$(S_\epsilon) \begin{cases} \epsilon \dot{p}_1 = h(p_1, q_x), \\ \dot{q}_x = g(p_1, q_x). \end{cases}$$

The parameter ϵ is a small positive scalar which serves to represent the different timescales of the two processes, where the velocity of the state process, $\dot{p}_1 = h(p_1, q_x)/\epsilon$, is fast when ϵ is small.

The theory of singular perturbed differential equations gives an easy way to solve an approximation of the system when $\epsilon \to 0$. We can consider the *quasi-steady-state-model* [157] by first solving in p_1 the transcendental equation $0 = h(p_1, q_x)$ and then rewriting the differential equation $\dot{q}$ as a function of the obtained roots. As the latter equation has a unique real solution $\bar{p}_1 := \pi_1(q_x)$, our system is in *normal form*. This allows us to solve the second differential equation called the quasi-steady-state equation:

$$\dot{q}_x = g(\pi_1(q_x), q_x). \tag{5.9}$$

As the assumption [157] $\frac{\partial h}{\partial p_1}(p_1, q_x) < 0$ is satisfied, the reduced model is a good approximation of the original system. The two-time-scale behavior of $p_1(t)$ and $q_x(t)$ has a geometric interpretation, as trajectories in $\mathbb{R}^2$. If we define the manifold sets $M_\epsilon := \{\phi \;\text{ s.t. }\; p_1 = \phi(q_x, \epsilon) \;\text{ and }\; \epsilon = h(q_x, \phi(q_x, \epsilon))\}$, it is possible to rewrite the problem in terms of invariant manifolds. When $\epsilon = 0$, the manifold M_0 corresponds to the expression of the quasi steady state model. When the condition $\frac{\partial h}{\partial p_1}(p_1, q_x) < 0$ is satisfied, we have that the equilibrium manifold M_0 is stable (attractive). Particularly, the important result is that the existence of a conditionally stable manifold M_0 for $\epsilon = 0$ implies the existence of an invariant manifold M_ϵ satisfying the following convergence for all $\epsilon \in [0, \epsilon^*]$:

$$\phi(\epsilon, q_x) \to \phi(0, q_x), \quad \text{and} \quad M_\epsilon \to M_0 \quad \text{as} \quad \epsilon \to 0.$$

The positive constant ϵ^* is determined such that the following manifold condition is satisfied:

$$\epsilon \frac{\partial \phi}{\partial x} g(\phi(q_x, \epsilon), q_x) = h(\phi(q_x, \epsilon), q_x),$$

for all q_x and $\epsilon \in [0, \epsilon^*]$. The attractiveness of the slow manifold M_0 is illustrated in the numerical illustrations section. Let us now compute the solution of the approximate system (S_0). We then consider the stationary regime of the individual state dynamics (expressed by Eq. (5.3)). By imposing $\dot{p}_1 = 0$, we obtain the following slow manifold $M_0 := \{\phi \ \text{ s.t. } \ p_1 = \phi(q_x, 0) \ \text{ and } \ 0 = h(q_x, \phi(q_x, 0))\}$:

$$\phi(q_x, 0) = \frac{\mu}{\mu + \mu_x q_x + \mu_y(1 - q_x)} := \pi_1(q_x). \tag{5.10}$$

The PbRE (5.5) can now be rewritten as:

$$\begin{aligned} \dot{q}_x(t) = q_x(t)(1 - q_x(t)) \times \ldots \\ \left[\bar{F}_x(\pi_1(q_x(t)), q_x(t)) - \bar{F}_y(\pi_1(q_x(t)), q_x(t))\right]. \end{aligned}$$

Proposition 5.22 *Considering the singular perturbations method with $\epsilon \to 0$, the solutions of the coupled differential equations (S_0) is the population profile $\xi^* = (p_1^*, q_x^*)$, such that:*

$$p_1^* = \frac{\mu - s^*(\mu_x - \mu_y)}{\mu + \mu_y} \quad \textit{and} \quad q_x^* = \frac{s^*(\mu + \mu_y)}{\mu - s^*(\mu_x - \mu_y)}, \tag{5.11}$$

where s^* is the equilibrium of the standard replicator dynamics (5.2) when considering payoff matrix (5.4):

$$s^* = \frac{d - b}{\Delta} \quad \textit{with} \quad \Delta = a - b - c + d.$$

Note that the rest point q_x^* of the PbRE (5.5) verifies:

$$q_x^* \pi_1(q_x^*) = s^*.$$

This result says that the equilibrium probability that any individual picked out randomly in the population, is playing action x is equal to s^*. This value is the mixed equilibrium of the standard matrix game. It means that, if we consider a state dependent action game, the equilibrium is obtained under conditional probability over the state.

We have the following necessary and sufficient condition under which the solution obtained is a strict interior point.

Lemma 5.3 *The solution q_x^* obtained in Proposition 5.22 is a strict interior point if and only if:*

$$\mu > \mu_x \frac{s^*}{1 - s^*}.$$

An important remark is that this condition does not depend on the rate μ_y.

In the next section, we present an alternative method based on rewriting our game problem into a matrix game considering only pure policies.

5.3.2 Matrix Approach

The two time-scales assumption implies that we can consider individuals in stationary state. We can thus rewrite our model as a matrix game in which individuals play stationary deterministic policies instead of actions. We get the following bimatrix game:

$$\begin{array}{cc} & \begin{array}{cc} u_y & \qquad\qquad u_x \end{array} \\ \begin{array}{c} u_y \\ u_x \end{array} & \begin{pmatrix} (J(u_y, u_y), J(u_y, u_y)) & (J(u_y, u_x), J(u_x, u_y)) \\ (J(u_x, u_y), J(u_y, u_x)) & (J(u_x, u_x), J(u_x, u_x)) \end{pmatrix} \end{array}$$

where $J(u_i, u_j)$ is the expected average fitness of an individual playing pure policy u_i against an individual using u_j, $i, j \in \{x, y\}$. The stationary distributions in states 1 and 0 are given by the following time ratios:

$$\bar{T}_1(i) = \frac{\frac{1}{\mu_i}}{\frac{1}{\mu} + \frac{1}{\mu_i}} = \frac{\mu}{\mu + \mu_i} \quad \text{and} \quad \bar{T}_0(i) = \frac{\mu_i}{\mu + \mu_i},$$

where $i \in \mathcal{A}$ denotes the choice of policy u_i. The expected average fitness $J(u_i, u_j)$ can thus be expressed as a function of the time ratios:

$$J(u_i, u_j) = \sum_{s,s' \in \mathcal{S}} \sum_{a,a' \in \mathcal{A}} T_s(a) T_{s'}(a') R(a, a'),$$

where $R(a, a')$ is the immediate fitness of a player using action a against an opponent playing a'. We consider the payoffs bimatix (5.4) and we thus obtain:

$$\begin{aligned} J(u_y, u_y) &= d, \quad J(u_x, u_y) = T_1(x)b + T_0(x)d, \\ J(u_y, u_x) &= T_1(x)c + T_0(x)d, \\ J(u_x, u_x) &= T_1(x)\left[T_1(x)a + T_0(x)b\right] \\ &\quad + T_0(x)\left[T_1(x)c + T_0(x)d\right]. \end{aligned}$$

If we consider this matrix game as a representation of a standard evolutionary game, we can write the replicator dynamics equation for this evolutionary game as:

$$\dot{\delta}(t) = \delta(t)(1 - \delta(t))(J(u_x, \delta(t)) - J(u_y, \delta(t))), \tag{5.12}$$

where $\delta(t)$ is the proportion of individuals in the population who play pure policy u_x at time t. The standard replicator dynamics equation for this matrix game, if it converges, converges to the ESS of the evolutionary game. The mixed equilibrium δ^* for this matrix game is obtained by solving the indifference principle equation: $J(u_y, \delta^*) = J(u_x, \delta^*)$, where $J(u_i, q) = (1 - q)J(u_i, u_y) + qJ(u_i, u_x)$ with $i \in \mathcal{A}$. We know from standard evolutionary game theory that the mixed equilibrium is expressed by:

$$\delta^* = \frac{J(u_y, u_y) - J(u_x, u_y)}{J(u_x, u_x) - J(u_y, u_x) + J(u_y, u_y) - J(u_x, u_y)}.$$

Thus, after some algebras, the ESS obtained by rewriting our game by considering stationary deterministic policies and average fitnesses is:

$$\delta^* = \frac{s^*}{\bar{T}_F(x)}. \tag{5.13}$$

5.3.3 Relations Between Equilibria

We Now compare the two equilibria we obtained by considering two different point of view of the problem. In Sect. 5.3.1, we supposed that each individual plays a deterministic policy u_x which consists in always choosing action x in state F and, by applying the singular perturbation method, we have been able to determine the equilibrium of such a game. In Sect. 5.3.2, we assumed that individuals are in their stationary states and we rewrite the game as a standard evolutionary game.

Proposition 5.23 *The relation between the equilibrium δ^* and the equilibrium q_x^* is the following:*

$$q_x^* < \delta^*.$$

We are able also to compare the two equilibria in terms of average fitness obtained by the population, i.e. $J(\delta^*, \delta^*)$ and $\bar{F}(q_x^*, p_1^*)$.

Proposition 5.24 *The average fitnesses of the population at the two equilibria points obtained with the two approaches are equals, i.e. $\bar{F}(q_x^*, p_1^*) = J(\delta^*, \delta^*)$.*

Finally, we prove that the two mixed strategies obtained with the two approaches are not equal, but are in the same equivalent class in terms of occupation measures. We denote by $\bar{T}_1(q)$ the average sojourn time in state F for an individual who plays mixed strategy q. This mixed strategy has two possibilities. First, it can be a mixed policy between the pure policies u_y and u_x, like proposed in the Sect. 5.3.2. Second, a mixed policy characterized by a probability q to play action x in state F, at each time an individual is in state F. This second point of view is proposed in Sect. 5.3.1. The two equilibria obtained by the singular perturbations method and the matrix game reformulation are in the same equivalent class in terms of the occupation measures. It means that they should satisfy: $\bar{T}_1(\delta^*) = \bar{T}_1(q_x^*)$.

We have for the case of pure policies:

$$\bar{T}_1(\delta^*) = \delta^* \frac{\mu}{\mu + \mu_x} + (1 - \delta^*) \frac{\mu}{\mu + \mu_y}. \tag{5.14}$$

For the case of mixed policies, we have:

$$\bar{T}_1(q_x^*) = \pi_1(q_x^*) = \frac{\mu}{\mu + \mu_x q_x^* + \mu_y(1 - q_x^*)}. \tag{5.15}$$

This important result is proved in the following proposition.

Proposition 5.25 *The mixed equilibrium δ^* over the pure policies and the equilibrium obtained by the singular perturbation approach yield to the same occupation measures, i.e.*

$$\bar{T}_1(\delta^*) = \bar{T}_1(q_x^*).$$

This two previous results show that we can define two equivalent classes for deterministic stationary policies that yield same average fitness and occupation measures. This also leads us to generalize to several states and actions.

5.4 Numerical Illustrations

We illustrate here the theoretical results obtained in previous sections with numerical solutions. We consider a first numerical example with the following transition rates: $\mu = 10$, $\mu_x = 1.5$ and $\mu_y = 1$. The fitnesses of the matrix game are: $a = -0.3$, $c = 0$, $b = 1$ and $d = 0.5$. Those values yield to the following equilibrium of the standard evolutionary game $s^* = \frac{5}{8} = 0.625$.

We plot on Fig. 5.1 the trajectories of the system (S_ϵ) of the coupled differential equations for different initial conditions and for $\epsilon = 0.01$. We simulate a discrete time version of the differential equations. We plot also the invariant manifold M_0 and we observe that it is an attractor of the trajectories.

More, based on Proposition 5.22 we have the following solution of the system, by considering the singular perturbation method based on the steady-state model:

$$q_x^* = 0.7097, \quad \text{and} \quad p_1^* = 0.8807.$$

This couple corresponds exactly to the attractor of the trajectories on Fig. 5.1.

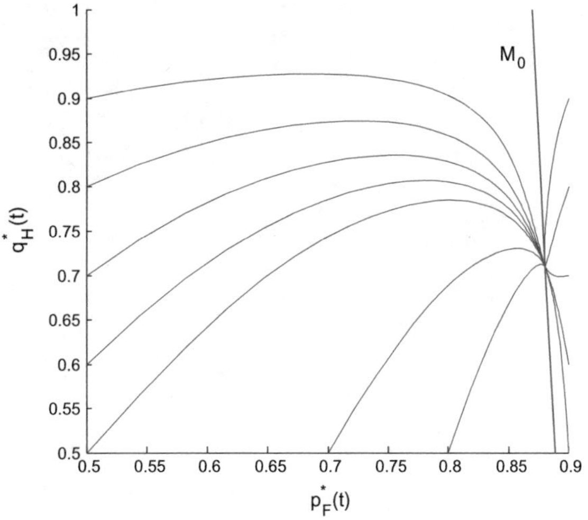

Fig. 5.1 Trajectories of the system (S_ϵ) from different starting points and the slow manifold M_0 with $\epsilon = 0.01$

5.5 Conclusions and Perspectives

In this chapter, we have considered a particular type of evolutionary game in which the action of the individual not only determines its immediate fitness but it also impacts his state. The aim is to describe the coupled dynamics of the individual states which is due to the direct control and the dynamics of policies which is determined by the replicator dynamics mechanism. Once we have introduced these combined dynamics, we have proved that any stable rest point corresponds to an equilibrium of the game. We have proposed two methods to obtain the rest points under the assumption that the two dynamics have different time scales. Finally, we discuss how to generalize our framework to any finite number of states and actions, which can lead to several applications of our framework in networking, social networks and complex systems. In perspective, we propose to further investigate the general case to extend our SPcD to the MDEG framework. We are also interested in studying the stability conditions for a rest point of the SPcD to be an ESS profile of our game.

Part II
Epidemics/Information Diffusion on Networks

Huijuan Wang

Epidemic spreading processes on networks have been widely used to model the diffusion of information, the propagation of computer worms and infectious diseases, the cascade of failures across infrastructures, the financial contagion in trade markets and the spread of social opinions etc.

One of the most studied epidemic models is the Susceptible-Infected-Susceptible (SIS) model. In the SIS model, the state of each node at any time t is a Bernoulli random variable, where $X_i(t) = 0$ if node i is susceptible and $X_i(t) = 1$ if it is infected. The recovery (curing) process of each infected node is an independent Poisson process with a recovery rate δ. Each infected agent infects each of its susceptible neighbors with a rate β, which is also an independent Poisson process. The ratio $\tau \triangleq \beta/\delta$ is the effective infection rate. A phase transition has been observed around a critical point τ_c in a single network. When $\tau > \tau_c$, a non-zero fraction of agents will be infected in the meta-stable state, whereas if $\tau < \tau_c$, infection rapidly disappears [46, 67].

The steady state of the SIS model is the susceptible state for all the nodes, which is the only absorbing state of the exact Markovian process. However, this absorbing state will be reached within an unrealistically long time for realistic sizes of networks [113]. We are interested in the meta-stable state in which the system stays for long and which will be reached fast and better characterizes real epidemics. The epidemic threshold τ_c and the average fraction of infected nodes in the meta-stable state for a given effective infection rate have been widely studied, especially their relation to properties of the network on which the SIS model is deployed.

We witness the progress of analytical approaches towards viral spreading models, which gradually take the heterogeneity in the network structure into account. Early compartment models, stemming from epidemiology assumes a homogenous population, i.e. at any time, each node is connected to the same number of nodes that are randomly selected. They predicted the phase transition around the epidemic threshold [39]. Moreno et al. [291] considered a network with heterogeneous degree distribution and Pastor-Satorras and Vespignani [213] studied epidemic spreading on scale-free networks. Although assuming still homogenous mixing, i.e. at any time, the neighbors of a node are selected randomly, ignoring the detailed connectivity of individuals, they revealed the importance of degree heterogeneity in influencing the behavior of viral spreading. Later, the heterogeneity in the connectivity of individuals was further captured by individual-based models, which studied viral spreading on a fixed social contact network. The N-Intertwined Mean field Approximation (NIMFA) is one of the mostly used individual based models [276, 272]. A substantial line of research has been devoted to understand the influence of the single network structure on the spreading dynamics.

In this part, we are going to present an overview of recent advances in viral spreading with respect to its relation to the underlying generic network structures including both single and interconnected networks (Chaps. 8 and 9) and its interplay with the underlying adaptive network (Chap. 10). Moreover, we will generalize and explore the classic immunization problem in the framework of game theory (Chap. 11).

Chapter 6
Community Networks with Equitable Partitions

Stefania Ottaviano, Francesco De Pellegrini, Stefano Bonaccorsi, Delio Mugnolo and Piet Van Mieghem

A community structure is an important non-trivial topological feature of a complex networks. Indeed community structures are a typical feature of social networks, tightly connected groups of nodes in the World Wide Web usually correspond to pages on common topics, communities in cellular and genetic networks are related to functional modules [46].

Thus, in order to investigate this topological feature, we consider, in this chapter, that the entire population is partitioned into communities (also called households, clusters, subgraphs, or patches). There is an extensive literature on the effect of network community structures on epidemics. Models utilizing this structure are commonly known as "metapopulation" models (see, e.g., [10, 125, 178]). Such models

---

Parts of this chapter have been written based on a thesis archived on Open Archives Initiative. http://eprints-phd.biblio.unitn.it/1684/.

S. Ottaviano · S. Bonaccorsi
University of Trento, Trento, Italy
e-mail: stefania.ottaviano@unitn.it

S. Bonaccorsi
e-mail: stefano.bonaccorsi@unitn.it

F. De Pellegrini (✉)
University of Avignon, Avignon, France
e-mail: francesco.de-pellegrini@univ-avignon.fr

D. Mugnolo
University of Hagen, Hagen, Germany
e-mail: delio.mugnolo@fernuni-hagen.de

P. Van Mieghem
Faculty of Electrical Engineering, Mathematics and Computer Science, Delft University of Technology, Delft, The Netherlands
e-mail: p.f.a.vanmieghem@tudelft.nl

E. Altman et al. (eds.), *Multilevel Strategic Interaction Game Models for Complex Networks*, https://doi.org/10.1007/978-3-030-24455-2_6

assume that each community shares a common environment or is defined by a specific relationship.

Some of the most common works on metapopulation regard a population divided into households with two level of mixing [31, 32, 220]. These models typically assume that contacts, and consequently infections, between nodes in the same group occur at a higher rate than those between nodes in different groups [35]. Thus, groups can be defined, e.g., in terms of spatial proximity, considering that between-group contact rates (and consequently the infection rates) depend in some way on spatial distance. In this type of heterogeneous contact networks each node can be theoretically infected by any other node. However, an underlying network contact structure, where infection can only be transmitted by node directly linked by an edge, may provide a more realistic approach for the study of the evolution of the epidemics, [35]. In turn, an important challenge is how to consider a realistic underlying structure and appropriately incorporate the influences of the network topology on the dynamics of epidemics [33, 34, 98, 212, 270, 283].

6.1 Equitable Partitions

We consider the diffusion of epidemics over an undirected graph $G = (V, E)$ with edge set E and node set V. The order of G, denoted N, is the cardinality of V, whereas the size of G is the cardinality of E, denoted L. Connectivity of the graph G is conveniently encoded in the $N \times N$ adjacency matrix A. We are interested in the case of networks that can be naturally partitioned into n communities: they are represented by a node set partition $\pi = \{V_1, ..., V_n\}$, i.e., a sequence of mutually disjoint nonempty subsets of V, called cells, whose union is V.

The dynamics of epidemic transmission is modeled by an SIS (susceptible-infected-susceptible) model, where we consider the continuous-time mean-field approximation of the exact Markovian SIS model, NIMFA [274]. We consider a curing rate δ equals for all nodes. Instead, compared to the homogeneous case, where the infection rate is the same for all pairs of nodes, in this framework we consider two infection rates: the *intra-community* infection rate β, for infecting individuals in the same community, and the *inter-community* infection rate $\varepsilon\beta$, i.e., the rate at which individuals among different communities get infected. We assume $0 < \varepsilon < 1$, the customary physical interpretation being that infection across communities occur at a much smaller rate. Indeed the presence of communities generates a strong mixing effect at local level (e.g., the rate of infection inside a community tends to be homogeneous) as opposed to the much lower speed of mixing (i.e., much larger inhomogeneity) within the whole population. Specifically, the contact newtork structure that we consider has an *equitable partition* of its node set. The original definition of equitable partition is due to Schwenk [234].

Definition 6.13 Let $G = (V, E)$ be a graph. The partition $\pi = \{V_1, ..., V_n\}$ of the node set V is called *equitable* if for all $i, j \in \{1, \ldots, n\}$, there is an integer d_{ij} such that

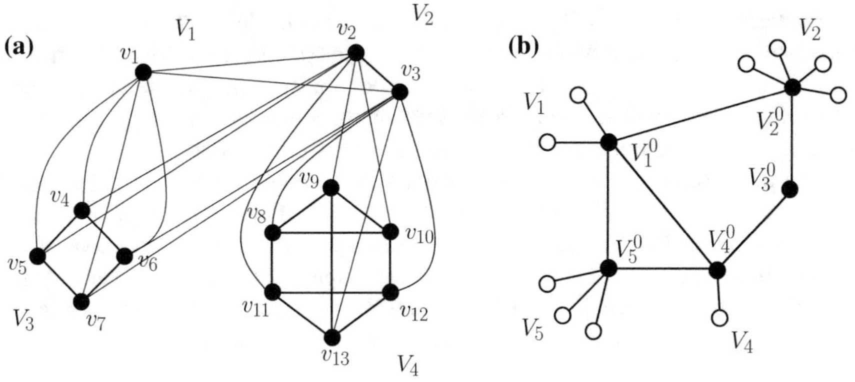

Fig. 6.1 A sample graphs with equitable partition. **a** $V = \{V_1, V_2, V_3, V_4\}$, **b** Interconnected star networks: $V = \{V_1^0, V_2^0, V_3^0, V_4^0, V_5^0, V_1, V_2, V_3, V_4, V_5\}$

$$d_{ij} = \deg(v, V_j) := \# \left\{ e \in E : e = \{v, w\}, w \in V_j \right\}.$$

independently of $v \in V_i$.

We shall identify the set of all nodes in V_i with the i-th *community* of the whole population. Basically this means that all nodes belonging to the same community have the same internal degree: the subgraph G_i of $G(V, E)$ induced by V_i is regular for all i's (recall that $\pi = \{V_1, ..., V_n\}$ is a partition of the node set V, which is assumed to be given a priori). Furthermore, for any two subgraphs G_i, G_j, each node in G_i is connected with the same number of nodes in G_j (see as examples Fig. 6.1).

Thus, a network with an equitable partition of its node set posses a sort of interesting symmetry properties, where with the word symmetry we refer to a certain structural regularity of the graph connectivity. Such kind of network structure can be observed, e.g., in the architecture of some computer networks where clusters of clients connect to single routers, whereas the routers' network has a connectivity structure with nodes' degree constrained by the number of ports. An other real-word circumstance, that can be model by an equitable partition of the population, is the epidemics transmission between e.g. households, classes in a school or work offices in the same department, i.e. small communities whose members know each other. This framework can be modeled representing the internal structure of each community by a complete graph. Moreover given two connected communities, all of their nodes can be considered mutually linked, indeed each member of those two communities may potentially come into contact.

Equitable partitions appears also in the study of synchrony and pattern formation in coupled cell networks [108, 250] where they are named "balanced" partitions. Equitable partitions have been used also to analyze the controllability of multi-agent systems, for the case of a multi-leader setting [216], and for the leader-selection controllability problem, in characterizing the set of nodes from which a given networked control system (NCS) is controllable/uncontrollable [7]. These works show interesting realistic scenarios for the use of equitable partitions. Since the size of some real networks might pose limitations in our ability to investigate their spectral properties, as we shall see, we can leverage on the structural regularity of network with equitable partition to reduce the dimensionality of our system.

The macroscopic structure of a network with an equitable partition of its node set can be described by a *quotient graph* G/π, which is a *multigraph* with cells $V_1, \ldots, V_n$ as vertices and $k_i d_{ij}$ edges between V_i and V_j. For the sake of explanation, in the following we will identify G/π with the (simple) graph having the same vertex set, and where an edge exists between V_i and V_j if at least one exists in the original multigraph. We shall denote by B the adjacency matrix of the graph G/π.

Remark 6.3 We use the notation lcm and gcd to denote the least common multiple and greatest common divisor, respectively. We can observe that the partition of a graph is equitable if and only if

$$d_{ij} = \alpha \frac{\mathrm{lcm}(k_i, k_j)}{k_i}$$

where α is an integer satisfying $1 \leq \alpha \leq \gcd(k_i, k_j)$ and k_i the number of nodes in V_i, for all $i = 1, ..., n$.

Example 6.1 Let us assume that the adjacency matrix B of the quotient graph is given and that, for any $i, j \in \{1, \ldots, n\}$, $b_{ij} \neq 0$ implies $d_{ij} = k_j$, i.e., each node in V_i is connected with every node inside V_j. We can explicitly write the adjacency matrix A in a block form. Let $C_{V_i} = (c_{ij})_{k_i \times k_i}$ be the adjacency matrix of the subgraph induced by V_i and $J_{k_i \times k_j}$ is an all ones $k_i \times k_j$ matrix; then

$$A = \begin{bmatrix} C_{V_1} & \varepsilon J_{k_1 \times k_2} b_{12} & .. & \varepsilon J_{k_1 \times k_n} b_{1n} \\ \varepsilon J_{k_2 \times k_1} b_{21} & C_{V_2} & .. & \varepsilon J_{k_2 \times k_n} b_{2n} \\ . & . & .. & . \\ . & . & .. & . \\ . & . & .. & C_{V_n} \end{bmatrix} \tag{6.1}$$

We observe that (6.1) represents a block-weighted version of the adjacency matrix A. The derivation of NIMFA for the case of two different infection rates, considered in this paper, results in the replacement of the unweighted adjacency matrix in the NIMFA system (6.6) with its weighted version.

A matrix smaller than the adjacency matrix A, that contains the relevant information for the evolution of the system, is associated with the quotient graph. Such a matrix is the *quotient matrix* Q of the equitable partition.

The quotient matrix Q can be defined for any equitable partition: in view of the internal structure of a graph with an equitable partition, it is natural to consider the cell-wise average value of a function on the node set, that is to say the projection of the node space into the subspace of cell-wise constant functions.

Definition 6.14 Let $G = (V, E)$ a graph. Let $\pi = \{V_i,\ i = 1, \ldots, n\}$ be any partition of the node set V, let us consider the $n \times N$ matrix $S = (s_{iv})$, where

$$s_{iv} = \begin{cases} \frac{1}{\sqrt{|V_i|}} & v \in V_i \\ 0 & \text{otherwise.} \end{cases}$$

The *quotient matrix* of G (with respect to the given partition) is

$$Q := SAS^T.$$

Observe that by definition $SS^T = I$.

In the case of equitable partitions, the expression for Q writes

$$Q = \text{diag}(d_{ii}) + (\sqrt{d_{ij}d_{ji}}\varepsilon b_{ij})_{i,j=1,\ldots n}.$$

A key feature of the model is that the spectral radius of this smaller quotient graph (which only captures the macroscopic structure of the community network) is all we need to know in order to decide whether the epidemics will go extinct in a reasonable time frame. Indeed, the spectral radius is related to the epidemic threshold of the system. NIMFA determines the epidemic threshold for the effective spreading rate β/δ as $\tau_c^{(1)} = \frac{1}{\lambda_1(A)}$, where $\lambda_1(A)$ is the spectral radius of A and the superscript (1) refers to the first-order meanfield approximation [50, 274]. Since Q and A have the same spectral radii [267, art. 62] we can compute the spectral radius of Q in order to estimate the epidemic threshold. This may lead to a significative computational advantage in the calculation of $\tau_c^{(1)}$, since the order of Q is smaller than that of A [50, Sect. 3.3].

6.1.1 Lower Bounds for the Epidemic Threshold

We can write $Q = D + \widehat{B}$, where $D = \text{diag}(d_{ii})$ and $\widehat{B} = (\sqrt{d_{ij}d_{ji}}\varepsilon b_{ij})_{i,j=1,\ldots n}$. By the Weyl's inequality [180] we have

$$\lambda_1(Q) \leq \lambda_1(D) + \lambda_1(\widehat{B}) = \max_{1 \leq i \leq n} d_{ii} + \lambda_1(\widehat{B}). \tag{6.2}$$

Since

$$\tau_c^{(1)} = 1/\lambda_1(A) = 1/\lambda_1(Q),$$

a lower bound for the epidemic threshold can be derived from (6.2)

$$\tau_c^{(1)} \geq \tau^\star = \min_i \frac{1}{d_{ii} + \lambda_1(\widehat{B})}, \tag{6.3}$$

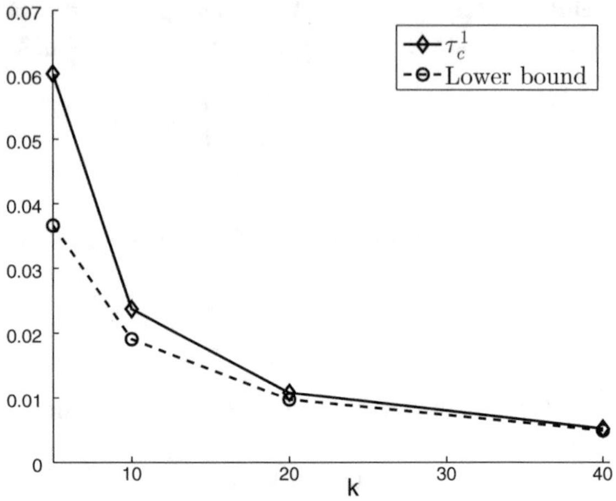

Fig. 6.2 Lower bound (6.3) versus epidemic threshold: comparison for different values of k in a 40-communities network. The internal structure of each community is a ring and $d_{ij} = 2$ for all $i, j = 1, \ldots, n$

Moreover let us note that $\lambda_1(\widehat{B}) \leq \max_i \sum_j \hat{b}_{ij}$ [181, pp. 24–26], hence

$$\tau_c^{(1)} \geq \frac{1}{\max_i(d_{ii} + \sum_j \hat{b}_{ij})}. \tag{6.4}$$

Figure 6.2 reports on the comparison of the lower bound (6.3) and the actual threshold value: it refers to the case of a sample equitable partition composed of interconnected rings for increasing values of the community order.

We observe that obtaining a lower bound for $\tau_c^{(1)}$ is meaningful because $\tau_c^{(1)}$ is itself a lower bound for the epidemic threshold τ_c of the exact stochastic model, i.e., $\tau_c = \alpha\tau_c^{(1)}$ with $\alpha \geq 1$ [61, 274]. In fact, smaller values of the effective spreading rate τ, namely $\delta > \beta/\tau_c^{(1)}$, correspond, in the exact stochastic model, to a region where the infectious dies out exponentially fast for sufficiently large times [106, 262, 269, 273]. Thus, in applications, when designing or controlling a network, τ^* (or the more conservative bound in (6.4)) can be adopted to determine a safety region $\{\tau \leq \tau^\star\}$ for the effective spreading rate that guarantees the extinction of epidemics in a reasonable time frame (above the threshold, the overall-healthy state is only reached after an unrealistically long time).

Equality can be attained in (6.3): consider for instance the graph described by the adjacency matrix A in (6.1). Furthermore, we may require that all V_i's have the same number of nodes $k_i = k$ and same internal degree $d_{ii} = d$, $i = 1, \ldots, n$. In this case $Q = d\,\mathrm{Id}_n + \widehat{B}$, where $\widehat{B} := (k\varepsilon b_{ij})_{i,j=1,\ldots n}$, and

$$\lambda_1(Q) = d + k\varepsilon\lambda_1(B),$$

which is the exact value of $\lambda_1(A)$ and consequently of $\tau_c^{(1)}$.

Remark 6.4 Let us underline that if we remove edges between the communities, or inside the communities, in a network whose set nodes has an equitable partition, the lower bound (6.3) still holds. This because the spectral radius of an adjacency matrix is monotonically non increasing under the deletion of edges.

6.1.2 *Infection Dynamics for Equitable Partitions*

The NIMFA model describes the process of diffusion of epidemics on a graph by expressing the time-change of the probability p_i that node i is infected.

Thus, node i obeys a following differential equation [274]

$$\frac{dp_i(t)}{dt} = \beta \sum_{j=1}^{N} a_{ij} p_j(t)(1 - p_i(t)) - \delta_i(t), \qquad i \in \{1, \ldots, N\} \tag{6.5}$$

The time-derivative of the infection probability of node i consists of two competing processes:

1. while healthy (with probability $1 - p_i(t)$), all infected neighbors, whose average number is $s_i(t)$, try to infect node i at rate β;
2. while node i is infected (with probability $p_i(t)$) it is cured at rate δ.

The following matrix representation of (6.5) holds

$$\frac{dP(t)}{dt} = (\beta A - \delta I)P(t) - \beta\, \text{diag}(p_i(t))AP(t). \tag{6.6}$$

where $P(t) = (\, p_1(t)\, p_2(t) \ldots p_N(t)\,)^T$ and $\text{diag}(p_i(t))$ is the diagonal matrix with elements $p_1(t), p_2(t), \ldots, p_N(t)$. Clearly we study the system for $(p_1, \ldots, p_N) \in I_N = [0, 1]^N$. It can be shown that the system (6.6) is positively invariant in I_N, i.e. if $P(0) \in I_N$ then $P(t) \in I_N$ for all $t > 0$ [163, Lemma 3.1].

The following theorem shows under which conditions the matrix Q can be used in order to express the epidemic dynamics introduced in (6.6). This allows us to describe the time-change of the infection probabilities by a system of n differential equations instead of N. For the proof we refer to [50].

Theorem 6.10 *Let $G = (V, E)$ a graph and $\pi = \{V_j,\ j = 1, \ldots, n\}$ an equitable partition of the node set V. Let G_j be the subgraph of $G = (V, E)$ induced by cell V_j. If $p_h(0) = p_w(0)$ for all $h, w \in G_j$ and for all $j = 1, \ldots, n$, then $p_h(t) = p_w(t)$ for all $t > 0$. In this case we can reduce the number of equations representing the time-change of infection probabilities using the quotient matrix Q.*

Basically the theorem shows that the following subset of I_N, defined by restricting nodes in the same community to have the same state

$$M = \{P \in [0,1]^N | p_1 = \ldots = p_{k_1} = \overline{p}_1, p_{k_1+1} = \ldots = p_{k_1+k_2} = \overline{p_2}, \\ \ldots, p_{(k_1+..k_{n-1}+1)} = \ldots = p_N = \overline{p}_n\}$$

is a positively invariant set for the system (6.6).

Thus, let us consider $P(0) \in M$ and $\overline{P} = (\overline{p}_1, \ldots, \overline{p}_n)$, we can write

$$\frac{d\overline{p}_j(t)}{dt} = \beta(1 - \overline{p}_j(t)) \sum_{m=1}^{n} \varepsilon b_{jm} d_{jm} \overline{p}_m(t) \\ + \beta d_j (1 - \overline{p}_j(t)) \overline{p}_j(t) - \delta \overline{p}_j(t), \qquad j = 1, \ldots, n \tag{6.7}$$

After some manipulations we arrive to the following matrix representation of (6.7)

$$\frac{d\overline{P}(t)}{dt} = \beta \left(\mathrm{I}_n - \mathrm{diag}(\overline{p}_j(t)) \right) \widetilde{Q} \overline{P}(t) - \delta \overline{P}(t), \tag{6.8}$$

where $\widetilde{Q} = \mathrm{diag}\left(\frac{1}{\sqrt{k_j}} \right) Q \, \mathrm{diag}(\sqrt{k_j})$. It is immediate to observe that $\sigma(Q) = \sigma(\tilde{Q})$.

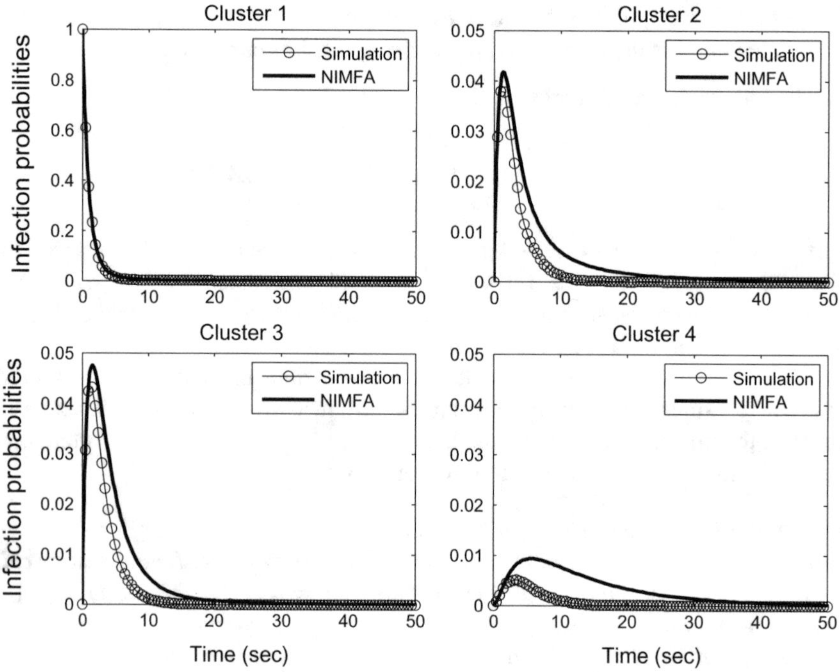

Fig. 6.3 Dynamics of infection probabilities for each community of the network in Fig. 6.1: simulation versus numerical solutions of (6.8); $\tau = \beta/\delta < \tau_c^{(1)} = 0.3178$, with $\beta = 0.29$ and $\delta = 1$, $\varepsilon = 0.3$. At time 0 the only infected node is node 1

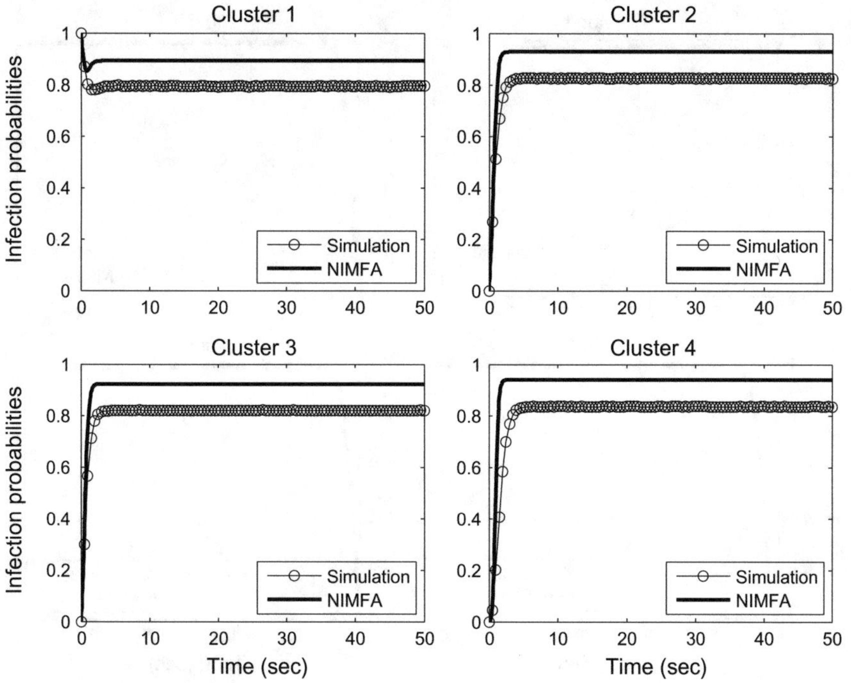

Fig. 6.4 Dynamics of infection probabilities for each community of the network in Fig. 6.1: simulation versus numerical solutions of (6.8); $\tau = \beta/\delta > \tau_c^{(1)} = 0.3178$, with $\beta = 1.5$ and $\delta = 0.3$, $\varepsilon = 0.3$; initial conditions as in Fig. 6.3

In Figs. 6.3 and 6.4 we provide a comparison between the solution of the reduced ODE system (6.8) for the graph in Fig. 6.1 and the averaged 50×10^4 sample paths resulting from a discrete event simulation of the exact SIS process. The discrete event simulation is based on the generation of independent Poisson processes for both the infection of healthy nodes and the recovery of infected ones. We observe that, as expected, NIMFA provides an upper bound to the dynamics of the infection probabilities.

Figure 6.5 depicts the same comparison in the case of a network with eighty nodes partitioned into four communities; each community is a complete graph and all nodes belonging to two linked communities are connected. The agreement between NIMFA and simulations improves compared to Fig. 6.4. This is expected, because the accuracy of NIMFA is known to increase with network order N, under the assumption that the nodes' degree also increases with the number of nodes. Conversely, it is less accurate, e.g., in lattice graphs or regular graphs with fixed degree not depending on N [264, 274].

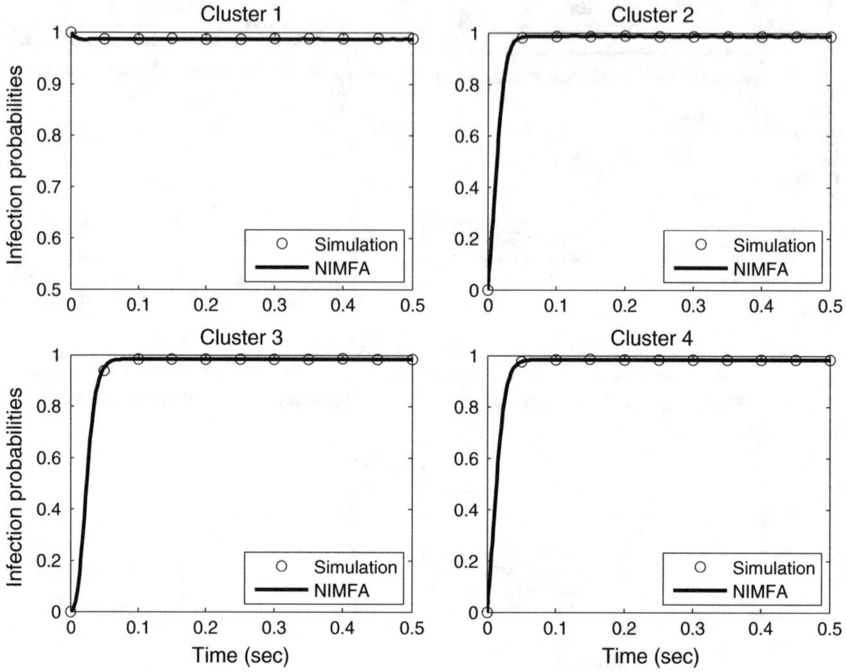

Fig. 6.5 Infection probabilities for each community in a network with $N = 80$, $d_{ii} = k_i - 1 = 19$ and $d_{ij} = 20$, for all $i, j = 1, .., 4$: simulation versus numerical solutions of (6.8); $\tau = \beta/\delta > \tau_c^{(1)} = 0.0348$, with $\beta = 5$ and $\delta = 2$, $\varepsilon = 0.3$; at time 0 all nodes of the 1-st community are infected

6.1.3 Steady State

Corollary 6.5 *When $\tau > \tau_c^{(1)}$ the metastable state P_∞ of the system* (6.6) *belongs to $M - \{0\}$.*

The result above is proved in [48]. Basically, Corollary 6.5 says that one can compute the $n \times 1$ vector, $\overline{P}_\infty$, of the reduced system (6.8) in order to obtain the $N \times 1$ vector, P_∞, of (6.6): indeed $p_{z\infty}, \ldots, p_{x\infty} = \overline{p}_{j\infty}$, for all $z, x \in G_j$ and $j = 1, \ldots, n$. This provides a computational advantage by solving a system of n equations instead of N. Moreover, since P_∞ is a globally asymptotically stable equilibrium in $I^N - \{0\}$, the trajectories starting outside M will approach those starting in $M - \{0\}$, as time elapses. The same holds clearly for trajectories starting in I^N and in M when $\tau \leq \tau_c^{(1)}$. Numerical experiments in Fig. 6.6 depict this fact.

We focus now on the computation of the steady-state $P_\infty = \left(p_{i\infty}\right)_{i=1,\ldots,N}$ of system (6.6). To this aim, by Corollary 6.5, we can compute the steady-state $\overline{P}_\infty = \left(\overline{p}_{j\infty}\right)_{j=1,\ldots,n}$ of the reduced system (6.8) and obtain

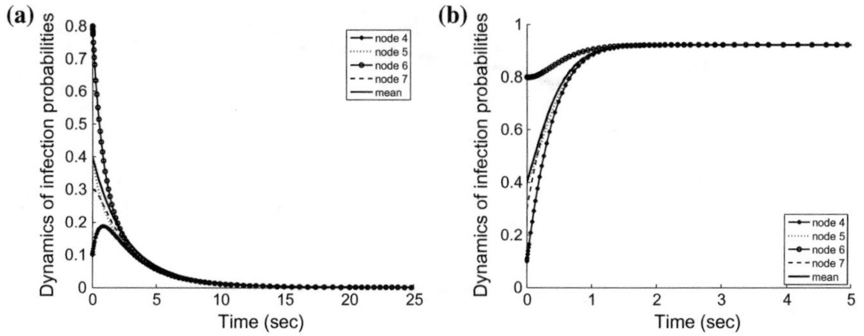

Fig. 6.6 Comparison between the dynamics of the original system (6.6) for each of the nodes belonging to V_3 in Fig. 6.1, for different initial conditions and the dynamics of the reduced system (6.8). In the latter case the initial conditions for each node are the mean value of the $p_i(0)$s. **a** case below the threshold: $\beta = 0.29$, $\delta = 1$, $\varepsilon = 0.3$ **b** case above the threshold: $\beta = 1.5$, $\delta = 0.3$, $\varepsilon = 0.3$

$$\beta(1-\overline{p}_{j\infty})\sum_{m=1}^{n}\left(\frac{k_j}{k_m}\right)^{-1/2} q_{jm}\overline{p}_{m\infty} - \delta\overline{p}_{j\infty} = 0, \qquad j = 1,\ldots,n$$

whence

$$\begin{aligned}\overline{p}_{j\infty} &= 1 - \frac{1}{1+\tau\sum_{m=1}^{n}\left(\frac{k_j}{k_m}\right)^{-1/2} q_{jm}\overline{p}_{m\infty}} \\ &= 1 - \frac{1}{1+\tau g_j\left(\overline{P}\right)}\end{aligned} \tag{6.9}$$

where

$$g_j\left(\overline{P}\right) := \left(d_{jj} + \varepsilon\sum_{m=1}^{n}\left(\frac{k_j}{k_m}\right)^{-1/2}\sqrt{d_{jm}d_{mj}}\right) - \sum_{m=1}^{n}\left(\frac{k_j}{k_m}\right)^{-1/2} q_{jm}(1-\overline{p}_{m\infty}).$$

By introducing $1-\overline{p}_{m\infty} = \frac{1}{1+\tau\sum_{z=1}^{n}\left(\frac{k_m}{k_z}\right)^{-1/2} q_{mz}\overline{p}_{z\infty}}$ in (6.9), we can express $\overline{p}_{j\infty}$ as a continued fraction iterating the formula

$$x_{j,s+1} = f(x_{1;s},..,x_{n;s}) = 1 - \frac{1}{1+\tau g_j(x_{1;s},..,x_{n;s})},$$

As showed in [274], after a few iterations of the formula above, one can obtain a good approximation of $\overline{p}_{j\infty}$, with a loss in the accuracy of the calculation around $\tau = \tau_c$.

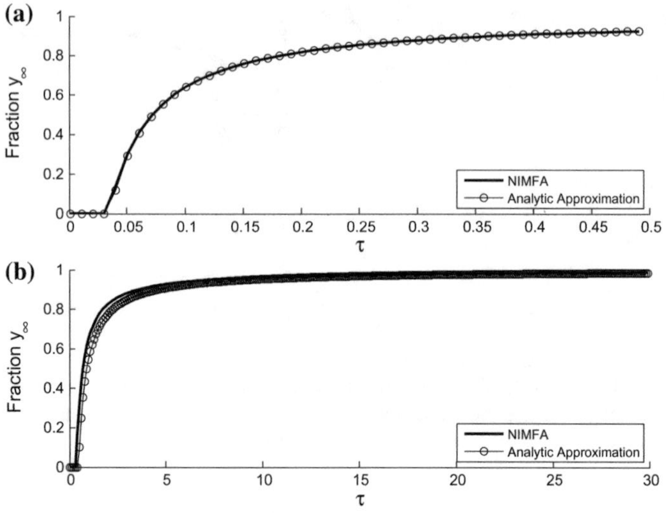

Fig. 6.7 Steady-state average fraction of infected nodes, for different values of τ: comparison between the approximation (6.10) and the exact computation through (6.6); **a** the graph is the one considered in Fig. 6.1a and **b** the one considered in Fig. 6.5

If we consider a regular graph where communities have the same number of nodes, then

$$\overline{p}_{j\infty} = 1 - \left(1/\tau \left(d_{jj} + \varepsilon \sum_{m=1}^{n} \left(\frac{k_j}{k_m}\right)^{-1/2} \sqrt{d_{jm} d_{mj}}\right)\right)$$

is the exact solution of (6.9).

Now let $r_j = d_{jj} + \varepsilon \sum_{m=1}^{n} \left(\frac{k_j}{k_m}\right)^{-1/2} \sqrt{d_{jm} d_{mj}}$ and $r(1) = \min_j r_j$; relying on the estimate $\overline{p}_{j\infty} \approx 1 - \left(1/\tau r_j\right)$ we can express the steady-state average fraction of infected nodes $y_\infty(\tau) = (1/N) \sum_{j=1}^{n} k_j p_{j\infty}(\tau)$ by

$$y_\infty(\tau) \approx 1 - \frac{1}{\tau N} \sum_{j=1}^{n} k_j \frac{1}{d_{jj} + \varepsilon \sum_{m=1}^{n} \left(\frac{k_j}{k_m}\right)^{-1/2} \sqrt{d_{jm} d_{mj}}}. \tag{6.10}$$

According to the analysis reported in [274], approximation (6.10) becomes the more precise the more the difference $r(2) - r(1)$ is small, where $r(2)$ is the second smallest of the r_j's (Fig. 6.7).

6.1.4 Clique Case

A *clique* of a graph is a set of vertices that induces a complete subgraph of that graph. Here we consider the specific case, analyzed in [49], where we have a *clique cover* of the graph, i.e., a set of cliques that partition its vertex set.

Thus, basically, all elements in a community are connected, i.e, $d_{ii} = k_i - 1$ for all $i = 1, ..., n$. Moreover we assume that all nodes belonging to two linked communities i and j are connected, i.e., $d_{ij} = k_j$ and $d_{ji} = k_i$. A sample graph is depicted in Fig. 6.8.

In [49] sufficient conditions for the extinction of epidemics have been found explicitly in terms of the dimension of the communities, their connectivity, and the parameters of the model. In the following we report the main results (for the derivation of the results see [49]).

Theorem 6.11 *Let $G = (V, E)$ be a graph with partition $\pi = \{V_i, i = 1, ..., n\}$, such that all V_i's induce a complete subgraph G_i of G, and all V_i's have the same order $k_i = k$. Moreover let us consider that whenever a node of G_i is connected with a node in G_j, then it is connected with all nodes in G_j. Therefore a sufficient condition for the uniqueness of the zero steady-state is the following:*

$$\frac{d_{\max}\varepsilon\beta + (1 - \frac{1}{k})\beta}{\delta} < \frac{1}{k},$$

where $d_{\max} = \max_i d_i$, and d_i is the number of communities with which the i-th is connected.

Theorem 6.12 *Let $G = (V, E)$ be a graph with partition $\pi = \{V_i, i = 1, ..., n\}$, such that all V_i's induce a complete subgraph G_i of G, each of arbitrary order k_i. Moreover let us consider that whenever a node of G_i is connected with a node in G_j, then it is connected with all nodes in G_j. Therefore a sufficient condition for the uniqueness of the zero steady state is the following:*

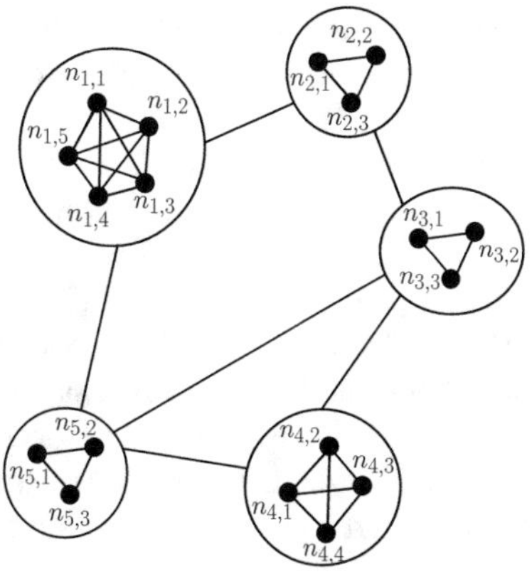

Fig. 6.8 Interconnected cliques. A link between two cliques means that each node in one clique is linked with all nodes in the other clique

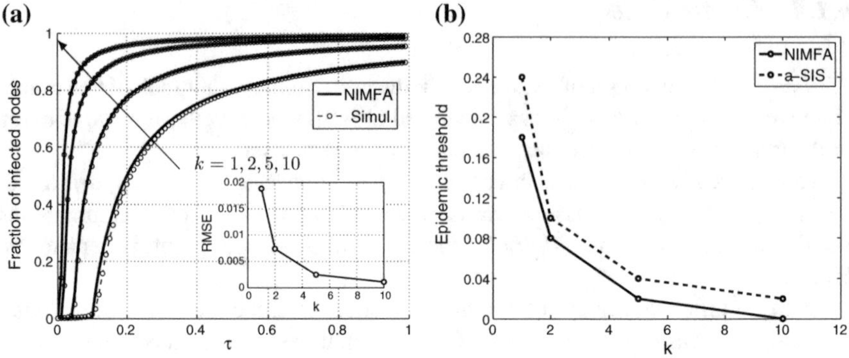

Fig. 6.9 **a** Fraction of infected nodes for different values of k as a function of $\tau = \beta/\delta$, with fixed the ratio $\varepsilon = 1/2$ and the value $\delta = 1$. The network of the communities is a regular graph with degree 10; the number of nodes is $N = 500$. The inserted plot represents the root mean square error between the simulated and the approximated fraction of infected nodes. **b** The corresponding value of the epidemic threshold for the NIMFA and the exact a-SIS model

$$\forall i = 1, \ldots, n : \qquad \frac{d_i\, \varepsilon\beta + (1 - \frac{1}{k_i})\beta}{\delta} < \frac{1}{k_i}.$$

Our NIMFA-like approximation is validated here by comparison with the exact SIS model. From the operative standpoint, we compare NIMFA with the a-SIS model [133, 271] where a nodal self-infection is allowed, at rate a. This model has no absorbing state and its stationary distribution, that can be computed for explicitly, can be made arbitrarily close to the quasi-stationary distribution of the original SIS model, by considering appropriate and small values of $a > 0$ [172, 271]. For a detailed explanation on the simulation process see [172].

Effect of community dimension. We depict first, in Fig. 6.9a, the impact of the community dimension k on the fraction of infected nodes in the steady-state, and compare the results of our model to the a-SIS model. The epidemic threshold of the a-SIS model is measured as the value of τ where the second derivative of the steady-state fraction of infected nodes equals zero. We consider a range for $\tau = \beta/\delta$, for constant ratio $\varepsilon = 1/2$ and fixed $\delta = 1$.

The sample network, representing the connections between the communities, has constant degree $d = 10$. The total number of nodes is $N = 500$. The number of elements k is the same for all communities: curves are drawn for increasing values of k ($k = 1, 2, 5, 10$), where $k = 1$ denotes the absence of local clusters. The threshold effect is well visible in the graphs depicted in Fig. 6.9a. As can be further observed, our model and the exact SIS model are in good agreement and the root mean square error between them decreases as k increases.

In Fig. 6.9b the corresponding value of the epidemic threshold for the NIMFA and the a-SIS model is reported. As expected from Theorem 6.11, the critical threshold above which a persistent infection exists decreases with the dimension of the com-

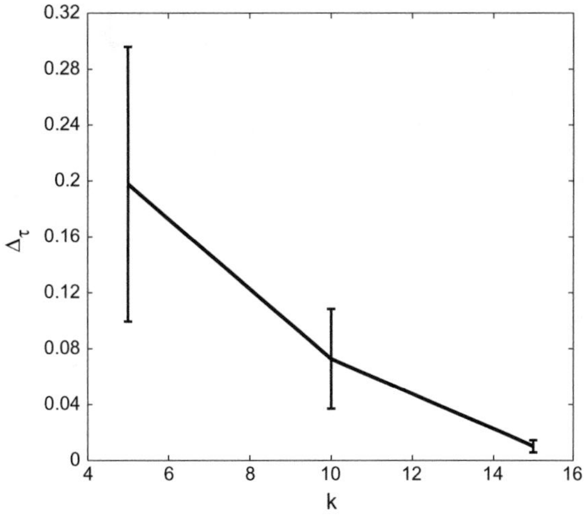

Fig. 6.10 Difference Δ_τ between the epidemic threshold in the case of homogeneous cluster distribution and inhomogeneous cluster distribution for different values of k (5, 10, 15), being fixed the ratio $\varepsilon = 1/8$. The difference was obtained averaging over 300 instances of tree graphs of 10 clusters, the level of confidence is set to 98%

munities. Thus, for large values of the community dimension, a very small value of τ is sufficient to cause epidemic outbreaks, irrespective of the actual network structure.

Effect of the heterogeneity of the community dimension.

One interesting question that concerns the two-scale epidemic model is the influence of the community dimension distribution onto the epidemic threshold. In general, it is not obvious whether, fixing all remaining system's parameters, a constant community dimension will lead to a lower or larger epidemic threshold for the same network.

In Fig. 6.10 we performed a test using a set of 300 sample tree graphs for depicting the connectivity of the communities. Each graph is the spanning tree of an Erdős-Rényi graph of order $n = 10$ and $p = 0.3$. The ratio ε is set to 1/8. The plot draws the difference Δ_τ, obtained averaging over the 300 sample graphs, between the epidemic threshold measured for homogeneous cluster distribution, and the epidemic threshold measured in the case of inhomogeneous cluster distribution.
In particular, for each sample tree, we considered different values of the average cluster dimension $k = 5, 10, 15$. In the case of heterogeneous cluster distribution half of the communities have dimension 2 and half of them have dimension $2k - 2$.

Figure 6.10 exemplifies that heterogeneity of communities' dimension lowers the epidemic threshold compared to the case of constant dimension. This observation agrees with the theory, indeed from the inequality [265, (3.34) on p. 47]:

$$\lambda_1 \geq \frac{2L}{N}\sqrt{1 + \frac{\mathrm{Var}[d]}{(\mathbb{E}[d])^2}},$$

where λ_1 is the spectral radius of a given graph with N nodes and L links, and d is the degree of a randomly chosen node in the graph, we have

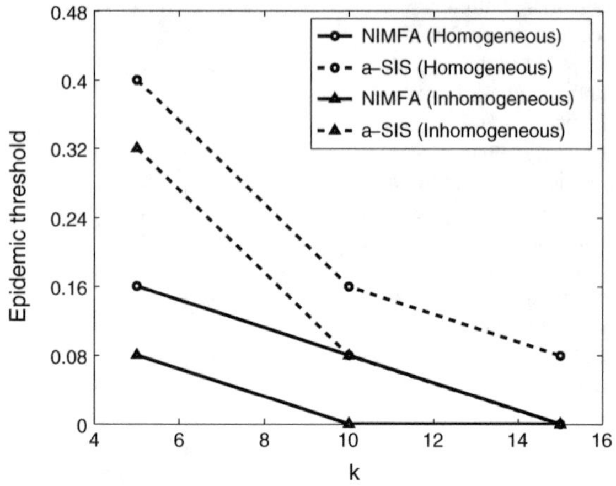

Fig. 6.11 The epidemic threshold in the case of homogeneous cluster distribution and inhomogeneous cluster distribution for different values of k, where the network of the communities is a spanning tree of an Erdős-Rényi graph of order $n = 10$ and $p = 0.3$. Both the NIMFA and the a-SIS thresholds are shown

$$\tau_c^{(1)} = \frac{1}{\lambda_1} \leq \frac{N}{2L} \frac{1}{\sqrt{1 + \frac{\text{Var}[d]}{(\mathbb{E}[d])^2}}}$$

implying that, the larger the variance in the degree d, the lower the NIMFA epidemic threshold $\tau_c^{(1)}$. Unfortunately, since $\tau_c^{(1)} \leq \tau_c$, we cannot conclude that an increase in Var[d] also always lowers the exact epidemic threshold τ_c.

Figure 6.11 shows the epidemic threshold measured for homogeneous community dimension and the epidemic threshold measured for inhomogeneous community dimension, by considering one instance of the previous set of spanning trees of an Erdős-Rényi graph. We report both the results obtained for our model and the results obtained for the a-SIS model: the NIMFA epidemic threshold well estimates the a-SIS epidemic threshold in both community dimension distributions.

6.2 Almost Equitable Partions

In this section we consider graphs where the partition of the vertex set is *almost equitable.*

Definition 6.15 The partition $\pi = \{V_1, ..., V_n\}$ is called *almost equitable* if for all $i, j \in \{1, \ldots, n\}$ with $i \neq j$, there is an integer d_{ij} such that for all $v \in V_i$, it holds

$$d_{ij} = deg(v, V_j) := \#\left\{e \in E : e = \{v, w\}, w \in V_j\right\}$$

independently of $v \in V_i$.

The difference between equitable and almost equitable partitions is that, in the former case, subgraph G_i of G induced by V_i has regular structure, whereas the latter definition does not impose any structural condition into G_i. Ideally we can think of a network $\tilde{G}$ whose node set has an almost equitable partition as a network G with equitable partition where links between nodes in one or more communities have been added or removed.

The objective is to obtain lower bounds on threshold $\tau_c^{(1)}$, useful in determining a safety region for the extinction of epidemics. We start assuming that links are added only.

To this aim, let us consider two graphs $G = (V, E)$ and $\tilde{G} = (V, \tilde{E})$ with the same partition $\{V_1, \ldots, V_n\}$, but different edge sets $E \subsetneq \tilde{E}$, and assume G to have an equitable partition but $\tilde{G}$ to have merely an almost equitable partition. Then if $\tilde{A}$ and A are the adjacency matrices of $\tilde{G}$ and G respectively it holds

$$\tilde{A} = A + R,$$

where $R = \text{diag}(R_1, \ldots, R_n)$; the dimension of R_i is $k_i \times k_i$ for $i = 1, ..., n$, as before k_i is the order of G_i and n is the number of the communities.

The theorem of Weyl can be applied to $\tilde{A} = A + R$ and then it yields

$$\lambda_1(\tilde{A}) \leq \lambda_1(A) + \lambda_1(R). \tag{6.11}$$

Proposition 6.26 *Let $G = (V, E)$ and $\tilde{G} = (V, \tilde{E})$ be two graphs and consider a partition $\{V_1, \ldots, V_n\}$ of the set of vertices V; we shall denote by $G_i = (V_i, E_i)$ and $\tilde{G}_i = (V_i, \tilde{E}_i)$ the subgraph of G and $\tilde{G}$ induced by the cell V_i, respectively, for $i = 1, ...n$. Assume this partition to be equitable for G and almost equitable for $\tilde{G}$. Let $E \subset \tilde{E}$ with*

$$\tilde{E} \setminus E = \bigcup_{i=1}^{n} (\tilde{E}_i \setminus E_i)$$

(i.e., the edge sets can only differ within cells) and denote by R the adjacency matrix corresponding to a graph with $\tilde{E} \setminus E$ as edge set. Finally, let us denote by G_i^C the graph with edge set $\tilde{E}_i \setminus E_i$ and whose node set is simply the set of endpoints of its edges (i.e., no further isolated nodes).

1. *If $\Delta(G_i^C)$ denotes the maximal degree in G_i^C, $i = 1, \ldots, n$, then*

$$\lambda_1(R) \leq \max_{1 \leq i \leq n} \min \left\{ \sqrt{\frac{2e_i(k_i - 1)}{k_i}}, \Delta(G_i^C) \right\},$$

   where e_i is the number of edges added to G_i, i.e., $e_i = (|\tilde{E}_i| - |E_i|)$, and k_i is the number of nodes in V_i.
2. *If additionally G_i^C is connected for each $i = 1, \ldots, n$, then*

$$\lambda_1(R) \leq \max_{1 \leq i \leq n} \min \left\{ \sqrt{2e_i - k_i' + 1}, \Delta(G_i^C) \right\},$$

where k_i' is the number of nodes of G_i^C.

Thus, by using estimate (6.2) and Proposition 6.26, we can derive a lower bound for the epidemic threshold, actually

$$\tau_c^{(1)} = \frac{1}{\lambda_1(\tilde{A})} \geq \tau^\star = \frac{1}{\max\limits_{1\leq i\leq n} \lambda_1(C_{V_i}) + \lambda_1(\widehat{B}) + \max\limits_{1\leq i\leq n} \min\left\{\sqrt{\frac{2e_i(k_i-1)}{k_i}}, \Delta(G_i^C)\right\}}. \tag{6.12}$$

Now let us consider the case where we remove edges, inside the communities, in a network whose set nodes has an equitable partition, thus because the spectral radius of an adjacency matrix is monotonically non increasing under the deletion of edges, we have

$$\lambda_1(\tilde{A}) \leq \lambda_1(A)$$

whence

$$\frac{1}{\lambda_1(\tilde{A})} \geq \frac{1}{\lambda_1(A)} \geq \min_i \frac{1}{d_{ii} + \lambda_1(\widehat{B})}.$$

The bounds developed so far support the design of community networks with a safety region for the effective spreading rate, that guarantees the extinction of epidemics. E.g. if we consider some G_i, $i = 1, \ldots, n$, it is possible to connect them such in a way to form a graph $\tilde{G} = (V, \tilde{E})$ with an almost equitable partition. Now, any subgraph obtained from $\tilde{G}$, by removing edges inside the communities, will have smaller spectral radius than $\tilde{G}$ and, consequently, a larger epidemic threshold. Thus the lower bound in (6.12) still holds.

6.3 Heterogenous SIS on Networks

Several analytic studies in the literature have determined the conditions for the appearance of endemic infectious states over a population under the assumptions of homogeneous infection and recovery rates.

However, in many real situations, e.g., in social, biological and data communications networks, homogeneity is a demanding assumption and it appears more appropriate to consider instead an heterogeneous setting [263]. A concise overview on the literature considering heterogeneous populations can be found in [215, 289].

To this aim, we report some results in [50], where heterogeneous infection and curing rates have been included.

Thus, hereafter, we denote by β_{ij} the infection rate of node j towards node i, and we exclude self-infection phenomena, i.e., $\beta_{ii} = 0$. Thus, we include the possibility that the infection rates depend on the connection between two nodes, covering a much more general case, than e.g. in [263], where a node i can infect all neighbors with the same infection rate β_i. Basically we allow for the epidemics to spread over a *directed weighted* graph. Moreover each node i recovers at rate δ_i, so that the curing rate is node specific.

As for the homogeneous case, the SIS model with heterogeneous infection and recovery rates is a Markovian process as well. The time for infected node j to infect any susceptible neighbor i is an exponential random variable with mean β_{ij}^{-1}. Also, the time for node j to recover is an exponential random variable with mean δ_j^{-1}. In the same way as in the homogeneous setting, we provide the NIMFA approximation. The NIMFA governing equation for node i in the heterogeneous setting writes as

$$\frac{dp_i(t)}{dt} = \sum_{j=1}^{N} \beta_{ij} p_j(t) - \sum_{j=1}^{N} \beta_{ij} p_i(t) p_j(t) - \delta_i p_i(t), \; i = 1, \ldots, N. \tag{6.13}$$

Let the vector $P = (p_1, \ldots, p_N)^T$ and let $\overline{A} = (\overline{a}_{ij})$ be the matrix defined by $\overline{a}_{ij} = \beta_{ij}$ when $i \neq j$, and $\overline{a}_{ii} = -\delta_i$; moreover let $F(P)$ be a column vector whose i-th component is $-\sum_{j=1}^{N} \beta_{ij} p_i(t) p_j(t)$. Then we can rewrite (6.14) in the following form:

$$\frac{dP(t)}{dt} = \overline{A} P(t) + F(P). \tag{6.14}$$

Let

$$r(\overline{A}) = \max_{1 \le j \le N} Re(\lambda_j(\overline{A}))$$

be the *stability modulus* [163] of $\overline{A}$, where $Re(\lambda_j(\overline{A}))$ denotes the real part of the eigenvalues of $\overline{A}$, $j = 1, \ldots, N$. We report a result from [163] that lead us to extend the stability analysis of NIMFA in [50] to the heterogeneous case (see [163, Theorem 3.1] for the proof).

Theorem 6.13 *If* $r(\overline{A}) \le 0$ *then* $P = 0$ *is a globally asymptotically stable equilibrium point in* $I_N = [0, 1]^N$ *for the system* (6.14)*, On the other hand if* $r(\overline{A}) > 0$ *then there exists a constant solution* $P^\infty \in I_N - \{0\}$*, such that* P^∞ *is globally asymptotically stable in* $I_N - \{0\}$ *for* (6.14).

Finally, in [205] we have defined the equitable partitions for the case of a directed weighted networks, and we have extended the analysis in [50] to this framework. For the purpose of modeling, nodes of the quotient graph can represent communities, e.g., villages, cities or countries. Link weights in the quotient graph in turn provide the strength of the contacts between such communities. In particular, the weight of a link may be (a non-negative) function of the number of people traveling per day between two countries; in fact, the frequency of contacts between them correlates with the propensity of a disease to spread between nodes.

Chapter 7
Epidemic Spreading on Interconnected Networks

Huijuan Wang

Most real-world networks are not isolated. In order to function fully, they are interconnected with other networks. Real-world power grids, for example, are almost always coupled with communication networks. Power stations need communication nodes for control and communication nodes need power stations for electricity. The interconnected network framework that more accurately represents real-world networks was proposed in 2010 [167, 210]. Since then, the influence of interconnected networks on diverse dynamic processes including cascading failures and opinion interactions has been widely studied [107, 141, 174, 206, 210, 301].

An interconnected networks scenario is essential when modeling epidemics because diseases spread across multiple networks, e.g., across multiple species or communities, through both contact network links within each species or community and interconnected network links between them [77, 81, 103, 229]. Dickison et al. [81] study the behavior of susceptible-infected-recovered (SIR) epidemics in interconnected networks. Depending on the infection rate in weakly and strongly coupled network systems, where each individual network follows the configuration model and interconnections are randomly placed, epidemics will infect none, one, or both networks of a two-network system. Mendiola et al. [229] show that in susceptible-infected-susceptible SIS model an endemic state may appear in the coupled networks even when an epidemic is unable to propagate in each network separately.

The interconnected network structure allows different infection rates within component networks and between component networks, a systematic way to gradually introduce the heterogeneity not only in network topology but also in dynamic features such as infection rates. In this chapter, we will explore how both the structural properties of each individual network and the interconnections between them determine

H. Wang (✉)
Multimedia Computing Group, Faculty of Electrical Engineering, Mathematics and Computer Science, Delft University of Technology,
Van Mourik Broekmanweg 6, Delft 2628 XE, The Netherlands
e-mail: H.Wang@tudelft.nl

E. Altman et al. (eds.), *Multilevel Strategic Interaction Game Models for Complex Networks*, https://doi.org/10.1007/978-3-030-24455-2_7

the epidemic threshold of two generic interconnected networks for SIS model. Moreover, many networks are spatially embedded such as transportation networks, so are the interconnected networks. In this case, the spatial constraints of the interconnection links may significantly influence the epidemic spreading. We will explore the effect of spatial constraints of interconnected lattices and small-world networks on the epidemic threshold as well as the average fraction of infected nodes in the meta-stable state.

7.1 Effect of the Interconnected Network Structure on the Epidemic Threshold

In order to represent two generic interconnected networks, we represent a network G with N nodes using an $N \times N$ adjacency matrix A_1 that consists of elements a_{ij}, which are either one or zero depending on whether there is a link between nodes i and j. For the interconnected networks, we consider two individual networks G_1 and G_2 of the same size N. When nodes in G_1 are labeled from 1 to N and in G_2 labeled from $N+1$ to $2N$, the two isolated networks G_1 and G_2 can be presented by a $2N \times 2N$ matrix $A = \begin{bmatrix} A_1 & \mathbf{0} \\ \mathbf{0} & A_2 \end{bmatrix}$ composed of their corresponding adjacency matrix A_1 and A_2 respectively. Similarly, a $2N \times 2N$ matrix $B = \begin{bmatrix} \mathbf{0} & B_{12} \\ B_{12}^T & \mathbf{0} \end{bmatrix}$ represents the symmetric interconnections between G_1 and G_2. The interconnected networks are composed of three network components: network A_1, network A_2, and interconnecting network B.

We consider the continuous SIS model on interconnected networks. We assume that the curing rate δ is the same for all the nodes, that the infection rate along each link of G_1 and G_2 is β, and that the infection rate along each interconnecting link between G_1 and G_2 is $\alpha\beta$, where α is a real constant ranging within $[0, \infty)$ without losing generality.

It has been proved in [283] that the epidemic threshold for β/δ in interconnected networks via NIMFA is $\tau_c = \frac{1}{\lambda_1(A+\alpha B)}$, where $\lambda_1(A+\alpha B)$ is the largest eigenvalue of the matrix $A + \alpha B$.

Expressing $\lambda_1(A+\alpha B)$ as a function of each network component A_1, A_2, and B and their eigenvalues/eigenvectors could indeed reveal the contribution of each component network to the epidemic threshold. This is, however, a significant mathematical challenge, except for special cases, e.g., when A and B commute, i.e., $AB = BA$ (see Sect. 7.1.1). The main progress has been made is that we could analytically derive for the epidemic characterizer $\lambda_1(A+\alpha B)$ (a) its perturbation approximation for small α, (b) its perturbation approximation for large α, and (c) its lower and upper bound for any α as a function of component network A_1, A_2, and B and their the largest eigenvalues/eigenvectors. Numerical simulations in Sect. 7.1.2 verify that these approximations and bounds well approximate $\lambda_1(A+\alpha B)$, and thus reveal the effect of component network features on the epidemic threshold of the whole system of interconnected networks, which provides essential insights into designing interconnected networks that are robust against the spread of epidemics (see Sect. 7.1.3).

Sahneh et al. [77] recently studied SIS epidemics on generic interconnected networks in which the infection rate can differ between G_1 and G_2, and derived the epidemic threshold for the infection rate in one network while assuming that the infection does not survive in the other. Their epidemic threshold was expressed as the largest eigenvalue of a function of matrices. Here, we explain further how the epidemic threshold of generic interconnected networks is related to the properties (eigenvalue/eigenvector) of each network component A_1, A_2, and B without any approximation on the network topology.

Graph spectra theory [266] and modern network theory, integrated with dynamic systems theory, can be used to understand how network topology can predict these dynamic processes. Youssef and Scoglio [298] have shown that a SIR epidemic threshold via NIMFA also equals $1/\lambda_1$. The Kuramoto synchronization process of coupled oscillators [217] and percolation [213] also features a phase transition that specifies the onset of a remaining fraction of locked oscillators and the appearance of a giant component, respectively. Note that a mean-field approximation predicts both phase transitions at a critical point that is proportional to $1/\lambda_1$. Thus we expect the results in this section to apply to a wider range of dynamic processes in interconnected networks.

7.1.1 Analytic Approach: $\lambda_1(A + \alpha B)$ in Relation to Component Network Properties

The spectral radius $\lambda_1(A + \alpha B)$ is able to characterize epidemic spreading in interconnected networks. We explore how $\lambda_1(A + \alpha B)$ is influenced by the structural properties of interconnected networks and by the relative infection rate α along the interconnection links. Specifically, we express $\lambda_1(A + \alpha B)$ as a function of the component network A_1, A_2, and B and their eigenvalues/eigenvectors.

Special cases

We start with some basic properties related to $\lambda_1(A + \alpha B)$. The spectral radius of a sub-network is always smaller or equal to that of the whole network. Hence,

Lemma 7.1

$$\lambda_1(A + \alpha B) \geq \lambda_1(A) = \max\left(\lambda_1(A_1), \lambda_1(A_2)\right)$$

Lemma 7.2

$$\lambda_1(A + \alpha B) \geq \alpha \lambda_1(B)$$

We summarize several special cases in which the relation between $\lambda_1(A + \alpha B)$ and the structural properties of network components A_1, A_2 and B are analytically tractable.

Lemma 7.3 *When G_1 and G_2 are both regular graphs with the same average degree $E[D]$ and when any two nodes from G_1 and G_2 respectively are randomly interconnected with probability p_I, the average spectral radius of the interconnected networks follows*

$$E[\lambda_1(A+\alpha B)] = E[D] + \alpha N p_I$$

if the interdependent connections are not sparse.

A dense Erdős-Rényi (ER) random network approaches a regular network when N is large. Lemma 7.3, thus, can be applied as well to cases where both G_1 and G_2 are dense ER random networks.

For any two commuting matrices A and B, thus $AB = BA$, $\lambda_1(A+B) = \lambda_1(A) + \lambda_1(B)$ [266]. This property of commuting matrices makes the following two special cases analytically tractable.

Lemma 7.4 *When $A + \alpha B = \begin{bmatrix} A_1 & \mathbf{0} \\ \mathbf{0} & A_1 \end{bmatrix} + \alpha \begin{bmatrix} \mathbf{0} & I \\ I & \mathbf{0} \end{bmatrix}$, i.e., the interconnected networks are composed of two identical networks, where one network is indexed from 1 to N and the other from $N+1$ to $2N$, with an interconnecting link between each so-called image node pair $(i, N+i)$ from the two individual networks respectively, its largest eigenvalue $\lambda_1(A+\alpha B) = \lambda_1(A) + \alpha$.*

Lemma 7.5 *When $A + \alpha B = \begin{bmatrix} A_1 & \mathbf{0} \\ \mathbf{0} & A_1 \end{bmatrix} + \alpha \begin{bmatrix} \mathbf{0} & A_1 \\ A_1 & \mathbf{0} \end{bmatrix}$, its largest eigenvalue λ_1 $(A + \alpha B) = (1+\alpha)\,\lambda_1(A_1)$.*

When A and B are not commuting, little can be known about the eigenvalues of $\lambda_1(A+\alpha B)$, given the spectrum of A and of B.

Lower bounds for $\lambda_1\,(A+\alpha B)$

We now denote matrix $A + \alpha B$ to be W. Applying the Rayleigh inequality [266, p. 223] to the symmetric matrix $W = A + \alpha B$ yields

$$\frac{z^T W z}{z^T z} \leq \lambda_1\,(W)$$

where equality holds only if z is the principal eigenvector of W.

Theorem 7.1 *The best possible lower bound $\frac{z^T W z}{z^T z}$ of interdependent networks W by choosing z as the linear combination of x and y, the largest eigenvector of A_1 and A_2 respectively, is*

$$\lambda_1\,(W) \geq \max\left(\lambda_1\,(A_1)\,,\lambda_1\,(A_2)\right) + \left(\sqrt{\left(\frac{\lambda_1\,(A_1) - \lambda_1\,(A_2)}{2}\right)^2 + \xi^2} - \left|\frac{\lambda_1\,(A_1) - \lambda_1\,(A_2)}{2}\right|\right) \tag{7.1}$$

where $\xi = \alpha x^T B_{12} y$.

When $\alpha = 0$, the lower bound becomes the exact solution $\lambda_1(W) = \lambda_L$. When the two individual networks have the same largest eigenvalue $\lambda_1(A_1) = \lambda_1(A_2)$, we have

$$\lambda_1(W) \geq \lambda_1(A_1) + \alpha x^T B_{12} y$$

Theorem 7.2 *The best possible lower bound* $\lambda_1^2(W) \geq \frac{z^T W^2 z}{z^T z}$ *by choosing* z *as the linear combination of* x *and* y*, the largest eigenvector of* A_1 *and* A_2 *respectively, is*

$$\lambda_1^2(W) \geq \frac{\left(\lambda_1^2(A_1) + \alpha^2 \left\| B_{12}^T x \right\|_2^2 + \lambda_1^2(A_2) + \alpha^2 \left\| B_{12} y \right\|_2^2\right)}{2} + \sqrt{\left(\frac{\lambda_1^2(A_1) + \alpha^2 \left\| B_{12}^T x \right\|_2^2 - \lambda_1^2(A_2) - \alpha^2 \left\| B_{12} y \right\|_2^2}{2}\right)^2 + \theta^2} \tag{7.2}$$

where $\theta = \alpha(\lambda_1(A_1) + \lambda_1(A_2)) x^T B_{12} y$.

Theorems 7.1 and 7.2 express the lower bound as a function of component network A_1, A_2 and B and their eigenvalues/eigenvectors, which illustrates the effect of component network features on the epidemic characterizer $\lambda_1(W)$.

Upper bound for $\lambda_1(A + \alpha B)$

Theorem 7.3 *The largest eigenvalue of interdependent networks* $\lambda_1(W)$ *is upper bounded by*

$$\lambda_1(W) \leq \max(\lambda_1(A_1), \lambda_1(A_2)) + \alpha\lambda_1(B) \tag{7.3}$$

$$= \max(\lambda_1(A_1), \lambda_1(A_2)) + \alpha\sqrt{\lambda_1(B_{12} B_{12}^T)} \tag{7.4}$$

This upper bound is reached when the principal eigenvector of $B_{12} B_{12}^T$ coincides with the principal eigenvector of A_1 if $\lambda_1(A_1) \geq \lambda_1(A_2)$ and when the principal eigenvector of $B_{12}^T B_{12}$ coincides with the principal eigenvector of A_2 if $\lambda_1(A_1) \leq \lambda_1(A_2)$.

Perturbation analysis for small and large α

We could further derive the perturbation approximation of $\lambda_1(W)$ for small and large α, respectively, as a function of component networks and their eigenvalues/eigenvectors.

We start with small α cases. The problem is to find the largest eigenvalue $\sup_{z \neq 0} \frac{z^T W z}{z^T z}$ of W, with the condition that

$$\begin{cases} (W - \lambda I) z = 0 \\ z^T z = 1 \end{cases}$$

When the solution is analytical in α, we express λ and z by Taylor expansion as

$$\lambda = \sum_{k=0}^{\infty} \lambda^{(k)} \alpha^k$$
$$z = \sum_{k=0}^{\infty} z^{(k)} \alpha^k$$

Let $\lambda_1(A_1)$ ($\lambda_1(A_2)$) and x (y) denote the largest eigenvalue and the corresponding eigenvector of $A_1(A_2)$ respectively. We examine two possible cases: (a) the non-degenerate case when $\lambda_1(A_1) > \lambda_1(A_2)$ and (b) the degenerate case when $\lambda_1(A_1) = \lambda_1(A_2)$ and the case $\lambda_1(A_1) < \lambda_1(A_2)$ is equivalent to the first.

Theorem 7.4 *For small α, in the non-degenerate case, thus when $\lambda_1(A_1) > \lambda_1(A_2)$,*

$$\lambda_1(W) = \lambda_1(A_1) + \alpha^2 (x^{(0)})^T B_{12} \left(\lambda_1(A_1) I - A_2\right)^{-1} B_{12}^T x^{(0)} + O(\alpha^3) \tag{7.5}$$

where $\left(z^{(0)}\right)^T = \left(x^T \ \mathbf{0}^T\right)$.

Theorem 7.5 *For small α, when the two component networks have the same largest eigenvalue $\lambda_1(A_1) = \lambda_1(A_2)$,*

$$\lambda_1(W) = \lambda_1(A_1) + \frac{1}{2}\alpha x^T B_{12} y + \alpha^2 (y^{(0)})^T B_{12}^T (\lambda^{(0)} I - A_1 + x^{(0)} (x^{(0)})^T)^{-1} \cdot \tag{7.6}$$
$$(B_{12} y^{(0)} - \lambda^{(1)} x^{(0)} + (x^{(0)})^T B_{12} (\lambda^{(0)} I - A_2 + x^{(0)} (x^{(0)})^T)^{-1} B_{12}^T x^{(0)} - \lambda^{(1)} y^{(0)}) + O(\alpha^3)$$

In the degenerate case, the first order correction is positive and the slope depends on B_{12}, y, and x.

Lemma 7.6 *For large α, the spectral radius of interconnected networks is*

$$\lambda_1(A + \alpha B) = \alpha \lambda_1(B) + v^T A v + O\left(\alpha^{-1}\right) \tag{7.7}$$

where v is the eigenvector belonging to $\lambda_1(B)$ and

$$\lambda_1(A + \alpha B) \le \lambda_1(A) + \alpha \lambda_1(B) + O\left(\alpha^{-1}\right)$$

Importantly, the analytical results derived so far are valid for arbitrary interconnected network structures.

7.1.2 Verification of Theoretical Results

Via numerical calculations, we are going to evaluate to what extent the perturbation approximation (7.5) and (7.6) for small α, the perturbation approximation (7.7) for large α, the upper (7.3) and lower bound (7.2) are close to the exact value

$\lambda_1(W) = \lambda_1(A + \alpha B)$. We would like to investigate the condition under which the approximations provide better estimates.

Although the analytical results derived earlier are valid for arbitrary interconnected network structures, we consider two classic network models as possible topologies of G_1 and G_2 in the simulations: (i) the Erdős-Rényi (ER) random network [47, 91, 201] and (ii) the Barabási-Albert (BA) scale-free network [88]. ER networks are characterized by a Binomial degree distribution $\Pr[D = k] = \binom{N-1}{k} p^k (1-p)^{N-1-k}$, where N is the size of the network and p is the probability that each node pair is randomly connected. In scale-free networks, the degree distribution is given by a power law $\Pr[D = k] = ck^{-\lambda}$ such that $\sum_{k=1}^{N-1} ck^{-\lambda} = 1$ and $\lambda = 3$ in BA scale-free networks.

We take the non-degenerate case as an example to illustrate the method. We consider the non-degenerate case in which G_1 is a BA scale-free network with $N = 1000$, $m = 3$, G_2 is an ER random network with the same size and link density $p_{ER} = p_{BA} \simeq 0.006$, and the two networks are randomly interconnected with link density p_I. We compute the largest average eigenvalue $E[\lambda_1(W)]$ and the average of the perturbation approximations and bounds mentioned above over 100 interconnected network realizations for each interconnection link density $p_I \in [0.00025, 0.004]$ such that the average number of interdependent links ranges from $\frac{N}{4}, \frac{N}{2}, N, 2N$ to $4N$ and for each value α that ranges from 0 to 10 with step size 0.05.

For a single BA scale-free network, where the power exponent $\beta = 3 > 2.5$, the largest eigenvalue is $(1 + o(1))\sqrt{d_{\max}}$ where $d_{\max}$ is the maximum degree in the network [72]. The spectral radius of a single ER random graph is close to the average degree $(N-1)p_{ER}$ when the network is not sparse. When $p_I = 0$, $\lambda_1(G) = \max(\lambda_1(G_{ER}), \lambda_1(G_{BA})) = \lambda_1(G_{BA}) > \lambda_1(G_{ER})$. The perturbation approximation is expected to be close to the exact $\lambda_1(W)$ only for $\alpha \to 0$ and $\alpha \to \infty$. However, as shown in Fig. 7.1a, the perturbation approximation for small α approximates $\lambda_1(W)$ well for a relative large range of α, especially for sparser interconnections, i.e., for a

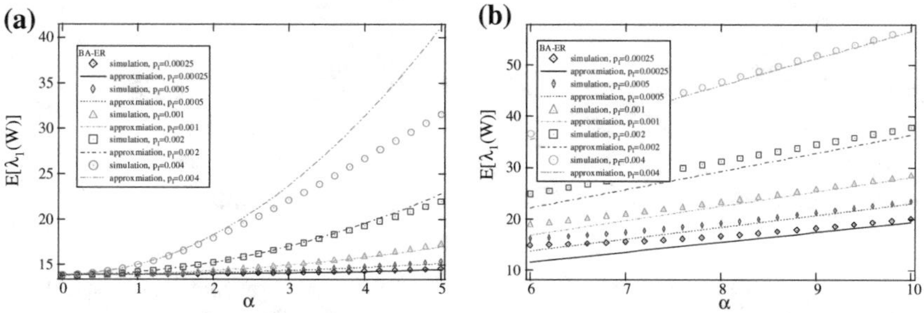

Fig. 7.1 A plot of $\lambda_1(W)$ as a function of α for both simulation results (symbol) and its **a** perturbation approximation (7.5) for small α (dashed line) and **b** perturbation approximation (7.7) for large α (dashed line). The interconnected network is composed of an ER random network and a BA scale-free network both with $N = 1000$ and link density $p = 0.006$, randomly interconnected with density p_I. All the results are averages of 100 realizations

smaller interconnection density p_I. Figure 7.1b shows that the exact spectral radius $\lambda_1(W)$ is already close to the large α perturbation approximation, at least for $\alpha > 8$.

As depicted in Fig. 7.2, the lower bound (7.2) and upper bound (7.3) are sharp, i.e., close to $\lambda_1(W)$ for small α. The lower and upper bounds are the same as $\lambda_1(W)$ when $\alpha \to 0$. For large α, the lower bound better approximates $\lambda_1(W)$ when the interconnections are sparser. Another lower bound $\alpha\lambda_1(B) \leq \lambda_1(W)$, i.e., Lemma 7.2, is sharp for large α, as shown in Fig. 7.3, especially for sparse interconnections. We do not illustrate the lower bound (7.1) because the lower bound (7.2) is always sharper or equally good. The lower bound $\alpha\lambda_1(B)$ considers only the largest eigenvalue of the interconnection network B and ignores the two individual networks G_1 and G_2. The difference $\lambda_1(W) - \alpha\lambda_1(B) = v^T A v + O\left(\alpha^{-1}\right)$ according to the large α perturbation approximation, is shown in Fig. 7.3 to be larger for denser interconnections. It suggests that G_1 and G_2 contribute more to the spectral radius of the interconnected networks when the interconnections are denser in this non-degenerate case. For large α, the upper bound is sharper when the interconnections are denser or when p_I is larger, as depicted in Fig. 7.2b. This is because $\alpha\lambda_1(B) \leq \lambda_1(W) \leq \alpha\lambda_1(B) + \max\left(\lambda_1(A_1), \lambda_1(A_2)\right)$. When the interconnections are sparse, $\lambda_1(W)$ is close to the lower bound $\alpha\lambda_1(B)$ and hence far from the upper bound. Besides the nondegenerate case illustrated here, we have observed the same in degenerate case that $\lambda_1(W)$ is well approximated by a perturbation analysis for a large range of small α, especially when the interconnections are sparse, and also for a large range of large α. The lower bound (7.2) and upper bound (7.3) are sharper for small α. Most real-world interconnected networks are sparse and non-degenerate, where our perturbation approximations are precise for a large range of α, and thus reveal well the effect of component network structures on the epidemic characterizer $\lambda_1(W)$.

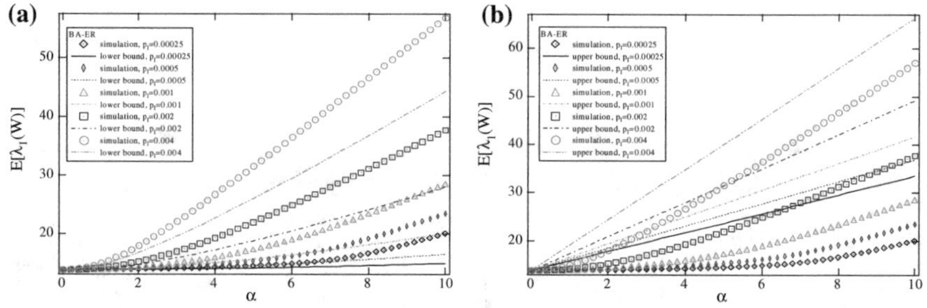

Fig. 7.2 Plot $\lambda_1(W)$ as a function of α for both simulation results (symbol) and its **a** its lower bound (7.2) (dashed line) and **b** upper bound (7.3) (dashed line). The interconnected network is composed of an ER random network and a BA scale-free network both with $N = 1000$ and link density $p = 0.006$, randomly interconnected with density p_I. All the results are averages of 100 realizations

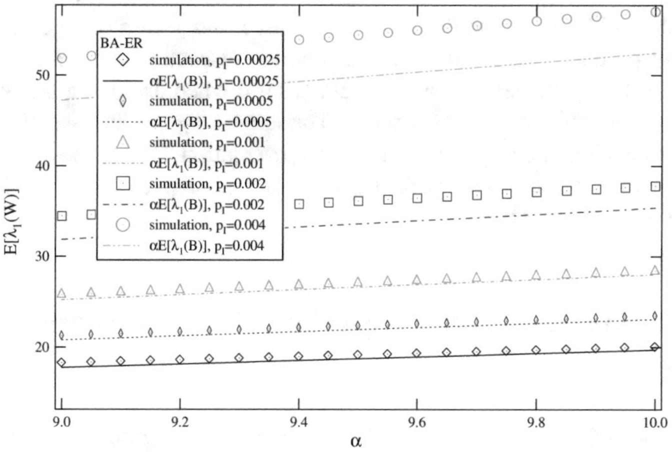

Fig. 7.3 Plot $\lambda_1(W)$ as a function of α for both simulation results (symbol) and its lower bound $\alpha\lambda_1(B)$ (dashed line). The interconnected network is composed of an ER random network and a BA scale-free network both with $N = 1000$ and link density $p = 0.006$, randomly interconnected with density p_I. All the results are averages of 100 realizations

7.1.3 Conclusion

We consider interconnected networks that are composed of two individual networks G_1 and G_2, and interconnecting links represented by adjacency matrices A_1, A_2, and B respectively. We focus on the SIS epidemic spreading in these generic coupled networks, where the infection rate within G_1 and G_2 is β, the infection rate between the two networks is $\alpha\beta$, and the recovery rate is δ for all agents. The epidemic threshold with respect to β/δ is shown to be $\tau_c = \frac{1}{\lambda_1(A+\alpha B)}$, where $A = \begin{bmatrix} A_1 & \mathbf{0} \\ \mathbf{0} & A_2 \end{bmatrix}$ is the adjacency matrix of the two isolated networks G_1 and G_2. The largest eigenvalue $\lambda_1(A + \alpha B)$ can thus be used to characterize epidemic spreading. This eigenvalue $\lambda_1(A + \alpha B)$ of a function of matrices seldom gives the contribution of each component network. Perturbation approximation for small and large α, lower and upper bounds for any α, of $\lambda_1(A + \alpha B)$ have been derived analytically as a function of component networks A_1, A_2, and B and their largest eigenvalues/eigenvectors. Numerical simulations verify that these approximations or bounds approximate well the exact $\lambda_1(A + \alpha B)$, especially when the interconnections are sparse and when the largest eigenvalues of the two networks G_1 and G_2 are different (the non-degenerate case), as is the case in most real-world interconnected networks. Hence, these approximations and bounds reveal how component network properties affect the epidemic characterizer $\lambda_1(A + \alpha B)$. Note that the term $x^T B_{12} y$ contributes positively to the perturbation approximation (7.6) and the lower bound (7.2) of $\lambda_1(A + \alpha B)$ where x and y are the principal eigenvector of network G_1 and G_2. This suggests that, given two isolated networks G_1 and G_2, the interconnected networks have a larger $\lambda_1(A + \alpha B)$

or a smaller epidemic threshold if the two nodes i and j with a larger eigenvector component product $x_i y_j$ from the two networks, respectively, are interconnected. This observation provides essential insights useful when designing interconnected networks to be robust against epidemics. The largest eigenvalue also characterizes the phase transition of coupled oscillators and percolation. Hence, these results apply to arbitrary interconnected network structures and are expected to apply to a wider range of dynamic processes.

7.2 Effect of the Spatial Constraints of Interconnected Networks

In the previous section, analytical and numerical results regarding to the epidemic threshold have been discussed for an arbitrary interconnected network structure. In reality, an interconnection link between two nodes from the two component networks respectively that are far away in location can be costly, thus, less likely to appear. For example, both the railway and bus transportation networks are spatially embedded and they are strongly coupled. A train station can be interconnected (share flow) with bus stations that are spatially close by. Such spatial spatial constraints may significantly influence the dynamical processes in the networks [76, 89, 90, 169, 173].

In this section, we explore the influence of the spatial constraints of interconnected networks on SIS epidemic spreading, including the epidemic threshold and the average fraction of infected nodes in the meta stable state [170, 175].

7.2.1 Epidemics on Interconnected Lattices

To address the effect of the spatial constraints in interconnected networks on epidemic spread, without loss of generality and for simplicity, we start with the interconnected lattices coupled with links of spatial length r and compare the results with those of interconnected ER networks (see Fig. 7.4). In this spatially embedded network, a node is only coupled with the nodes at length r in the other network. On average, a node has q interconnected links.

Interconnected network construction

Interconnected lattices with spatial constraint r can be generated as follows: generate two square lattices A and B with the same number of the nodes $N = L \times L$. L is the linear size of lattice. Choose a node i located at (x_i, y_i) in network A, and randomly choose another node j located at (x_j, y_j) in network B with the condition:

$$\begin{aligned} &\left|x_i - x_j\right| = r \text{ and } \left|y_i - y_j\right| \le r \text{ or} \\ &\left|x_i - x_j\right| \le r \text{ and } \left|y_i - y_j\right| = r \end{aligned} \tag{7.8}$$

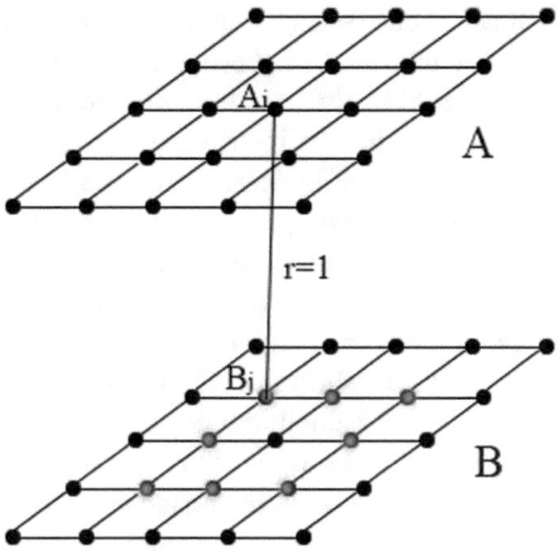

Fig. 7.4 Interconnected lattices A and B, where one node in network A is randomly interconnected with a node in network B with a spatial length $r = 1$, as defined in Eq. (7.8). There are 8 nodes (gray) meeting the condition for A_i when the spatial length $r = 1$, thus B_i is randomly chosen to connect to A_i

Then between the two nodes an interconnected link is formed. In total, $q \times N$ interconnected links will be added between network A and network B. Multiple links between the same pair are allowed, which represents internal coupling strength between two nodes. Randomly interconnected ER networks are generated and will be compared with interconnected lattices with respect to their influence on the SIS epidemic spreading. Firstly, we generate two ER networks, whose average degree is k and the number of nodes is N. The network nodes have similar coordinates as lattices, while links within each ER are randomly placed. A node in network A is randomly selected and interconnected to a randomly selected node in network B. In this way, $q \times N$ interconnection links are added.

Epidemic spreading model

The discrete time SIS model is considered. At each time step, a susceptible node has a probability β to be infected due to each infected neighbor either in the same network or in the other component network. At each time step, each infected node can be recovered and become susceptible again with probability δ. The assumption that the infection probabilities within a component network and between two networks are the same allows us to further explore the epidemic threshold τ_c for the effective infection rate $\tau = \beta/\delta$, above (below) which the epidemic spreads (die) out in the meta-stable state.

Results

Simulations have been performed to obtained the average fraction of infected nodes ρ in the meta-stable state as a function of the effective infection rate τ for different spatial lengths of interconnected links in interconnected lattices, which is compared with that in interconnected ER networks. Initially, we randomly infect 10% of the nodes in network A. The results are averaged over 100 realizations.

As shown in Fig. 7.5, the epidemic threshold τ_c in interconnected ER networks is found smaller than that in interconnected lattices. When the effective infection rate τ

is small, it can be seen that the infection density of the interconnected ER networks is larger than that of the interconnected lattices. When λ is increased and larger than a certain value, the infection density in interconnected ER networks becomes smaller than that in interconnected lattices. This is because there are two effects of interconnected lattices. This is because of two effects of interconnection spatial constraints: the clustering effect due to strong spatial constraints (e.g., $r = 1$) and the short cut effect due to weak spatial constraints (e.g., $r = \infty$). For a small r, epidemics spread more easily in a local region between networks due to the clustering of infected neighbors within spatial constraints. For a large r, epidemics can spread to distinct sites due to the short cut effect of relatively longer interconnected links. As shown in Fig. 7.6, the epidemic threshold decreases as the spatial length of interconnected links increases regardless of the finite-size effect, which also agrees with the results in Fig. 7.5. The epidemic threshold decreases slowly when the interconnected length r is large. The epidemic threshold starts to increase due to the finite-size effect since $r = 40$ when the minimal epidemic threshold is obtained.

Conclusion

We show that spatial constraints in interconnected networks, will strongly influence the epidemics outbreak and spreading. In this initial study, we consider the epidemics in interconnected lattices, which are compared with interconnected ER networks. It shown that the epidemic threshold decreases as interconnected length increases, regardless of the system size. When the infection rate is small, the disease is limited by the spatial constraints, where the infection density in interconnected lattices is lower than that in interconnected ER networks. However, the contrary is observed

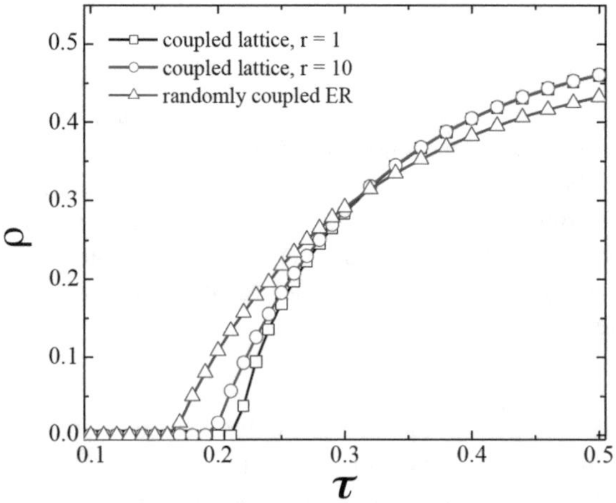

Fig. 7.5 Average fraction ρ of infected nodes as a function of the effective infection rate τ in interconnected lattices with different spatial length of interconnected links ($r = 1$ (square), and $r = 10$ (circle)) and randomly interconnected ER networks (up triangle). The density of interconnected links is $q = 1$. Initially 10% of nodes in network A are infected randomly. The average degree of the ER network is $k = 4$. The network size is $N = 10000$

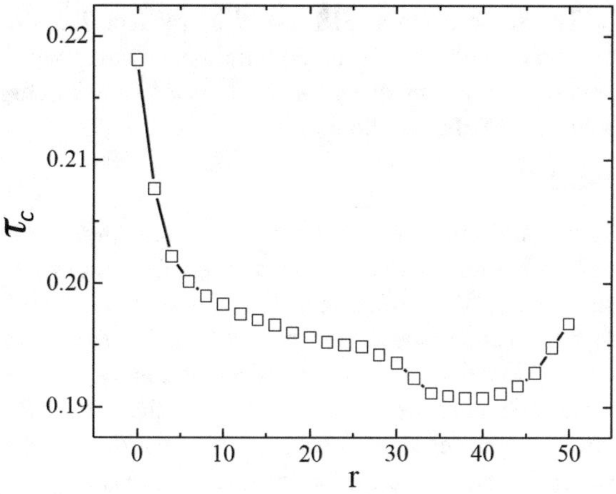

Fig. 7.6 Epidemic threshold τ_c as a function of spatial length r of interconnected links in interconnected lattices with the density of the coupling links $q = 1$. Initially 10% of nodes in network A are infected randomly. Network size $N = 10000$

when the infection rate is large enough. More investigations about the spreading process over time and space can be found in [170]. These findings will help to develop mitigation strategy for epidemics, which is spatially embedded in critical transportation infrastructures.

7.2.2 *Epidemics on Interconnected Small-World Networks*

The smallworld model [286] proposed by Watts and Strogatz captures the features of high clustering and small average path length, which have been widely observed in real-world networks. The small-world model can be constructed from a regular lattice. In such a network, each existing link is randomly rewired with a rewiring probability p, which tunes the nature of the network between that of a regular network ($p = 0$) and that of a random network ($p = 1$). We are going to investigate how do the rewiring probability and the spatial constraint of interconnected small-world networks affect the viral spreading process.

Interconnected small-world network generation

Firstly, A and B, two identical square lattices of linear size L and with $N = L \times L$ nodes are generated. For each node $n_i = n_0, \ldots n_{N-1}$ with lattice coordinates (x_i, y_j) and its neighbor $n_j(x_j, y_j)$ in network A, remove the lattice link (n_i, n_j) that satisfies $x_i < x_j$ or $y_i < y_j$ and add a link between n_i and n_k with probability p_A, where n_k is randomly chosen among all possible nodes, avoiding self-loops and duplicate links. The same process is also applied to network B with rewiring probability p_B. In this paper, we consider $p_A = p_B = p$.

The interconnection links with spatial constraint R are added as described in Sect. 7.2.1. The coupling density is q. Hence, $q \times N$ interconnection links are constructed between network A and network B.

The same discrete SIS model as in Sect. 7.2.1 is considered. The same infection probability β is assumed within each component network and between the two networks. The recovery probability is δ for each node. The SIS process is thus characterized by the effective infection rate $\tau = \beta/\delta$.

Results

Two basic properties that characterize the epidemic spreading will be explored: the epidemic threshold τ_c and the average fraction of infected nodes in the meta-stable state, also called infection density. As shown in Fig. 7.7, for a fixed R, the epidemic threshold decreases as p increases. This is because, as p increases, more links in the original lattices A and B are rewired, reducing the average distance of both networks, which enhances the spreading of epidemic. And as R increases, while the number of interconnections stays the same, the interconnections can bridge distinct locations, which facilitates the epidemics spreading. When p is small, the impact of the spatial length of the interconnections on the epidemic threshold is significant. When p is arge, however, the epidemic threshold is barely influenced by this spatial length. Both a high rewiring probability p and a large spatial length R contribute to the heterogeneity of the connections formed in the interconnected networks, which is beneficial to the spreading of epidemics. As shown in Fig. 7.8, interconnected networks with a larger rewiring probability p have a smaller epidemic threshold, confirming the results presented in Fig. 7.7. When the infection rate τ (larger than τ_c) is small, the average infection density ρ of interconnected networks with a higher rewiring

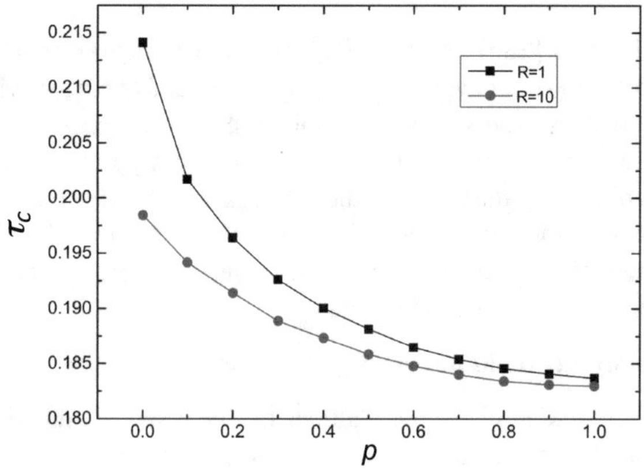

Fig. 7.7 Epidemic threshold τ_c as a function of the rewiring probability for different spatial constraints R of the interconnection links. For a given rewiring probability p and a given spatial constraint R, we gradually increase the infection rate τ and find the corresponding infection density in the meta-steady state for each infection rate. We consider τ_c as the first τ value corresponding to a non-zero infection density in the meta-steady state. Each component network has a small-world topology with a rewiring probability p. The density of the interconnection links is $q = 1$. Initially 10% of nodes in network A are infected randomly. Network size is $N = 10000$. The results have been averaged over 100 realizations

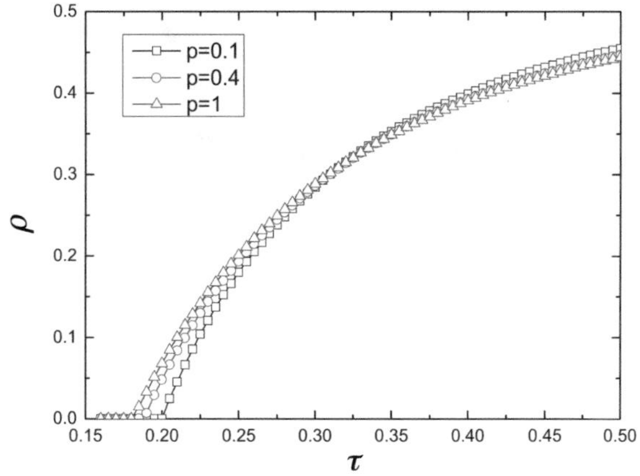

Fig. 7.8 Density ρ of infected nodes as a function of the infection rate τ for various rewiring probabilities. The density ρ is the average of the infection density ρ_A and ρ_B. The density of the interconnection links is $q = 1$, and the spatial constraint is $R = 1$

probability is larger than that of interconnected networks with a small rewiring probability. As the infection rate is further increased, however, the infection density of the interconnected networks is hardly influenced by the rewiring probability.

More investigations about the spreading process over time and space can be found in [175].

Conclusion

In summary, we studied the spread of epidemics in interconnected small-world networks with spatial constraints. We found that the rewiring probability of the small-world networks strongly affects the epidemic spreading behavior. We demonstrated that the epidemic threshold decreases as the rewiring probability increases. When the infection rate is low, the steady-state infection density varies with the rewiring probability. However, when the infection rate is sufficiently high, the infection density does not differ considerably for different rewiring probabilities.

While previous studies have focused on the epidemics spreading on single networks, recent work on viral spreading in interconnected and multilayer networks reveals new phenomena that cannot be captured in a single network. We deem the development of immunization and vaccination strategies based on the realistic interconnected and multilayer networks as the promising and challenging future work.

Chapter 8
Adaptive Networks

Huijuan Wang, Stojan Trajanovski, Dongchao Guo and Piet Van Mieghem

We have so far concentrated on networks, which do not change over time. In reality, a network may change over time in an independent process from the epidemic spread. Such networks, where the topology changes according to some rule or pattern, are known as evolving networks. The epidemic threshold in evolving networks has been studied in the past [214, 280]. Adaptive networks possess more complex properties than evolving networks, such that the topology is modified based on epidemic processes.

An adaptive model over the standard SIS model has been considered by Gross et al. [116]. This model is based on fixed probability of an infected nodes to infect and incident susceptible node and a fixed recovery probability of an infected node. Similarly, a link between a susceptible and an infected node is broken with a fixed probability and subsequently, a connection is established between the susceptible node and another susceptible node at random, which is an example of a *rewiring process*. Moreover, Gross et al. [116] have found a bifurcation pattern between the healthy, endemic and bi-stable states in their model. Related model to the model of Gross et al. have been studied by Marceau et al. [177] and Risau-Gusmán [299], while Lagorio et al. [162] have considered a discrete variant of Susceptible-Infected-Recovered (SIR) model in a combination with a rewiring process.

H. Wang (✉) · P. Van Mieghem
Faculty of Electrical Engineering, Mathematics and Computer Science,
Delft University of Technology, Delft, The Netherlands
e-mail: H.Wang@tudelft.nl

S. Trajanovski
Data Science Department, Philips Research, Eindhoven, The Netherlands

D. Guo
School of Computer Science, Beijing Information Science and Technology University,
Beijing, China

E. Altman et al. (eds.), *Multilevel Strategic Interaction Game Models for Complex Networks*, https://doi.org/10.1007/978-3-030-24455-2_8

Separate link-activation and link-deactivation strategies, different from link-rewiring, involving SIR model has been studied by Valdez et al. [261]. This model is discrete and an infected node can infect a susceptible neighbor with a certain probability, otherwise a link is broken for a constant time period. After this time, the link is established again. Following the SIR concept, an infected node becomes susceptible after a fixed time. The model of Valdez et al. [261] differs from the global link rewiring concept of Gross et al. [116] that its dynamic relies on local information and the infectious state of the neighbors of a node. The existence of epidemic threshold was discovered in the model of Valdez et al. [261]. A related model based on the SIS was proposed by Tunc et al. [260].

The majority of these models assume mean-field approximations, thus neglecting the high-order correlations and the local connectivity. The outlook of the final network topology when a meta-stable state is achieved, mean degree [218], degree distribution [116, 290] or concentration of susceptible and infected nodes into loosely connected clusters have not been thoroughly studied.

In this chapter, we will introduce two adaptive spreading processes on networks: the Adaptive Susceptible-Infected-Susceptible (SIS) epidemic model (ASIS) and the adaptive information diffusion (AID) model. The epidemic dynamic in the two models is the some, while the topology dynamic is opposite. In the former, an existing link is broken if one of its end-nodes is infected and the other susceptible; while in the later a link is established between an infected and a susceptible node. Furthermore, a link is established between two susceptible nodes in ASIS model, while an existing link is broken between two such nodes in AID model. ASIS models a process of isolation and distancing from infected nodes, while straightening the susceptible part of the networks, while AID aims to capture the spreading in information and social networks where nodes tend to connect with the information hubs, while the interest in less popular or information lacking nodes diminishes. We will firstly focus on the ASIS model, using both analytical and numerical results to reveal the epidemic threshold, the prevalence and topological features in the metastable-state in relation to ASIS dynamics. Afterwards, we will compare these two models showing that the models have different, but surprisingly not opposite characteristics.

8.1 Adaptive SIS Model

Assuming infection rate β and recovery rate δ, the continuous SIS model drives the epidemic dynamic in ASIS model. The link-dynamic is determined by the adjacency matrix $A(t)$ at time t. The existence of a link between two nodes i and j is specified by $a_{ij}(t) \in \{0, 1\}$ of this adjacency matrix. Each $a_{ij}(t)$ is a Bernoulli random variable, such that $a_{ij}(t) = 1$ with probability $\Pr[a_{ij}(t) = 1]$, while a link absence ($a_{ij}(t) = 0$) happens with probability $1 - \Pr[a_{ij}(t) = 1]$. The link-breaking and -creating (Fig. 8.1a), visualized in Fig. 8.1a and b, processes are based on viral states of the involved nodes and they are independent from one another. First, if exactly one of nodes i and j is infected and the other susceptible and a link is present between

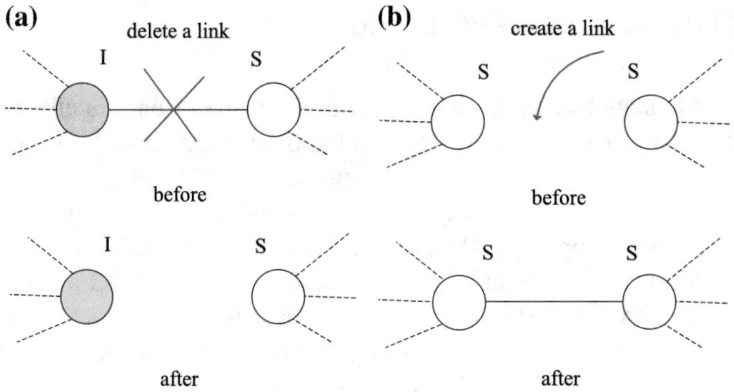

Fig. 8.1 (Color online) Changes of the link states based on the viral states of two nodes. **a** link breaking between a susceptible and an infected node; **b** link creation between two susceptible nodes

them ($a_{ij}(0) = 1$), the link can be deactivated with a Poisson rate ζ. Second, a non existing link can be created between two susceptible nodes i and j with a Poisson rate ξ.

For simplicity and we introduce the following notation:

$$\tilde{t} = t\delta, \quad \tilde{\zeta} = \frac{\zeta}{\delta}, \quad \tilde{\xi} = \frac{\xi}{\delta}, \quad \tau = \frac{\beta}{\delta}, \quad \omega = \frac{2\zeta}{\xi} \tag{8.1}$$

such that τ and ω are the effective infection and link-breaking rates, respectively, while the variable $\tilde{t}$ is the time t scaled by the curing rate δ. For simplicity, in what follows, we will drop ˜ notation and continue with these dimensionless parameters. The governing equation of the ASIS dynamics is the following

$$\frac{d}{dt}E[X_i] = E\left[-X_i + (1 - X_i)\tau \sum_{j=1}^{N} a_{ij} X_j\right] \tag{8.2}$$

$$\frac{d}{dt}E[a_{ij}] = a_{ij}(0) \cdot E\left[-\zeta a_{ij}\left(X_i - X_j\right)^2 + \xi\left(1 - a_{ij}\right)(1 - X_i)(1 - X_j)\right] \tag{8.3}$$

How this general model is related to previously introduced ones can be found in [118]. We restrict our analysis in the complete graph K_N in the starting moment $t = 0$, because only for a complete graph K_N, an exact analysis is possible.

8.1.1 The Metastable State of ASIS

The metastable state The meta-stable state is the value of empirically determined time point of the plateau of the average number of infected nodes [62]. The obstacle of this approach is the presence of uncertainty in the choice of the time moment to calculate the value of the meta-stable state value, which depends on the spreading rates and the network topology. On the other hand, the metastable state can be determined by ε-SIS model [272] and finding its stable state. The ε-SIS model [272] is a generalized version of the SIS model, introducing a small self-infection rate $\varepsilon < \frac{\delta}{N}$. This assumption contributes to diminishing the absorbing state and a steady-state is always present for a positive ε. If $\varepsilon = 0$, ε-SIS model boils down to the SIS model. Extending the ε-SIS model with appropriate link dynamics as defined before leads to adaptive ε-SIS model (ε-ASIS model). In such a model and for a small ε, it is possible to calculate the average steady-state values for many metrics, including the number of infected nodes on average or the number of links.

It has been shown [118] that the steady-state in ε-ASIS model resembles the metastable state of the ASIS model.

8.1.2 The Average Metastable-State Fraction of Infected Nodes

The fraction of infected nodes is defined as $Z = \frac{1}{N}\sum_i X_i$, while the average fraction of infected nodes in the metastable state is denoted as $y = E[Z^*]$. Here, we employ the notion of Z^* for the fraction of infected nodes and subsequently, similar is done for other metrics. Assuming a complete graph as an initial topology, we have the following Theorem 8.1.

Theorem 8.1 *For a complete graph K_N as an initial topology and the average metastable-state of infected nodes $y = E[Z^*]$, using (8.2) and (8.3), the following quadratic equation holds*

$$y^2 - 2Vy + H = 0 \tag{8.4}$$

such that

$$V = 1 - \frac{1}{2N} + \frac{\omega - 1}{2\tau N} \tag{8.5}$$

and

$$H = 1 - \frac{1}{N} + Var\left[Z^*\right] - E\left[\frac{1}{N^2}\sum_{j=1}^{N} d_j^*\left(1 - X_j^*\right)\right] \tag{8.6}$$

The solution of the quadratic equation (8.4) is given by

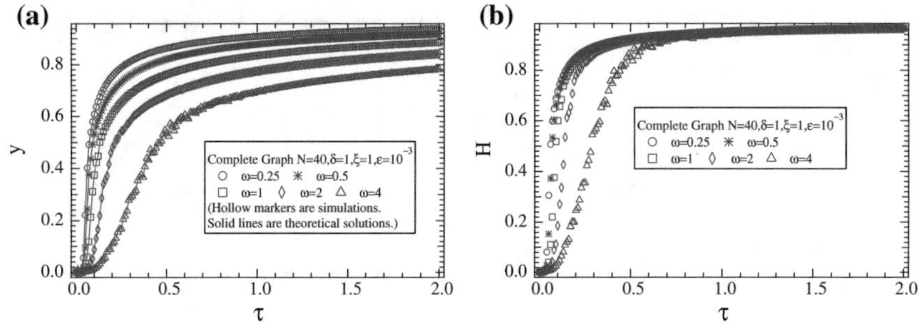

Fig. 8.2 **a** Numerically determined values (in the solid red lines) from (8.7) compared to the simulation results (with blue markers) show a good agreement for the average fraction of infected nodes y in the metastable state as a function of τ. The initial topology is a complete graph. **b** The corresponding values of H in (8.6)

$$y = \left(1 - \frac{1}{2N} + \frac{\omega - 1}{2\tau N}\right)\left(1 \pm \sqrt{1 - \frac{1 - \frac{1}{N} + Var[Z^*] - E\left[\frac{1}{N^2}\sum_{j=1}^{N} d_j^*\left(1 - X_j^*\right)\right]}{\left(1 - \frac{1}{2N} + \frac{\omega-1}{2\tau N}\right)^2}}\right). \quad (8.7)$$

$Var[Z^*]$ is the variance of the fraction of infected nodes, while nodal degree of j is denoted by d_j^*, such that $\frac{d}{dt}E\left[\frac{2L}{\xi} - \frac{(\omega-1)N}{\beta}Z\right] = 0$.

Although the Eq. (8.7) formally has two solution, only one is physically possible. In a case of $\tau \to \infty$ with finite ω, the solution with plus sign in (8.7) is physically valid, while in the opposite case, the solution with minus sign in (8.7) is relevant. If there is no link dynamics i.e. $\omega \to 0$, (8.4) boils down to an equation for a complete graph [60] independent from the time t.

In addition, for several values of effective link-breaking rates ω and given ink-creating rate ξ, Theorem 8.1 has been confirmed by simulations. The solution, given in (8.7), is calculated numerically by applying the values of Var$[Z^*]$ and $E\left[\sum_j d_j^* X_j^*\right]$ that are taken from the simulations. Figure 8.2a show that this solution of (8.7), obtained numerically, is in accordance with the simulation results for multiple cases. The values of H in (8.6) is smaller than 1, and this has been verified in Fig. 8.2b. Figure 8.3a and b depict the behavior of y and H as a function of the rate ω. Additionally, Theorem 8.1 and the fact that $H < 1$ are reaffirmed again. The average fraction of infected nodes in the metastable state decreases as a function of the effective link-breaking rate ω, thus the topology adaptation contributes to the suppression of the virus spread.

Epidemic Threshold

Theorem 8.2 *In ASIS model on K_N, for the epidemic threshold holds*

$$\tau_c(\omega; \xi) = \frac{\omega - 1}{N\left(h(\omega; \xi) - 2 + \frac{1}{N}\right)} \quad (8.8)$$

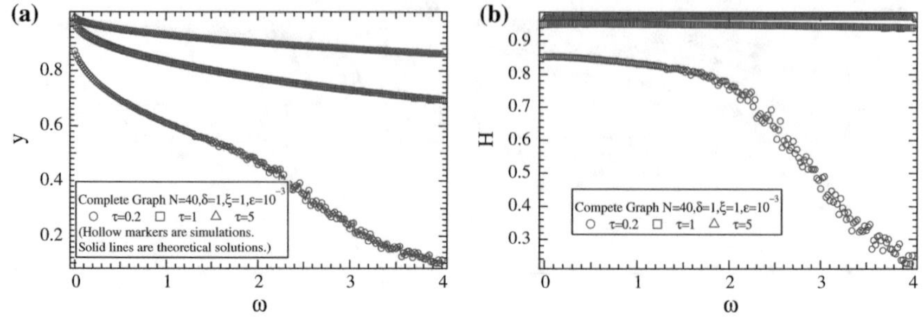

Fig. 8.3 (Color online) **a** Numerically determined values (in the solid red lines) from (8.7) compared to the simulation results (with blue markers) show a good agreement for the average fraction of infected nodes y in the metastable state as a function of ω. The initial topology is a complete graph. **b** The corresponding values of H in (8.6)

such that $h(\omega; \xi) = \lim_{y \downarrow 0} \frac{H}{y}$ *is a positive, but slowly changing function such that*

$$1 \leq h(\omega; \xi) \leq 2 + \frac{1}{N} \left(\frac{1}{\left. \frac{\partial \tau_c(\omega;\xi)}{\partial \omega} \right|_{\omega \to \infty}} - 1 \right)$$

for all $\omega > 0$ *and* $h(1; \xi) = 2 - \frac{1}{N}$.

According to Theorem 8.2, the epidemic threshold τ_c behaves as a linear function in ω since the function $h(\omega; \xi)$ changes very slowly in ω. This trend is mostly noticeable for large ω.

The function $h(\omega; \xi) = \frac{H(\tau_c)}{y_c}$ and the epidemic threshold are obtained experimentally. Figure 8.4b explains that $h(\omega; \xi)$ is slowly changing in ω. In particular, the inset of Fig. 8.4b shows that $h(\omega; \xi)$ is stable and close to a constant for large ω, while Fig. 8.4a depicts that τ_c is close to a linear function in ω. These two observations are in accordance to Theorem 8.2.

8.1.3 Metastable-State Topology

Impact of the Disease Dynamics on the Metastable-State Topology

In this section, we study several topological metrics such as: the modularity (as expressed in [275]), the assortativity [276], the connectivity (expressed as a probability of the that the graph being connected), the average number of components, the biggest component size and the number of links in the metastable state of ASIS. We consider $E[2L^*]/(N(N-1))$ the average number of links in the metastable state scaled by the maximum number of links ($\frac{N(N-1)}{2}$ in a complete

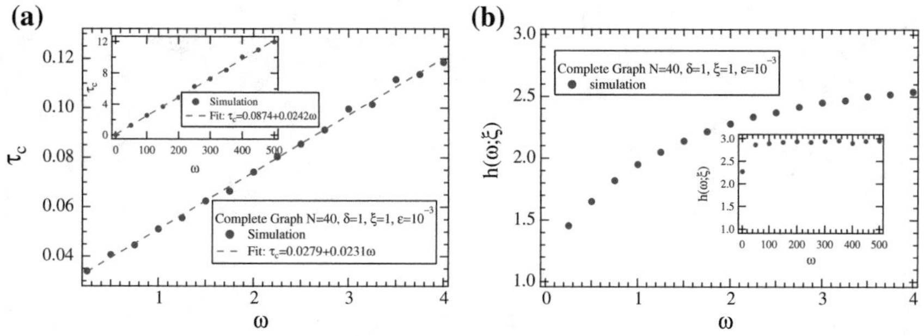

Fig. 8.4 (Color online) **a** Epidemic threshold τ_c as a function of ω for an initial complete graph with 40 nodes. The trend of τ_c for larger values of ω is shown in the inset. Based on the method from Sect. 8.1.2, simulations are conducted and are shown with blue circles. **b** The corresponding function of $h(\omega; \xi)$ in ω. The trend of $h(\omega; \xi)$ for larger values of ω is shown in the inset

graph). $E[2L^*]/(N(N-1)$ in the metastable state has a small value for small τ as shown in Fig. 8.5a. The reason behind is that effective infection rate contributes for the links to break. However, for $E[2L^*]/(N(N-1))$ slowly increases with τ (e.g., $\tau \in [0, 1500]$) as shown in the inset of Fig. 8.5a. The value of $E[2L^*]/(N(N-1))$ could easily reach a maximum value of 1 if τ is very high, because all the nodes will be infected very fast, there will be not enough time for healing or link breaking and subsequently no link will be broken once all are infected. There is also a strong correlation with the number of links with different metrics like the size of the biggest component or the connectivity. For example, the connectivity is shown in Fig. 8.5b, where if $\omega = 2\zeta/\xi > 1$ is high, the network is likely to be disconnected. Moreover for high enough $\omega > 1$, a common phenomena is that the network is partition into one big cluster (component) with almost all the nodes and few components with very small number of nodes. Gross et al. [116] reported that the inclusion of some moderate link-dynamics can introduce correlation in the network, which is also observed in this work. Figure 8.5f shows that the modularity and assortativity demonstrates a strong correlation, which was earlier observed by Van Mieghem et al. [275] in different networks. The process of breaking links contributes to a network separation into two weakly inter-connected components, namely a component of predominantly susceptible node (named as S component) and the other of mostly infected nodes (named as I component). On the other hand, the process of link-creation contributes to strengthening the S component and the connectivity between its nodes. The effect is opposite for the infection and curing rates, where both try to destroy the separation into S and I components. The epidemic and link dynamics are in "persistent competition", for example once the assortativity achieves a maximum value, it starts decreasing due to the increase of the infection rates. This can be observed in Fig. 8.5e. Something similar happens with the modularity (Fig. 8.5f). The presence of S and I components that weakly connected has also been observed in other models [116].

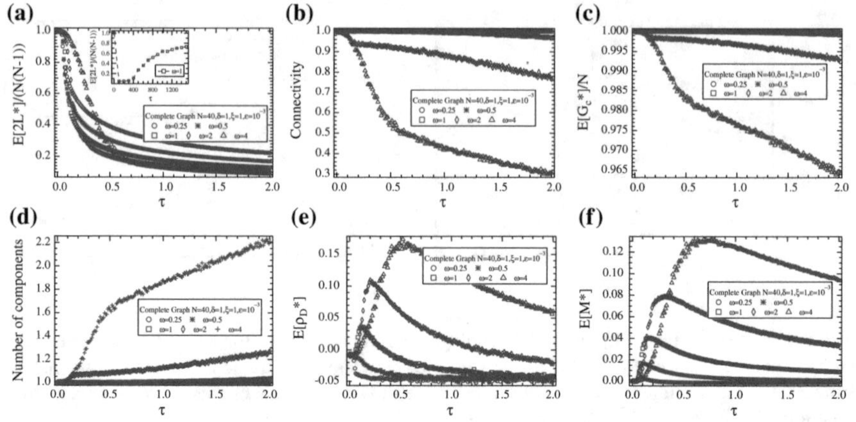

Fig. 8.5 (Color online) The effect of the effective infection rate τ on the topology in the metastable state. **a** $\frac{E[2L^*]}{N(N-1)}$ the average number of links in the metastable state scaled by the maximum number of links ($\frac{N(N-1)}{2}$ in a complete graph) as a function of τ. The effect of large values of τ is shown in the inset. **b** The probability of the graph being connected in the metastable state as a function of τ. **c** $\frac{E[G_c^*]}{N}$ the normalized average size of the biggest component in the metastable state as a function of τ. **d** The average number of components in the metastable state as a function of τ. **e** Assortativity value on average $E[\rho_D^*]$ in the metastable state as a function of τ. **f** Modularity value on average $E[M^*]$ in the metastable state as a function of τ

The Effect of the Link Dynamics on the Topology in the Metastable State

In a similar way, all these metastable stable network properties can be shown in relation to the effective link-breaking rate ω, which characterizes the link dynamics. It has been observed that as ω increases, the network becomes sparser, disconnected with a higher probability into more components, both the assortativity and modularity increase first and decline afterwards. More details can be found in [118].

Structure of the Topology in the Metastable State

Figure 8.6b presents an overview of the modularity as a function in effective rates τ and ω for $\delta = 1$ and $\xi = 1$. The modularity in the ASIS model is high and it appears to show "half-open", "elliptical-like" curves. The explanation behind high modularity lies in the fact (a) there is a clear separation between the susceptible and infected nodes and (b) the sizes of the S and I components are similar. This can be explained as follows. A high modularity means (i) that the infected nodes and the susceptible nodes are well separated, and (ii) that the I component is comparable in size to the S component. Such high values of modularity can be achieved for moderate values of τ and ω such that the epidemic can be spread it fast enough (the role of high enough/moderate τ) and it will not be suppressed (the role of moderate ω) and this can be achieved for several values of τ and ω thus forming "half-open", "elliptical-like" curves. On the other hand, similar low values of modularity can be achieved for either small τ with high ω or high τ with small ω. In the former, the epidemic spread is

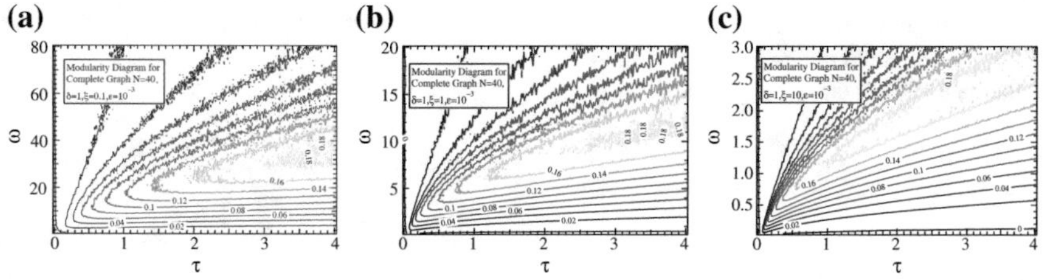

Fig. 8.6 (Color online) The contour lines of the modularity contours depending on τ and ω for three different values of link-creating rate ξ: **a** $\xi = 0.1$, **b** $\xi = 1$, and **c** $\xi = 10$

negligible and will be suppressed fast enough, thus leaving noticeable S component; while in the later the epidemic is spread very fast and cannot be suppressed, thus forming significant I component. Both cases lead to similar low modularity values also forming wider "half-open", "elliptical-like" curves. The connectivity also shows very similar "half-open", "elliptical-like" curves in a plane with τ and ω dependence, but opposite to modularity i.e. high connectivity leads to low modularity and *vice versa*. The degree distribution for the all nodes and separately for the infected and susceptible nodes have been discussed in more details in [118]. It has been shown that when the infection process is faster than the link dynamics, the final degree distribution is binomial-like, while in the opposite case there are multiple peaks in the degree distribution.

Determing the Bi-Stability in the ASIS Model

We explore the distribution $\Pr[Z^*]$ of the fraction Z^* of infected nodes in the metastable-state instead of the average $y = E[Z^*]$. Figure 8.7 shows the $\Pr[Z^*]$ for diverse effective infection rates τ and fixed link dynamic rates ξ and ζ. When $\tau = 0.15$ is low, the meta-stable state approaches the healthy state. When $\tau = 3$ is high, the meta-stable state is the endemic state. The fraction Z^* of infected nodes in the meta-stable state is either close to 0 or a non-zero positive value for some other cases (for example, the $\tau = 1$ case in Fig. 8.7). When $\tau = 1$, the probability $\Pr[Z^* = 0]$ approximates the probability $\Pr[Z^* = c]$), where c is positive and dependent on τ. This implies that the metastable state is likely stable at the two dramatically different infection states, a seemingly bi-stability phenomenon. Such phenomenon in epidemic spreading on adaptive networks was reported by Gross et al. in [116]. The bistable state is a metastable state where the infection persists or there is no infection in the ASIS model.

A bifurcation-like behavior is illustrated in Fig. 8.8. The probability $\Pr[Z^* = 0]$ is comparable with the probability $\Pr[Z^* = c]$ for a certain τ in value. The metastable state of the ASIS model is possibly stable in either of the two states. It seems that the metastable state changes from the healthy state, to the bi-stable state and to the endemic state as τ increases.

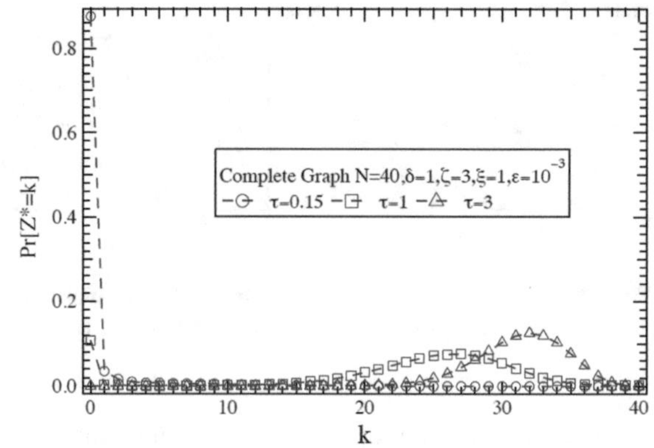

Fig. 8.7 (Color online) The fraction of infected nodes Z^* in the metastable state

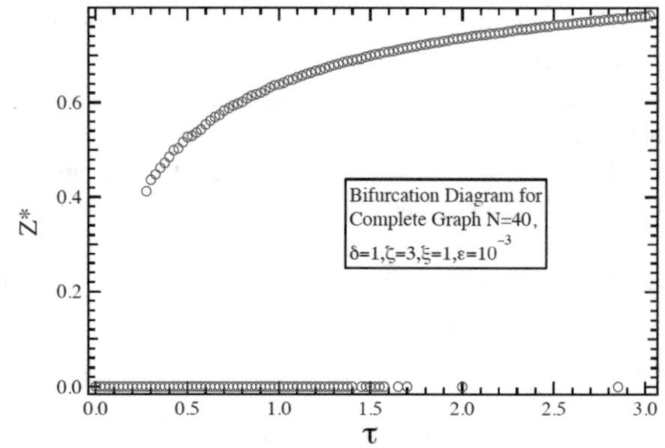

Fig. 8.8 (Color online) The bifurcation diagram of the fraction of infected nodes in metastable state. The metastable state is the healthy state when the effective infection rate $0 \leq \tau \leq 0.3$), the bi-stable infection state when $0.3 \leq \tau \leq 1.6$ and the endemic state when $\tau \geq 1.6$ in sequence as τ increases

8.1.4 Summary

We proposed an adaptive network model *ASIS* to characterize the interplay and co-evolution between the dynamics on a network (e.g. disease spreading) and the dynamics of the network (i.e. dynamics of the link state). This model includes a Poissonion link-breaking process with rate ζ and a Poissonion link-creating process with rate ξ in the classic Susceptible-Infected-Susceptible (SIS) model. When the initial topology of an adaptive network is a complete graph, the average fraction of infected nodes in metastable state has been derived (see Theorem 8.1). Moreover, we have proved and illustrated a linear law between the epidemic threshold τ_c and the effective link-breaking rate $\omega = 2\zeta/\xi$ We have also verified experimentally (see Theorem 8.2) that the phase transition that a disease can persist in the presence of link dynamics for the effective infection rate $\tau > \tau_c$, and the linear function $\tau_c(\omega)$.

Our simulations point out how the co-evolution of the disease and link dynamics promotes the emergent features of the adaptive network with respect to the connectivity, the number of links, the biggest component size, the associativity and modularity . Nodes group into two loosely inter-connected clusters according to their viral states, i.e. the I (infectious) component and the S (susceptible) component, based on which the modularity is calculated. When the disease dynamics is faster than the link-breaking process and the link-creating process in rate, the network evolves towards no apparent community structure and disassortative-mixing. When the epidemic spreading is slower than the link dynamics, the topology becomes slightly but clearly modular and assortative. A universal contour-line pattern can be observed in the modularity diagram as a function of τ and ω. A high link-breaking rate ω or a low infection rate τ may lead to disassortative networks with low modularity. In contrast, a low connectivity tend to contribute to a high modularity in network topologies.

Finally, our investigation on the distribution of the fraction of infected nodes in the metastable state shows that between the healthy state and the endemic state, a bi-stable state may exist where the fraction of infected nodes is stable either around 0 (the healthy state) or around a positive non-zero value (the endemic state).

8.2 Comparison of the ASIS and AID Model

8.2.1 The AID Model

Both the ASIS and AID are based on the SIS epidemic spreading model. However, the dynamics of the topology evolution in these two coevolution models are opposite. The Poissonian *link-breaking* and *link-creating* processes with rates ζ and ξ respectively, govern the evolution of the network topology. In the AID model, a link is created between a node pair when only one node but not both has the information. An existing link is removed between a node pair, when both nodes do not have the information, and if the two nodes were not connected in the original network. For the link existence probability $E\left[a_{ij}(t)\right] = \Pr\left[a_{ij}(t) = 1\right]$, we have the following governing equation

$$\begin{aligned}\frac{d}{dt}E[a_{ij}] =&(1 - a_{ij}(0))E\Big[- \zeta a_{ij}(1 - X_i)(1 - X_j) \\ &+ \xi(1 - a_{ij})\left(X_i - X_j\right)^2\Big]. \end{aligned} \tag{8.9}$$

We consider the simple case where the initial network is an empty graph with N isolated nodes and without any link. When both i and j have the information ($X_i = X_j = 1$), the link is preserved, i.e. $\frac{dE[a_{ij}]}{dt} = 0$. The link dynamics, thus, tend to increase (decrease) the degree of a node with (without) information.

The AID model has be verified to be realistic by using the Facebook wall posts dataset [117].

Table 8.1 Comparison of AID and ASIS models

Property/model	ASIS	AID
Metastable state	Always stable	Unstable (τ, ω) regions
Threshold $\tau_c(\omega)$	Linear	(mostly) constant
Topological metrics	"half-elliptical"	Rotated "half-elliptical"

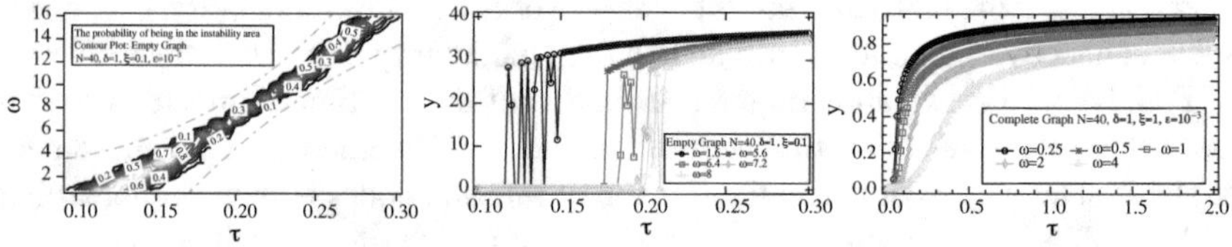

(a) Instability region for the AID model.

(b) The prevalence y in the AID model.

(c) The prevalence y the ASIS model.

Fig. 8.9 **a** and **b** demonstrate the instability in AID. **c** demonstrates the stability in ASIS

8.2.2 *Comparison*

We are going to illustrate the striking difference that emerges from AID and ASIS. The most important difference is that instability and non-existence of the metastable state are observed in the AID but not the ASIS model. Key differences between the two models are given in Table 8.1, which will be further explained.

8.2.3 *The Prevalence*

Exact expressions for the fraction of infected nodes and the epidemic threshold for the AID model has as well been derived in [117]. Although these relations are not of closed-form, they can well explain the existence of the metastable state and the stability of the prevalence for both models. We denote the prevalence in the metastable state by $Z^* = \frac{1}{N}\sum_{j=1}^{N} X_j^*$ and its average by $y = E[Z^*]$ where N is the number of nodes in the network. We further denote $T(N) = \frac{E\left[\sum_{i=1}^{N} d_i^*(1-X_i^*)\right]}{N^2}$, which is bounded by

$$0 \le T(N) \le \frac{E\left[\sum_{j=1}^{N} d_j^*\right]}{N^2} = \frac{E[2L^*]}{N^2} \le \frac{N(N-1)}{N^2} < 1.$$

In the AID model,

$$y = \frac{1}{2}\left(1 + \frac{\omega - 2}{2\tau N}\right)\left(1 \pm \sqrt{1 - \frac{4\text{Var}\left[Z^*\right] + 2\omega T(N)}{\left(1 + \frac{\omega-2}{2N\tau}\right)^2}}\right), \tag{8.10}$$

where Var[Z^*] is the variance of the prevalence and d_j^* is the degree of node j. Importantly, the argument under the square root in (8.10) is possibly negative, leading to the **non-existence** of the metastable state. Consider a large network, where $N \to \infty$. In this case, (8.10) can be simplified to

$$y = \frac{1}{2}\left(1 \pm \sqrt{1 - \left(2\omega T_\infty + 4\text{Var}\left[Z^*\right]\right)}\right). \tag{8.11}$$

The metastable state does not exist, if $4\text{Var}\left[Z^*\right] + 2\omega T_\infty > 1$. Therefore,

$$\text{Var}\left[Z^*\right] > \frac{1}{4}$$

is sufficient to lead to the non-existence of the metastable state. Furthermore, an upper bound for the link-breaking rate can be derived from (8.11):

$$\omega \leq \frac{1 - 4\text{Var}\left[Z^*\right]}{2T_\infty} \leq \frac{1}{2T_\infty},$$

otherwise, a metastable state solution does not exist. These findings in theory are confirmed by simulations. As shown in Fig. 8.9a and b, the metastable state does not exist in certain regions of (τ, ω). The **instability area** reveals a "sand clock" shape: as τ and ω increase, the area narrows first and then widens. The area vanishes for large enough τ and ω.

In contrast, the metastable state always exists in the ASIS according to (8.7).

Consider the combination of all the four parameters of the AID model. When the link breaking rate is higher than the creating rate but both are large and the spreading rate is small, a small fraction of nodes possessing the information are unlikely to stay long nor can be considered as a metastable state. In this case, both the number of links and infected nodes change dramatically over time, as shown in Fig. 8.10a. Whereas in other combinations of the parameters, there is usually a critical mass of links and nodes that possess the information, forcing of the epidemic to reach an equilibrium, i.e. the metastable state (see Fig. 8.10b).

8.2.4 *Epidemic Threshold τ_c*

We have shown that the epidemic threshold (8.8) in the ASIS model is linear in ω.

The threshold in the AID model is, however, the quotient of two linear functions, which approaches a constant if ω is large,

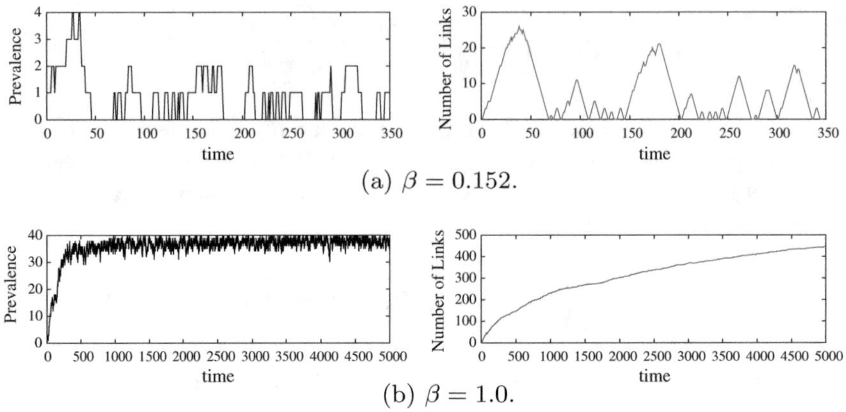

Fig. 8.10 (Color online) The numbers of links and infected nodes as functions of time in the AID model, where$N = 40$, $\zeta = 0.32$, $\xi = 0.1$, $\delta = 1$, $\varepsilon = 10^{-3}$ and different spreading rates β are considered. The points of instability/stability are in accordance to Fig. 8.9a

Fig. 8.11 (Color online) Threshold τ_c versus effective link-breaking rate ω for $N = 40$ (the inset: large range of ω)

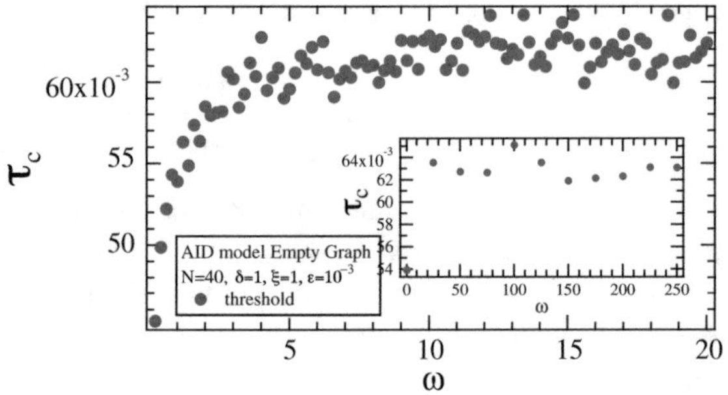

$$\tau_c(\omega; \xi) = \frac{\omega - 2}{2N\left(h_{\text{AID}}(\omega; \xi) - 1\right)}, \tag{8.12}$$

where $h_{\text{AID}}(\omega; \xi) \leq 1 + \max\{1, 1 + \frac{\omega-2}{2Na}\}$ and $a = \lim_{\omega\to\infty} \frac{\partial h_{\text{AID}}(\omega;\xi)}{\partial \omega}$ is approximately a constant. If $\omega > 2$, $h_{\text{AID}}(\omega; \xi)$ is almost a linear function of ω, obeying $h_{\text{AID}}(2; \xi) = 1$.

Figure 8.11 illustrates the relation between the epidemic threshold and ω and the epidemic threshold is almost a constant when ω is large. The epidemic threshold in Fig. 8.11 is relatively noisy, a fingerprint of the instability in the AID model.

8.2.5 Topological Properties

Figure 8.12 depicts the contour plot of the network modularity in the metastable state in the (τ, ω)-plane, for both ASIS and AID models. Interestingly,for a given effective

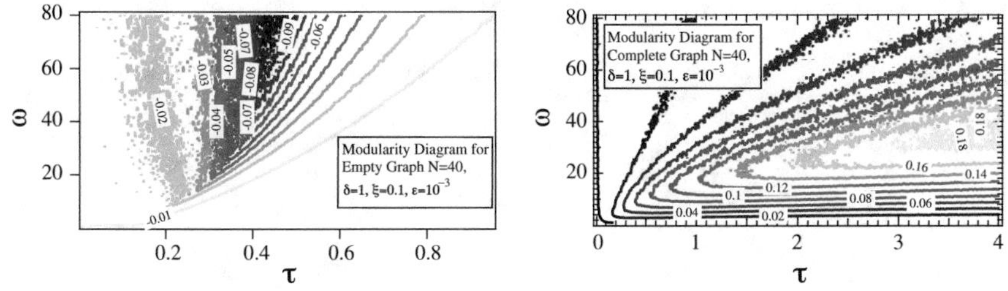

(a) AID model. (b) ASIS model.

Fig. 8.12 (Color online) Modularity in (τ, ω)-plane in the stable region for $\xi = 0.1$

infection rate τ, small and large effective link-breaking rates ω may lead to the same modularity. The ASIS and AID models differ in the order of the contour lines: "the inner contour" lines show higher (lower) modularity in ASIS (AID) but have similar shape, though rotated, in the contour lines.

In the (τ, ω)-plane, the instability area, which has "sand-clock" shape (Fig. 8.9a), exists only The instability area for the AID model, is close to the center of the coordinate system in the (τ, ω)-plane and bellow the "half-ellipses" extremal node.

The *metastable state* (when it exists) topologies in the AID model are random graphs. In the metastable state of the ASIS model, the networks have, however, two clusters that are sparsely connected. One cluster is composed of the susceptible nodes, which are almost fully connected and the other is composed of the infected nodes, connected like a random graph.

8.3 Conclusion

The two process and network coevolving models ASIS and AID share the same epidemic spreading process but different topology dynamics. Via our theoretical analysis and extensive simulation, we have observed and explained their differences from the following perspectives:

1. Instability of the metastable state exists in the AID but not the ASIS model.
2. The epidemic threshold τ_c tends to be independent of the effective link-breaking rate ω when ω is large in the AID model, whereas linearly increases with ω in the ASIS model.
3. Topological features such as the modularity of both models exhibit concentric half-ellipses in the (τ, ω)-plane. The two models differ in the order rotation of the contours.

Part III
Networking Games

Eitan Altman and Konstantin Avrachenkov

Routing games were brought into the community of Telecommunication Networks by Orda et al. [ORS-1993] on 1993. The special feature that differentiates this reference from previous work on RGs (in other communities) is that it considers finitely many decision makers, each of which can split its traffic among several paths (we shall present more details later). It is of interest for the telecommunication community since it models well competition between service providers that can control routes for traffic to or from their subscribers. An important motivation to study routing games is the well known Braess paradox where performance of all users may deteriorate (at equilibrium) when one adds links or when one adds capacity to some existing links. The paradox implies in particular that the well known upgrade policy of adding capacity to the most congested links may have quite opposite results of increasing the congestion.

Routing Games (RG) are concerned with choosing routes in networks with several traffic classes, each characterized by a source destination pair and a demand constraint. This branch developed mainly within the community of road traffic engineering, starting with the work of Wardrop [Wa-1952]. Below is a classification of routing games.

Population (non-atomic) games: These are games with a continuum (infinite) set of non-atomic players. Non-atomic means that the action of a single player has a negligible impact on the utilities of other players. These games which had an important impact on road traffic engineering are known there as the Traffic Assignment Problem, formalized by Wardrop [Wa-1952] to model the choice of routes of cars where each driver, modeled as a non-atomic player, minimizes his expected travel delay. This game was solved in [BMW-1956] by showing that it can be transformed into a global optimization problem. The cost function that appears in the global optimization problem is called a potential of the game. The theory mainly treats additive costs: the cost (delay) over a path is the sum of delays over each of its links. The link cost is assumed to be a function of the total amount of its flow. Some research [Pa-1991] has been devoted to multimodal traffic where one considers different traffic types (pedestrian, cars, bicycles, trucks etc.). The link cost may differ from one type to another, and may depend explicitly on the flows of each type through the link.

In both cases a potential does not exist anymore which renders the problems much harder to solve [Pa-1991].

Atomic non-splitable games: The setting is the same as in the previous model except that there are finitely many drivers. The link cost depends in general on the number of drivers using it and are again additive along paths. Congestion Games introduced by Rosenthal [Rl-1993] introduced these games on1978 and solved them game using a potential similar to [BMW-1956] assuming that the link cost (or delay) depends only on the sum of flows traversing the link and is the same for all users. The equilibrium is obtained through an integer linear program and can be achieved also in a decentralized way in which players update their decisions (using a best response) in an asynchronous way (one at a time). Again, the potential disappears once we allow costs to depend on the player. Crowding Games, introduced by Milchtaich [Mi-1996] (and references therein) are a special class of congestion games in which the cost is allowed to be user specific. They too are non-splitable discrete routing games. It extends congestion games in allowing the link cost to depend explicitly on the flow of each player that traverses it rather than on the sum of flows. It is more restrictive than congestion games in that two different paths do not have common links. Under some assumptions on the costs, some properties of congestion games still hold, even when a potential no more exists.

Non-atomic routing games as well as non-splitable atomic games are concerned with decisions of individuals. The theory serves in predicting the congestion as a function of the topology of the network and the capacity of its links. It is thus useful for a network manager or network owners. In contrast, routing games as developed in Electrical Engineering departments are often concerned with the Internet Service Providers (ISPs) each controlling the routes taken by many users (subscribers).

The deregulation of telecom industry in Europe during the 1990–2000 and the opening of national markets to competition between ISPs triggered research on a (relatively) new class of routing games called Competitive Routing (Splitable atomic games). They are concerned with a finite (or more generally discrete) number of decision makers (ISPs) whose decisions concern a continuum (infinite number) of non-atomic individuals. Their systematic study started with the pioneering work by Orda et al. [ORS-1993]. Yet, some previous work on this framework had already in the context of road traffic earlier. We refer to [HM-1985] who showed that the equilibrium in these types of games converges to the Wardrop equilibrium used in road traffic, as the number of players increases to infinity.

The first and the third chapters of Part III describe further developments in routing games and their application to telecommunication systems. Then in the forth chapter of Part III we apply the routing game models to discuss the issue of network neutrality.

For an extensive overview of networking games in road traffic we recommend the book [214]; for the crossover between games in road traffic and in telecommunication, we refer to the whole special issue that we edited on that topic in the journal Networks and Spacial Economics, 4(1), March 2004. The contribution of the community of Algorithmic games to RGs appears in [277].

Network formation games are nowadays a consolidated branch of game theory [87, 92, 118, 153]. They study which networks' structures arise when the nodes are selfish rational players, who can sever or create some links in order to increase the utility they perceive from the network. In particular, it is usually assumed that each node can unilaterally sever a link to one of its neighbors, while the creation of a new link requires the approval of both the participating nodes. This idea has lead to the concept of pairwise-stable networks [156], i.e., networks for which every existent link is beneficial to both the connected nodes and every inexistent link is not beneficial to at least one of the two nodes it would connect. Different dynamics for links' creation/destruction have been studied. In the second chapter of Part III we generalize the setting of [156] to coalitional better-response dynamics when several agents can form a coalition and to seek a move which benefits the coalition as a whole. Similarly to [295] we also admit the presence of random mistakes. In fact, we first present the general results for finite games and then apply the general results to the network formation games.

Chapter 9
"Beat-Your-Rival" Routing Games

Gideon Blocq and Ariel Orda

To date, game theoretic models have been employed in virtually all networking contexts. These include control tasks at the network layer, such as flow control and routing (e.g., [14, 36, 161, 191, 202, 224] and references therein), as well as numerous studies on control tasks at the link and MAC layers. A fundamental assumption in all of these referenced studies is that the selfish agents compete over resources in the network and aim to optimize their own performance; agents do not care (either way) about the performance of their competitors. However, and typically in the context of routing, scenarios exist in which this assumption is not warranted.

For example, consider the scenario where two content providers, A and B, offer video-on-demand services in a network. Both A and B compete over the network resources, however only content provider A aspires to minimize its own latency. Due to business considerations, content provider B aims at offering its clients a performance that is equal or better than A's performance. Thus, the objective of B is not solely to maximize its performance.

In light of examples like the one above, previous research in routing games has extended the classical model of "performance-maximizing" or "selfish" agents, and focused on different scenarios, e.g., settings where certain agents may act *maliciously* towards other agents [28, 45, 222]. Such malicious behavior could be due to a range of reasons, e.g., hackers or rivaling companies that aim to degrade network

Parts of this chapter have been published in 8th International Symposium, SAGT 2015, Saarbrücken, Germany, September 28–30, 2015. Proceedings.

G. Blocq (✉) · A. Orda
Viterbi Faculty of Electrical Engineering, Technion - Israel Institute of Technology, 3200003 Haifa, Israel
e-mail: gideon@alumni.technion.ac.il

A. Orda
e-mail: ariel@ee.technion.ac.il

E. Altman et al. (eds.), *Multilevel Strategic Interaction Game Models for Complex Networks*, https://doi.org/10.1007/978-3-030-24455-2_9

quality. In contrast, other studies in routing games consider agents to have an altruistic component to their objective [27, 57, 70, 134].

In order to best model real-life scenarios, each agent's objective should lie somewhere in the range between *malicious*, *selfish* and *altruistic*, as depicted in Fig. 9.1. This direction has been proposed in [67], where each agent i has a parameter that captures how important the social performance is to i. In this setting, a malicious agent aims to minimize the social performance, an altruistic agent aims to maximize it and a selfish agent does not take the social performance into account at all. However, [67] focuses on a non-atomic game, i.e., a game with an infinite amount of agents, where each agent controls a negligible amount of flow. Following a similar course, in [27, 57, 65, 134], agents are of finite size, and their objectives are parameterized to lie somewhere between *selfish* and *altruistic*, yet *malicious* objectives are not taken into account.

In this study, we intend to investigate agents of finite size whose objectives lie in the range between *malicious* and *selfish*. Per agent i, we parameterize this trade-off through a coefficient $\alpha^i \in [0, 1]$, where $\alpha^i = 1$ corresponds to a selfish agent and $\alpha^i = 0$ to a malicious agent. However, unlike [67], we represent agent i's cost as a combination of its own performance and that of its *rival*. We define the rival of an agent i as the agent $j \neq i$ with the current best performance in the system. Note that an agent's rival is not fixed, but is dependent on the current performance of all the agents in the system. In our setting, a totally malicious agent aims to minimize the performance of its rival, while a totally selfish agent does not take its rival's performance into account.

We consider two types of routing games based on the structure of the agents' performance objectives. The first game considers agents with *bottleneck objectives* (also known as Max-Min or Min-Max objectives), i.e., their performance is determined by the worst component (link) in the network [36, 54, 73, 129]. *Bottleneck routing games* have been shown to emerge in many practical scenarios. For example, in wireless networks, the weakest link in a transmission is determined by the node with the least remaining battery power. Hence, each agent would route traffic so as to maximize the smallest battery lifetime along its routing topology. Additionally, bottleneck routing games arise in congested networks where it is desirable to move traffic away from congested hot spots. For further discussion and additional examples see [36]. The second type of games considers agents with *additive* performance measures, e.g., delay or packet loss. Much of the current literature on networking games has focused on such games, e.g., [14, 128, 159, 161, 191, 202, 224], albeit in the traditional setting of selfish agents.

In [36] and [202], the existence of a Nash equilibrium has been established (respectively) for bottleneck and additive routing games with selfish agents. We note that a

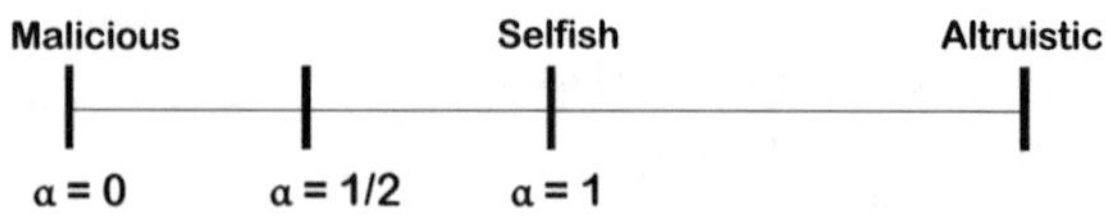

Fig. 9.1 The range of agents' objectives

major complication in proving the existence of a Nash equilibrium for agents with a malicious component, i.e, $\alpha^i < 1$, is the inherent lack of convexity of the objective functions. Thus, we cannot rely on the existence proofs from the referenced studies, and need to establish proofs of our own that do not require (quasi-) convexity of the performance functions.

For both types of games, many studies have attempted to bound the Price of Anarchy (PoA) [159] and the Price of Stability (PoS) [19]. The PoA and PoS quantify the deficiency of the network from a social perspective, at the worst and best Nash equilibrium (respectively). The concepts of PoA and PoS have also been extended to capture *Strong Nash equilibria (SNE)* [17, 22]. Specifically, The Strong Price of Anarchy (SPoA) and Strong Price of Stability (SPoS) quantify the deficiency of the network from a social perspective, at the worst and best Strong Nash equilibrium (respectively). Due to the ever-growing work in this context, it is beyond the scope of this article to do justice and present an exhaustive survey of previous work on routing games with selfish agents. We refer the reader to the above cited studies and to the references therein for a broader review of the literature.

9.1 Solution Concept

We focus our study on the atomic splittable routing model [36, 202], in which each agent sends its non-negligible demand to its destination by splitting it over a set of paths in the network. All agents share the same source and destination, and each agent i has a coefficient α^i, which captures the importance of its rival's performance. We first consider agents with *bottleneck performance measures*, and for which $\alpha^i \in [1/2, 1]$. Intuitively, this range of α^i implies that they care more about their own performance than that of their rivals'. We prove that any system-optimal flow profile is also a Strong Nash equilibrium, i.e., the Strong Price of Stability is equal to 1.[1] Moreover, we prove that the Price of Anarchy is unbounded. Furthermore, in a network with fully selfish agents, i.e., for all i, $\alpha^i = 1$, we establish that the Strong Price of Anarchy is equal to 1. Thus, for fully selfish agents, a flow profile is a Strong Nash equilibrium if and only if it is system-optimal.

We then consider agents with *additive* performance objectives and focus on the fundamental load balancing game of routing over parallel links. Beyond being a basic framework of routing, this is the *generic framework of load balancing* among servers in a network. Accordingly, it has been the subject of numerous studies in the context of non-cooperative networking games, e.g., [128, 159, 161, 202, 223, 281], to name a few. We consider agents that view their own performance and that of their rivals with equal importance, i.e., for all i, $\alpha^i = 1/2$. We establish the existence of a Nash equilibrium and show that the Wardrop equilibrium (which necessarily exists and is unique [152]) is also a Nash equilibrium. Moreover, for a system with two agents, we prove the Nash equilibrium's uniqueness, while for the general case of N

[1]Since any SNE is a Nash equilibrium, $PoS \leq SPoS$ and the *PoS* also equals 1.

agents, we provide an example of its non-uniqueness. Finally, we present an example of a network with agents for which $\alpha^i \in [0, 1]$ and show that for both *bottleneck* and *additive* routing games, no Nash equilibrium necessarily exists.

9.2 Model and Game Theoretic Formulations

9.2.1 Model

We consider a set $\mathcal{N} = \{1, 2, \ldots, N\}$ of selfish "users" (or, "players", "agents"), which share a communication network modeled by a directed graph $G(V, E)$ (Fig. 9.2). We denote by $\mathcal{P}$ the set of all paths in the network. Each user $i \in \mathcal{N}$ has a traffic demand r^i and all users share a common source S and common destination T. Denote the total demand of all the users by R, i.e., $R = \sum_{i \in \mathcal{N}} r^i$. For every i, we denote by $-i$ the set of all users in the system, excluding i. A user ships its demand from S to T by splitting it along the paths in $\mathcal{P}$, i.e., user i decides what fraction of r^i should be sent on through each path. We denote by f_p^i the flow that user $i \in \mathcal{N}$ sends on path $p \in \mathcal{P}$. User i can fix any value for f_p^i, as long as $f_p^i \geq 0$ (non-negativity constraint) and $\sum_{p \in \mathcal{P}} f_p^i = r^i$ (demand constraint); this assignment of traffic to paths, $\mathbf{f^i} = \{f_p^i\}_{p \in \mathcal{P}}$ shall also be referred to as the *routing strategy* of user i. The *(routing strategy) flow profile* $\mathbf{f}$ is the vector of all user routing strategies, $\mathbf{f} = (\mathbf{f^1}, \mathbf{f^2}, \ldots, \mathbf{f^N})$. We say that a flow profile $\mathbf{f}$ is feasible if it is composed of feasible routing strategies and we denote by $\mathbf{F}$ the set of all feasible flow profiles. Turning our attention to a path $p \in \mathcal{P}$, let f_p be the total flow on that path i.e., $f_p = \sum_{i \in \mathcal{N}} f_p^i$; also denote by f_e^i the flow that i sends on link $e \in E$, i.e., $f_e^i = \sum_{p | e \in p} f_p^i$. Similarly, the total flow on link $e \in E$ is denoted by $f_e = \sum_{i \in \mathcal{N}} f_e^i$. We associate with each link a performance function $T_e(\cdot)$, which corresponds to the *cost per unit of flow* through link e and only depends on the total flow f_e. Furthermore, we impose the following assumptions on $T_e(f_e)$:

A1 $T_e : [0, \infty) \to [0, \infty]$.
A2 $T_e(f_e)$ is continuous and strictly increasing in f_e.

Fig. 9.2 Communication network

The performance measure of a user $i \in \mathcal{N}$ is given by a cost function $H^i(\mathbf{f})$, which we shall refer to as the *selfish* cost of i. In bottleneck routing games, $H^i(\mathbf{f})$ corresponds to the performance of the worst-case link, and in additive routing games it corresponds to the sum of all link performances in the system. We define the *rival* of i at $\mathbf{f}$, as the user with the lowest selfish cost at $\mathbf{f}$, i.e., $\min_{j \neq i} H^j(\mathbf{f})$. The aim of each user is to minimize the weighted difference between its own cost and the cost of its *rival* in the network. Thus, the aim of i is to minimize

$$J^i(\mathbf{f}) \triangleq \alpha^i H^i(\mathbf{f}) - (1 - \alpha^i) \min_{j \neq i}\{H^j(\mathbf{f})\}. \tag{9.1}$$

Note that $J^i(\mathbf{f})$ is not necessarily convex in its user flows. Moreover, note that a user's rival is not fixed in our game, but is dependent on the routing strategy of all users.

9.2.2 Bottleneck Routing Cost Function

Following [36], we define the bottleneck of a user $i \in \mathcal{N}$, $b^i(\mathbf{f})$, as the worst performance of any link in the network that i sends a positive amount of flow on,

$$b^i(\mathbf{f}) = \max_{e \in E | f_e^i > 0} T_e(f_e).$$

The *selfish cost* of user i is equal to its bottleneck, $H^i(\mathbf{f}) = b^i(\mathbf{f}) = \max_{e \in E | f_e^i > 0} T_e(f_e)$. Thus, we consider users whose cost functions are of the following form,

$$J^i(\mathbf{f}) = \alpha^i \max_{e \in E | f_e^i > 0} \{T_e(f_e)\} - (1 - \alpha^i) \min_{j \neq i} \max_{l \in E | f_l^j > 0} \{T_l(f_l)\}. \tag{9.2}$$

In other words, user i aims to minimize the weighted difference between its bottleneck and that of its best-off competitor. We define the bottleneck of a path $p \in \mathcal{P}$ with $f_p > 0$ as $b_p(\mathbf{f}) = \max_{e \in p} T_e(f_e)$ and we define the bottleneck of the system as

$$b(\mathbf{f}) = \max_{e \in E | f_e > 0} T_e(f_e).$$

We equate the "welfare" of the system to its bottleneck and denote by $\mathbf{f}^* = (\mathbf{f}^*)_{e \in E}$, the optimal vector of link flows. Thus, the system-optimal cost equals $b(\mathbf{f}^*) = \min_{\mathbf{f} \in \mathbf{F}} b(\mathbf{f})$.

9.2.3 Additive Routing Cost Functions

Another important class of problems is when users are interested in additive performance measures, e.g., delay or packet loss. In this case, T_e may correspond to the total delay of link e. For *additive routing games*, we consider the framework of routing in a "parallel links" network (Fig. 9.3). Thus, $G(V, E)$ corresponds to a graph with parallel "links" (e.g., communication links, servers, etc.) $\mathcal{L} = \{1, 2, \ldots, L\}$, $L > 1$, and a users ships its demand by splitting it over the links $\mathcal{L}$. As observed in [158], it constitutes an appropriate model for seemingly unrelated networking problems. For example, in a QoS-supporting network architecture, bandwidth may be separated among different virtual paths, resulting effectively in a system of parallel and noninterfering links between the source and destination. Additionally, one can consider a corporation or organization that receives service from a number of different network providers. The corporation can split its total flow over the various network facilities (according to performance and cost considerations), each of which can be represented as a link in the parallel link model. More generally, the problem of routing over parallel links is, essentially, the generic problem of load balancing among several servers, and it has been the subject of numerous studies, including the seminal paper [159] and many others, e.g., [128, 161, 202, 223, 281]. In particular, we consider users whose selfish cost functions are of the following form:

$$H^i(\mathbf{f}) = \frac{1}{r^i} \sum_{l \in \mathcal{L}} f_l^i T_l(f_l). \tag{9.3}$$

Thus, $H^i(\mathbf{f})$ corresponds to the average sum of the link costs. From (9.1) we get that

$$J^i(\mathbf{f}) = \alpha^i \sum_{l \in \mathcal{L}} \frac{f_l^i}{r^i} T_l(f_l) - (1 - \alpha^i) \min_{j \neq i} \left\{ \sum_{n \in \mathcal{L}} \frac{f_n^j}{r^j} T_n(f_n) \right\}. \tag{9.4}$$

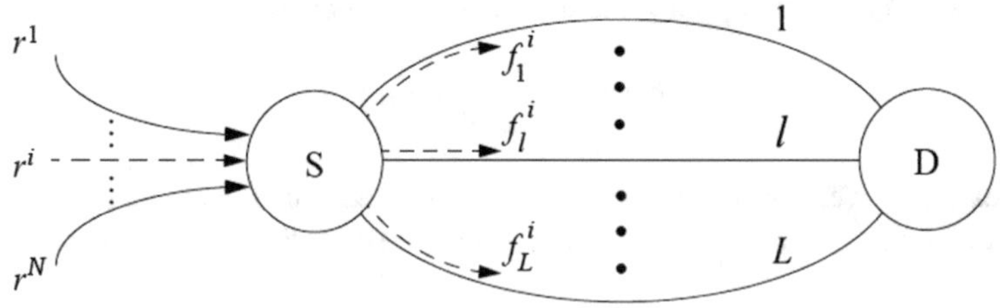

Fig. 9.3 Parallel links network

9.2.4 (Strong) Nash Equilibrium

A flow profile $\mathbf{f}$ is said to be a Nash equilibrium if, given $\mathbf{f^{-i}}$, no user finds it beneficial to deviate from its routing strategy $\mathbf{f^i}$. More formally, $\mathbf{f}$ is a Nash equilibrium if, for all $i \in \mathcal{N}$ and any feasible routing strategy $\bar{\mathbf{f}}^{\mathbf{i}} \neq \mathbf{f^i}$, the following condition holds

$$J^i(\mathbf{f^i}, \mathbf{f^{-i}}) \leq J^i(\bar{\mathbf{f}}^{\mathbf{i}}, \mathbf{f^{-i}}). \tag{9.5}$$

A flow profile $\mathbf{f}$ is said to be a Strong Nash equilibrium (SNE) [22] if no group of users can jointly change their strategy in a way which will decrease every user's cost in the group. Denote the routing strategies of a subset of users $S \subseteq \mathcal{N}$ as $\mathbf{f^S}$. Formally, $\mathbf{f}$ is a Strong Nash equilibrium if, for any coalition $S \subseteq \mathcal{N}$ and any feasible $\bar{\mathbf{f}}^{\mathbf{S}}$, given $\mathbf{f}^{\mathcal{N}\setminus \mathbf{S}}$, there exists some user $i \in S$ that does not decrease its cost, i.e.,

$$J^i(\mathbf{f^S}, \mathbf{f}^{\mathcal{N}\setminus \mathbf{S}}) \leq J^i(\bar{\mathbf{f}}^{\mathbf{S}}, \mathbf{f}^{\mathcal{N}\setminus \mathbf{S}}). \tag{9.6}$$

Note that any Strong Nash equilibrium is also a Nash equilibrium.

In order to quantify the degradation of a Nash equilibrium, we turn towards the *Price of Anarchy* [159] (*Price of Stability* [19]), which is defined as the ratio between the *worst* (*best*) Nash equilibrium, and the social optimum. Moreover, for bottleneck routing games we will also consider the *Strong Price of Anarchy* (*Strong Price of Stability*) [17], which is defined as the ratio between the *worst* (*best*) Strong Nash equilibrium, and the social optimum.

9.3 Bottleneck Routing Games

We start by establishing the existence of a Strong Nash equilibrium in our bottleneck routing game. Note that the user cost function in (9.2) is not continuous, as pointed out in [36]. Moreover, $J^i(\mathbf{f})$ is not necessarily quasi-convex in f_l^i. Consequently, we need to construct an existence proof that does not rely on the continuity or the quasi-convexity of the cost functions. We establish the existence of a Strong Nash equilibrium by constructing a feasible flow profile for all users, such that no coalition of users benefits by jointly deviating from its routing strategy. We first provide the following definition.

Definition 9.1 A flow profile, $\mathbf{f}$, is referred to as *balanced*, if for any two paths $p_1, p_2 \in \mathcal{P}$ with $f_{p_1} > 0$, it holds that $b_{p_1}(\mathbf{f}) \leq \max_{e \in p_2}\{T_e(f_e)\}$.

Thus, at a *balanced* flow profile, for any two paths $p_1, p_2 \in \mathcal{P}$ with positive flow, their bottlenecks are equal, $b_{p_1}(\mathbf{f}) = b_{p_2}(\mathbf{f})$. In order to construct a feasible SNE, we first establish the following lemma.

Lemma 9.1 *Consider a bottleneck routing game. Any system-optimal flow profile is balanced.*

Proof Consider a system-optimal flow profile $\mathbf{f}^*$. Let the set $\mathcal{P}_{\mathbf{f}^*}$ contain all paths with positive flow, whose bottlenecks are equal to the bottleneck of the system. Thus, for any path $p \in \mathcal{P}_{\mathbf{f}^*}$,

$$b_p(\mathbf{f}^*) = b(\mathbf{f}^*).$$

Now assume by contradiction that the system-optimal flow profile $\mathbf{f}^*$ is not balanced. Therefore, the set $\mathcal{P}\backslash\mathcal{P}_{\mathbf{f}^*}$ is not empty, and for any $q \in \mathcal{P}\backslash\mathcal{P}_{\mathbf{f}^*}$ and for all $p \in \mathcal{P}_{\mathbf{f}^*}$,

$$b_p(\mathbf{f}^*) > b_q(\mathbf{f}^*). \tag{9.7}$$

Consider a path $q \in \mathcal{P}\backslash\mathcal{P}_{\mathbf{f}^*}$. We construct a different feasible routing strategy $\mathbf{f}$ by sending a small amount of flow, $\epsilon > 0$ from all the paths in $\mathcal{P}_{\mathbf{f}^*}$ to path q. Specifically, for all $p \in \mathcal{P}_{\mathbf{f}^*}$, $f_p = f_p^* - \epsilon$, $f_q = f_q^* + |\bar{\mathcal{P}}| \cdot \epsilon$, for some small $\epsilon > 0$, and for all other paths, $p \in \mathcal{P}\backslash\{\mathcal{P}_{\mathbf{f}^*} \cup q\}$, $f_p = f_p^*$. If we consider a small enough ϵ, it follows from the strict inequality of (9.7) that for all $q \in \mathcal{P}\backslash\mathcal{P}_{\mathbf{f}^*}$ and for all $p \in \mathcal{P}_{\mathbf{f}^*}$

$$b_p(\mathbf{f}^*) > b_p(\mathbf{f}) > b_q(\mathbf{f}) > b_q(\mathbf{f}^*). \tag{9.8}$$

In other words, for a small enough ϵ,

$$b(\mathbf{f}^*) > b(\mathbf{f}). \tag{9.9}$$

By constructing a new routing strategy $\mathbf{f}$ we are able to lower the bottleneck of the system, which is a contradiction to the optimality of $\mathbf{f}^*$. □

We continue to construct a feasible flow profile, which is also an SNE. Specifically, we focus on any flow profile that is system-optimal.

Theorem 9.1 *Consider a bottleneck routing game where, for each user i, $\alpha^i \in [1/2, 1]$. Any system-optimal flow profile is a Strong Nash equilibrium.*

Proof First consider a system-optimal flow profile $\mathbf{f}$. We will establish that $\mathbf{f}$ is an SNE. As a result of Lemma 9.1, $\mathbf{f}$ is balanced, thus for all $k \in \mathcal{N}$, $b^k(\mathbf{f}) = b(\mathbf{f})$. Therefore, for any user k,

$$J^k(\mathbf{f}) = \alpha^k b^k(\mathbf{f}) - (1 - \alpha^k) \min_{j \neq k}\{b^j(\mathbf{f})\} = (2\alpha^k - 1)b(\mathbf{f}). \tag{9.10}$$

Now, assume by contradiction that $\mathbf{f}$ is not a Strong Nash equilibrium. In other words, there exists some coalition of users $S \subseteq \mathcal{N}$ that are jointly able to change their strategy to $\bar{\mathbf{f}}^{\mathbf{S}}$ such that, for all $k \in S$,

$$J^k(\bar{\mathbf{f}}) = \alpha^k b^k(\bar{\mathbf{f}}) - (1 - \alpha^k) \min_{j \neq k}\{b^j(\bar{\mathbf{f}})\} < (2\alpha^k - 1)b(\mathbf{f}) = J^k(\mathbf{f}), \tag{9.11}$$

where $\bar{\mathbf{f}} \triangleq (\bar{\mathbf{f}}^{\mathbf{S}}, \mathbf{f}^{\mathcal{N}\backslash\mathbf{S}})$. Denote the bottleneck link of the system at $\bar{\mathbf{f}}$ as n and suppose that $\sum_{k \in S} \bar{f}_n^k = 0$, hence $\sum_{k \notin S} \bar{f}_n^k > 0$. Since $\mathbf{f}$ is system-optimal it is clear that

$T_n(\bar{f}_n) = b(\bar{\mathbf{f}}) \geq b(\mathbf{f})$. However,

$$b(\bar{\mathbf{f}}) = T_n(\sum_{k\in S} \bar{f}_n^k + \sum_{k\notin S} \bar{f}_n^k) = T_n(\sum_{k\notin S} f_n^k) \leq T_n(f_n) \leq b(\mathbf{f}).$$

Therefore, $b(\bar{\mathbf{f}}) = b(\mathbf{f})$, and $\bar{\mathbf{f}}$ is also a system-optimal flow profile. As a result of (9.11), there exists some user $i \in \mathcal{N}$ for which $b^i(\bar{\mathbf{f}}) \neq b^i(\mathbf{f})$. Since $b^i(\mathbf{f}) = b(\mathbf{f})$ it must be that $b^i(\bar{\mathbf{f}}) < b(\bar{\mathbf{f}})$. However, this implies that there exists some path p for which $\bar{f}_p^i > 0$ and $b_p(\bar{\mathbf{f}}) < b(\bar{\mathbf{f}})$. This contradicts that $\bar{\mathbf{f}}$ is balanced.

Therefore, $\sum_{k\in S} \bar{f}_n^k > 0$. Consequently, there exists some user $i \in S$, for which $\bar{f}_n^i > 0$, i.e., $b^i(\bar{\mathbf{f}}) = b(\bar{\mathbf{f}})$. Hence, for user i we get that,

$$b^i(\bar{\mathbf{f}}) = b(\bar{\mathbf{f}}) \geq \min_{j\neq i}\{b^j(\bar{\mathbf{f}})\} \tag{9.12}$$

and, as a result of (9.12) it follows that

$$\begin{aligned} J^i(\bar{\mathbf{f}}) &= \alpha^i b^i(\bar{\mathbf{f}}) - (1-\alpha^i)\min_{j\neq i}\{b^j(\bar{\mathbf{f}})\} \\ &= \alpha^i b(\bar{\mathbf{f}}) - (1-\alpha^i)\min_{j\neq i}\{b^j(\bar{\mathbf{f}})\} \\ &\geq (2\alpha^i - 1)b(\bar{\mathbf{f}}) \\ &\geq (2\alpha^i - 1)b(\mathbf{f}) \\ &= J^i(\mathbf{f}), \end{aligned}$$

which is a contradiction to (9.11). Thus, any system-optimal flow profile is a Strong Nash equilibrium. □

Theorem 9.1 illustrates that, in any bottleneck routing game where, for each user i, $\alpha^i \in [1/2, 1]$, there exists a Strong Nash equilibrium. Moreover, there always exists a Strong Nash equilibrium, which is system optimal.[2] This brings us to the following conclusion.

Corollary 9.1 *Consider a bottleneck routing game, where for each user i, $\alpha^i \in [1/2, 1]$. The Price of Stability and the Strong Price of Stability are equal to 1.*

Through Theorem 9.1 and Corollary 9.1 we established that any flow profile that is optimal from a system's perspective is also stable against coalitional deviations. However, even though Theorem 9.1 establishes the existence of desirable (strong) equilibria from a system's perspective, there might also exist equilibria at which the system performance is substantially degraded. To quantify the potential degradation of the system performance, we investigate the Strong Price of Anarchy and the Price of Anarchy. We first consider network scenarios where users are not able to cooperate

---

[2] In [36] a similar theorem was proven for a Nash equilibrium and more general topology. However, they only considered selfish users (i.e, for all i, $\alpha^i = 1$).

and jointly change their strategies. In such scenarios, the lack of communication between the users can substantially degrade the system. This deficiency is captured by the Price of Anarchy.

Theorem 9.2 *Consider a bottleneck routing game where, for each user i, $\alpha^i \in [1/2, 1]$. The Price of Anarchy is unbounded.*

Proof We establish the theorem through the following example. □

Example 1 Consider the network $G = (V, E)$ as depicted in Fig. 9.4. Further, consider two users i and j, each with a flow demand of $r^i = r^j = \frac{R}{2}$ and $\alpha^i = \alpha^j \triangleq \alpha \in [1/2, 1]$. For any edge $e \in E$, the cost per unit of flow is equal to $T_e(f_e) = e^{f_e} - 1$. We focus on a specific flow profile $\mathbf{f}$, in which user i sends its total demand on a single path, namely $\{S, A, B, E, F, I, J, T\}$, and, user j sends its demand on the path $\{S, D, C, F, E, H, G, T\}$. The labels on the edges in Fig. 9.4 correspond to the portion of the total flow that transverses on that edge at $\mathbf{f}$, i.e., f_e/R. Thus,

$$J^i(\mathbf{f}) = \alpha b^i(\mathbf{f}) - (1-\alpha) b^j(\mathbf{f}) = (2\alpha - 1) \cdot (e^{R/2} - 1).$$

It is straightforward that $J^i(\mathbf{f}) = J^j(\mathbf{f})$. Now assume by contradiction that $\mathbf{f}$ is not a Nash equilibrium. Hence, there exists a different routing strategy for user i, $\bar{\mathbf{f}}^{\mathbf{i}} \neq \mathbf{f}^{\mathbf{i}}$, at which user i can decrease its cost. If i places a positive flow on either (S, D), (S, C) or (A, G), it is immediate that $b^i(\bar{\mathbf{f}}^{\mathbf{i}}, \mathbf{f}^{\mathbf{j}}) = b^j(\bar{\mathbf{f}}^{\mathbf{i}}, \mathbf{f}^{\mathbf{j}}) > b^i(\mathbf{f})$ and $J^i(\bar{\mathbf{f}}^{\mathbf{i}}, \mathbf{f}^{\mathbf{j}}) > J^i(\mathbf{f})$.

Thus, if i wishes to refrain from increasing its cost, it will send all its flow on (B, E) and its bottleneck will be at least $T_{(B,E)}(\frac{R}{2}) = e^{R/2} - 1$. It follows that at $(\bar{\mathbf{f}}^{\mathbf{i}}, \mathbf{f}^{\mathbf{j}})$, there cannot exist an edge on which both i and j send a positive amount of flow, otherwise i increases its cost. Thus, the bottleneck of j stays the same. Hence,

$$\begin{aligned} J^i(\bar{\mathbf{f}}^{\mathbf{i}}, \mathbf{f}^{\mathbf{j}}) &= \alpha b^i(\bar{\mathbf{f}}^{\mathbf{i}}, \mathbf{f}^{\mathbf{j}}) - (1-\alpha) b^j(\bar{\mathbf{f}}^{\mathbf{i}}, \mathbf{f}^{\mathbf{j}}) \\ &= \alpha b^i(\bar{\mathbf{f}}^{\mathbf{i}}, \mathbf{f}^{\mathbf{j}}) - (1-\alpha) b^j(\mathbf{f}) \\ &\geq (2\alpha - 1) \cdot (e^{R/2} - 1) \\ &= J^i(\mathbf{f}), \end{aligned}$$

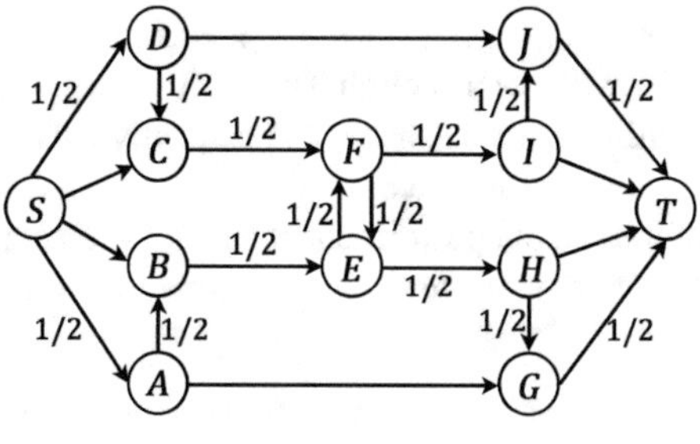

Fig. 9.4 Example of a network with an unbounded PoA

which is a contradiction. Because the users i and j are symmetric, the above analysis also holds for j. Therefore $\mathbf{f}$ is a Nash equilibrium and the bottleneck of the system at $\mathbf{f}$ is equal to $b(\mathbf{f}) = e^{R/2} - 1$.

On the other hand, at the system optimum, $\mathbf{f}^*$, an amount of flow, $R/4$, is sent through the following four paths: **1**: $\{S, A, G, T\}$, **2**: $\{S, B, E, H, T\}$, **3**: $\{S, C, F, I, T\}$, **4**: $\{S, D, J, T\}$. The system bottleneck at $\mathbf{f}^*$ is equal to $b(\mathbf{f}^*) = e^{R/4} - 1$. As a result, the Price of Anarchy in our example is lower-bounded by

$$PoA = \frac{e^{R/2} - 1}{e^{R/4} - 1} \geq \frac{e^{R/2} - 1}{e^{R/4}} = e^{R/4} - \frac{1}{e^{R/4}} \geq e^{R/4} - 1.$$

Since R can be arbitrarily large, the PoA is unbounded. □

Theorem 9.2 illustrates that when users are not able to cooperate and jointly change their strategies, the worst-case Nash equilibrium can be far from optimal. We continue to quantify the degradation of the system when users are able to jointly change their strategies. Specifically, we consider users that are fully selfish, i.e., for all i, $\alpha^i = 1$. First, we provide the following definition as an extension of Definition 9.1.

Definition 9.2 For any user $i \in \mathcal{N}$, a flow profile $\mathbf{f}$ is referred to as *balanced* in $\mathbf{f^i}$, if for any two paths $p_1, p_2 \in \mathcal{P}$ with $f^i_{p_1} > 0$, it holds that $b_{p_1}(\mathbf{f}) \leq \max_{e \in p_2} T_e(f_e)$.

Before we establish a bound on the SPoA, with the help of Definitions 9.1 and 9.2, we establish the following lemma.

Lemma 9.2 *Consider a bottleneck routing game, where for each user* i, $\alpha^i = 1$. *Any Nash equilibrium is balanced.*

Proof Consider a Nash equilibrium $\mathbf{f}$ and a user $i \in \mathcal{N}$. For any fixed $\mathbf{f^{-i}}$, from the perspective of i, each link $e \in E$ has an offset of f^{-i}_e. As a result, when i sends its flow according to its best-response strategy $\mathbf{f^i}$, it sends its demand as a single optimizing user in a network with an offset of f^{-i}_e on each link $e \in E$. Thus, according to Lemma 9.1, for any user $i \in \mathcal{N}$, $(\mathbf{f^i}, \mathbf{f^{-i}})$ is balanced in $\mathbf{f^i}$.

We now continue to prove that the Nash equilibrium $\mathbf{f}$ is balanced. Assume by contradiction that $\mathbf{f}$ is not balanced. Thus, there exist paths p, p' with $f_p > 0$, and

$$b_p(\mathbf{f}) > \max_{e \in p'} T_e(f_e). \tag{9.13}$$

Therefore, there exists some user j for which $f^j_p > 0$. As a result of (9.13) and Definition 9.2, $\mathbf{f} \triangleq (\mathbf{f^j}, \mathbf{f^{-j}})$ is not balanced in $\mathbf{f^j}$, which is a contradiction. Hence, any Nash equilibrium flow profile $\mathbf{f}$ is balanced. □

Theorem 9.3 *Consider a bottleneck routing game, where for each user* i, $\alpha^i = 1$. *The Strong Price of Anarchy equals 1.*

Proof Consider an system-optimal flow profile $\mathbf{f}$ and assume by contradiction that there exists an SNE $\hat{\mathbf{f}}$ for which $b(\hat{\mathbf{f}}) > b(\mathbf{f})$. It follows from Lemmas 9.1 and 9.2 that both $\mathbf{f}$ and $\hat{\mathbf{f}}$ are balanced, thus for any $i \in \mathcal{N}$,

$$J^i(\hat{\mathbf{f}}) = \alpha^i b^i(\hat{\mathbf{f}}) - (1 - \alpha^i) \min_{j \neq i}\{b^j(\hat{\mathbf{f}})\} = (2\alpha^i - 1)b(\hat{\mathbf{f}}) > (2\alpha^i - 1)b(\mathbf{f}) = J^i(\mathbf{f}).$$

In other words, at $\hat{\mathbf{f}}$ the cost of all users is strictly higher than at $\mathbf{f}$. Therefore, it is beneficial for all users to jointly change their strategy to any system-optimal flow profile. This is in contradiction to $\hat{\mathbf{f}}$ being an SNE. Hence, any Strong Nash Equilibrium is a system-optimal flow profile. □

Theorems 9.1 and 9.3 illustrate that, in any bottleneck routing game where, for each user i, $\alpha^i = 1$, the set of Strong Nash equilibria is equal to the set of system-optimal flow profiles. Moreover, for such fully selfish users there exists a wide gap between the Strong Price of Anarchy and the Price of Anarchy. This indicates that the degradation of the system at the worst Nash equilibrium results from the lack of cooperation of the users and not from their selfishness. Indeed, if they were allowed to jointly change their strategy they would bring the system to its optimum, while strictly benefiting themselves.

9.4 Additive Routing Games

In this section we consider additive performance measures, such as delay, jitter and packet loss. Moreover, we consider the specific case where, for all $i \in \mathcal{N}$, $\alpha^i = 1/2$,[3] and we focus on the fundamental load balancing game of routing over parallel links (Fig. 9.3). Similar to bottleneck routing games, we first need to prove the existence of a Nash equilibrium. As mentioned in Sect. 9.3, due to the lack of quasi-convexity we cannot rely on well known existence proofs for convex-games, such as the one given in [219]. We therefore establish the existence of a Nash equilibrium by constructing a feasible flow profile for all users, such that no user wishes to unilaterally deviate from its routing strategy. In other words, each user views its own performance and that of its rival, with equal importance. From (9.3), the cost of user i turns into

$$J^i(\mathbf{f}) = \sum_{l \in \mathcal{L}} \frac{f_l^i}{r^i} T_l(f_l) - \min_{j \neq i} \left\{ \sum_{n \in \mathcal{L}} \frac{f_n^j}{r^j} T_n(f_n) \right\}. \tag{9.14}$$

[3] An existence and uniqueness proof for fully selfish users is given in [202].

Note that we disregard $\alpha^i \triangleq \alpha = 1/2$ from our equilibrium analysis, since it multiplies all users' costs by the same constant. We now provide the following definition from [284].

Definition 9.3 A flow profile, $\mathbf{f}$, is a Wardrop equilibrium if for any two links $l, n \in \mathcal{L}$ with $f_l > 0$, $T_l(f_l) \leq T_n(f_n)$.

In any additive routing game, there exists a Wardrop equilibrium. Moreover, it is unique with respect to the aggregated link flows f_l, [152, 284]. We focus on a specific Wardrop equilibrium, which is also proportional in its user flows.

Definition 9.4 A flow profile, $\mathbf{f}$, is referred to as proportional, if for any path $p \in \mathcal{P}$, and for any user $i \in \mathcal{N}$, $f_p^i = \frac{r^i}{R} f_p$.

To demonstrate that a proportional flow profile is feasible, it needs to satisfy **(i)** the non-negativity constraint and **(ii)** the demand constraint of all users. Consider a user $i \in \mathcal{N}$. It follows that $f_p^i = \frac{r^i}{R} f_p \geq 0$, thus the non-negativity constraint is satisfied. Furthermore, $\sum_{p\in\mathcal{P}} f_p^i = \frac{r^i}{R} \sum_{p\in\mathcal{P}} f_p = r^i$, thus the demand constraint is also satisfied. In the following theorem we focus on the proportional Wardrop equilibrium. According to Definitions 9.3 and 9.4, this flow profile is unique with respect to the individual user flows.

Theorem 9.4 *Consider an additive routing game where, for each user i, $\alpha^i = 1/2$. There exists a Nash equilibrium. In particular, it is equal to the proportional Wardrop equilibrium.*

Proof We consider the unique proportional Wardrop equilibrium, $\mathbf{f}$, and prove that no user wishes to unilaterally deviate from $\mathbf{f}$. Assume by contradiction that $\mathbf{f}$ is not a Nash equilibrium. Hence, there exists a user i and a routing strategy $\bar{\mathbf{f}}^{\mathbf{i}} \neq \mathbf{f}^{\mathbf{i}}$ such that $J^i(\bar{\mathbf{f}}^{\mathbf{i}}, \mathbf{f}^{-\mathbf{i}}) < J^i(\mathbf{f})$. We split the set of links $\mathcal{L}$, into three subsets: $\mathcal{L}^+ = \{l \in \mathcal{L} | \bar{f}_l^i > f_l^i\}$, $\mathcal{L}^- = \{l \in \mathcal{L} | \bar{f}_l^i < f_l^i\}$ and $\mathcal{L}^0 = \{l \in \mathcal{L} | \bar{f}_l^i = f_l^i\}$. Since $\bar{\mathbf{f}}^{\mathbf{i}} \neq \mathbf{f}^{\mathbf{i}}$, it follows that $\mathcal{L}^+$ and $\mathcal{L}^-$ are not empty. For any link $l \in \mathcal{L}^+$, denote $\epsilon_l \equiv \bar{f}_l^i - f_l^i$ and for any link $l \in \mathcal{L}^-$, denote $\delta_l \equiv f_l^i - \bar{f}_l^i$. Since r^i is constant, the differences in $\mathcal{L}^+$ and $\mathcal{L}^-$ are equal and $\sum_{l\in\mathcal{L}^+} \epsilon_l = \sum_{l\in\mathcal{L}^-} \delta_l$. Because $\mathbf{f}$ is a proportional flow profile, it holds that, for any two users $i, k \in \mathcal{N}$ and for any link $l \in \mathcal{L}$, $f_l^i/r^i = f_l^k/r^k$. Thus, for any link $l \in \mathcal{L}^0$ and any user $k \in \mathcal{N}$,

$$\left[\frac{\bar{f}_l^i}{r^i} - \frac{f_l^k}{r^k}\right] = \left[\frac{f_l^i}{r^i} - \frac{f_l^k}{r^k}\right] = 0. \tag{9.15}$$

Equation (9.15) holds for any $k \in \mathcal{N}$, hence also for i's rival at $(\bar{\mathbf{f}}^{\mathbf{i}}, \mathbf{f}^{-\mathbf{i}})$. Denote i's rival at $(\bar{\mathbf{f}}^{\mathbf{i}}, \mathbf{f}^{-\mathbf{i}})$ as j. Combining (9.15) with (9.14), we get

$$J^i(\bar{\mathbf{f}}^{\mathbf{i}}, \mathbf{f}^{-\mathbf{i}}) = \sum_{l\in\mathcal{L}^+}\left[\frac{\bar{f}_l^i}{r^i} - \frac{f_l^j}{r^j}\right]T_l(\bar{f}_l^i + f_l^{-i}) + \sum_{l\in\mathcal{L}^-}\left[\frac{\bar{f}_l^i}{r^i} - \frac{f_l^j}{r^j}\right]T_l(\bar{f}_l^i + f_l^{-i}) \tag{9.16}$$

$$+ \sum_{l\in\mathcal{L}^0}\left[\frac{\bar{f}_l^i}{r^i} - \frac{f_l^j}{r^j}\right]T_l(\bar{f}_l^i + f_l^{-i})$$

$$= \sum_{l\in\mathcal{L}^+}\left[\frac{f_l^i + \epsilon_l}{r^i} - \frac{f_l^j}{r^j}\right]T_l(f_l + \epsilon_l) + \sum_{l\in\mathcal{L}^-}\left[\frac{f_l^i - \delta_l}{r^i} - \frac{f_l^j}{r^j}\right]T_l(f_l - \delta_l)$$

$$= \sum_{l\in\mathcal{L}^+}\frac{\epsilon_l}{r^i}T_l(f_l + \epsilon_l) - \sum_{l\in\mathcal{L}^-}\frac{\delta_l}{r^i}T_l(f_l - \delta_l) > \sum_{l\in\mathcal{L}^+}\frac{\epsilon_l}{r^i}T_l(f_l) - \sum_{l\in\mathcal{L}^-}\frac{\delta_l}{r^i}T_l(f_l).$$

The last inequality follows from Assumption A2. Since $\mathbf{f}$ is a Wardrop equilibrium, we make two observations, namely

(1): $\forall l \in \mathcal{L}^-$, $\delta_l > 0$, thus $f_l > 0$. Therefore, from Definition 9.3, for any two links $l, n \in \mathcal{L}^-$, $T_l(f_l) = T_n(f_n)$.

(2): From Definition 9.3 it follows that for any link $l \in \mathcal{L}^+$ and any link $n \in \mathcal{L}^-$, $T_l(f_l) \geq T_n(f_n)$.

Consider any link $e \in \mathcal{L}^-$. Consequently, Eq. (9.16) turns into

$$J^i(\bar{\mathbf{f}}^{\mathbf{i}}, \mathbf{f}^{-\mathbf{i}}) > \sum_{l\in\mathcal{L}^+}\frac{\epsilon_l}{r^i}T_e(f_e) - \sum_{l\in\mathcal{L}^-}\frac{\delta_l}{r^i}T_e(f_e) = \frac{1}{r^i}T_e(f_e)\cdot\left[\sum_{l\in\mathcal{L}^+}\epsilon_l - \sum_{l\in\mathcal{L}^-}\delta_l\right] = 0.$$

On the other hand, because $\mathbf{f}$ is proportional, it follows from (9.14) that

$$J^i(\mathbf{f}) = \sum_{l\in\mathcal{L}}\frac{f_l^i}{r^i}T_l(f_l) - \min_{k\neq i}\{\sum_{n\in\mathcal{L}}\frac{f_n^k}{r^k}T_n(f_n)\} = \sum_{l\in\mathcal{L}}\frac{f_l^i}{r^i}T_l(f_l) - \sum_{n\in\mathcal{L}}\frac{f_n^i}{r^i}T_n(f_n) = 0.$$

Thus, $J^i(\bar{\mathbf{f}}^{\mathbf{i}}, \mathbf{f}^{-\mathbf{i}}) > J^i(\mathbf{f})$, which is a contradiction. Hence, $\mathbf{f}$ is a Nash equilibrium. □

Now that we have proven the existence of a Nash equilibrium, we continue to investigate its uniqueness. We focus on a special case in which the network has two users, i.e., $N = 2$, and we denote them as i and j. It follows from (9.14) that $J^j(\mathbf{f}) = -J^i(\mathbf{f})$. In order to prove the Nash equilibrium's uniqueness, we make use of the following lemma.

Lemma 9.3 *Consider an additive routing game as described in Sect. 9.2, where $N = 2$ and $\alpha^i = \alpha^j = 1/2$. At any Nash equilibrium $\mathbf{f}$, $J^i(\mathbf{f}) = J^j(\mathbf{f}) = 0$.*

Proof Assume by contradiction that $J^i(\mathbf{f}) > 0$. Consider a different routing strategy $\bar{\mathbf{f}}^{\mathbf{i}}$ in which for any link $l \in \mathcal{L}$, i sends its flow according

$$\bar{f}_l^i = \frac{r^i}{r^j}f_l^j.$$

$\bar{\mathbf{f}}^{\mathbf{i}}$ is a feasible routing strategy for i, since **(1)** $\forall l \in \mathcal{L}$, $\bar{f}_l^i \geq 0$ and **(2)** $\sum_{l\in\mathcal{L}} \bar{f}_l^i = \sum_{l\in\mathcal{L}} \frac{r^i}{r^j} f_l^j = r^i$. Moreover, from (9.14) it follows that $J^i(\bar{\mathbf{f}}) = 0$, which contradicts the fact that $\mathbf{f}$ is a Nash equilibrium. Thus, at any Nash equilibrium $\mathbf{f}$, $J^k(\mathbf{f}) \leq 0$, for $k = i, j$.

Now assume by contradiction that $J^i(\mathbf{f}) < 0$. Thus, $J^j(\mathbf{f}) > 0$, which is a contradiction. □

We are now ready to prove the following.

Theorem 9.5 *For $N = 2$, the proportional Wardrop equilibrium is the unique Nash equilibrium, i.e., the Nash equilibrium is unique in the users' individual flows.*

Proof Denote the proportional Wardrop equilibrium as $\hat{\mathbf{f}}$ and assume by contradiction that there exists another Nash equilibrium $\mathbf{f} \neq \hat{\mathbf{f}}$. From Lemma 9.3 it follows that $J^i(\mathbf{f}) = J^j(\mathbf{f}) = 0$. Consider a new routing strategy for user i, $\bar{\mathbf{f}}^{\mathbf{i}} \neq \mathbf{f}^{\mathbf{i}}$. At $\bar{\mathbf{f}}^{\mathbf{i}}$, user i sends its flow according $\bar{f}_l^i = \frac{r^i}{r^j} f_l^j$ for any $l \in \mathcal{L}$. Denote the link flow of link l at flow profile $(\bar{\mathbf{f}}^{\mathbf{i}}, \mathbf{f}^{\mathbf{j}})$ as $\bar{f}_l$. Thus,

$$\bar{f}_l = \bar{f}_l^i + f_l^j = f_l^j \left[1 + \frac{r^i}{r^j}\right] = f_l^j \frac{R}{r^j}.$$

In other words, $f_l^j = \frac{r^j}{R} \bar{f}_l$ and $\bar{f}_l^i = \frac{r^i}{R} \bar{f}_l$, i.e., $(\bar{\mathbf{f}}^{\mathbf{i}}, \mathbf{f}^{\mathbf{j}})$ is a proportional flow profile. From (9.14) and Lemma 9.3, it follows that $J^i(\bar{\mathbf{f}}^{\mathbf{i}}, \mathbf{f}^{\mathbf{j}}) = J^i(\mathbf{f}^{\mathbf{i}}, \mathbf{f}^{\mathbf{j}}) = 0$, hence i does not increase its cost by changing its strategy to $\bar{\mathbf{f}}^{\mathbf{i}}$. At the new routing strategy flow profile $(\bar{\mathbf{f}}^{\mathbf{i}}, \mathbf{f}^{\mathbf{j}})$, either one of the two cases holds:

Case 1: $(\bar{\mathbf{f}}^{\mathbf{i}}, \mathbf{f}^{\mathbf{j}})$ is not a Wardrop equilibrium.
Case 2: $(\bar{\mathbf{f}}^{\mathbf{i}}, \mathbf{f}^{\mathbf{j}})$ is a Wardrop equilibrium.

Consider Case 1. There exist two links, $l, n \in \mathcal{L}$ such that $\bar{f}_l > 0$ and $T_l(\bar{f}_l) > T_n(\bar{f}_n)$. Moreover, since $\bar{f}_l^i = \frac{r^i}{R} \bar{f}_l$, it follows that $\bar{f}_l^i > 0$. We now construct a new strategy for i, at which i sends a small amount of flow, $\epsilon > 0$, from link l to n. Denote this new strategy as $\tilde{\mathbf{f}}^{\mathbf{i}}$. At $(\tilde{\mathbf{f}}^{\mathbf{i}}, \mathbf{f}^{\mathbf{j}})$, i's cost is equal to:

$$\begin{aligned} J^i(\tilde{\mathbf{f}}^{\mathbf{i}}, \mathbf{f}^{\mathbf{j}}) = & \left[\frac{\bar{f}_l^i - \epsilon}{r^i} - \frac{f_l^j}{r^j}\right] T_l(\bar{f}_l - \epsilon) \\ & + \left[\frac{\bar{f}_n^i + \epsilon}{r^i} - \frac{f_n^j}{r^j}\right] T_n(\bar{f}_n + \epsilon) \\ & + \sum_{e\in\mathcal{L}\setminus\{l,n\}} \left[\frac{\bar{f}_e^i}{r^i} - \frac{f_e^j}{r^j}\right] T_e(f_e). \end{aligned} \tag{9.17}$$

Thus, (9.17) turns into:

$$J^i(\tilde{\mathbf{f}}^{\mathbf{i}}, \mathbf{f}^{\mathbf{j}}) = \frac{\epsilon}{r^i}\left[T_n(f_n + \epsilon) - T_l(f_l - \epsilon)\right], \tag{9.18}$$

which is negative for a small enough ϵ. Hence, we constructed a feasible strategy in which user i decreases its cost, which is a contradiction to $\mathbf{f}$ being a Nash equilibrium.

Now consider Case 2. It follows that $(\bar{\mathbf{f}}^{\mathbf{i}}, \mathbf{f}^{\mathbf{j}})$ is the unique proportional Wardrop equilibrium. However, in Theorem 9.4 it is proven that any unilateral deviation of user i from the proportional Wardrop equilibrium causes a strict increase in cost to user i. In other words,

$$J^i(\hat{\mathbf{f}}^{\mathbf{i}}, \hat{\mathbf{f}}^{\mathbf{j}}) = J^i(\bar{\mathbf{f}}^{\mathbf{i}}, \mathbf{f}^{\mathbf{j}}) < J^i(\mathbf{f}^{\mathbf{i}}, \mathbf{f}^{\mathbf{j}}),$$

which is a contradiction. □

An immediate consequence of Theorem 9.5 is that the PoA and PoS of two-user systems are bounded by well known bounds on the Wardrop equilibrium, e.g., see [191, 224].

Although Theorem 9.5 holds for a network with two users, in the general case of N players, it does not hold. Indeed, the following example shows that, in games with $N > 2$ multiple equilibria may exist.

Example 2 Consider a network with four users i, j, k, h and with two parallel links, for which $T_1(f_1) = f_1$, $T_2(f_2) = f_2$. The demand of each individual user is equal to 1 and $\alpha^z = 1/2$ for $z = i, j, k, h$. Consider the flow profile, $\mathbf{f}$, at which $f_1^i = f_1^j = f_2^k = f_2^h = 1$ and $f_2^i = f_2^j = f_1^k = f_1^h = 0$. The selfish cost of the users is equal to

$$H^z(\mathbf{f}) = \sum_{l\in\mathcal{L}} \frac{f_l^z}{r^z} T_l(f_l) = \sum_{l\in\mathcal{L}} T_l(f_l) = 2$$

for $z = i, j, k, h$. Thus, $J^z(\mathbf{f}) = 2 - 2 = 0$ for $z = i, j, k, h$.

Now consider user i and a different flow profile $\bar{\mathbf{f}} = (\bar{\mathbf{f}}^{\mathbf{i}}, \mathbf{f}^{-\mathbf{i}})$, at which i sends an amount of $0 < \epsilon \leq 1$ to link 2. It follows that

$$H^i(\bar{\mathbf{f}}) = (1-\epsilon)(2-\epsilon) + \epsilon(2+\epsilon),$$

$H^j(\bar{\mathbf{f}}) = 2 - \epsilon$ and $H^k(\bar{\mathbf{f}}) = H^h(\bar{\mathbf{f}}) = 2 + \epsilon$. Thus,

$$J^i(\bar{\mathbf{f}}) = H^i(\bar{\mathbf{f}}) - H^j(\bar{\mathbf{f}}) = 2\epsilon^2 > 0.$$

Therefore, i increases its cost at $\bar{\mathbf{f}}^{\mathbf{i}}$. Since the example is symmetric, this holds for all users, which proves that $\mathbf{f}$ is a Nash equilibrium. However, $\mathbf{f}$ is not a proportional flow profile, thus, according to Theorem 9.4, there exists another Nash equilibrium.

In Theorems 9.4 and 9.5 we focused on users with $\alpha = 1/2$. The next example considers a more general case. Namely, we show that, in a network of users with

$\alpha^i \in [0, 1]$, a Nash equilibrium need not exist for either bottleneck routing games or additive routing games.

Example 3 Consider either a bottleneck or an additive routing game. Moreover, consider a network with two parallel links, two users with demands $r^1 = 1, r^2 = 1/2$, $\alpha^1 = 1$ and $\alpha^2 = 0$. Both links have the following cost per unit of flow, for $l = 1, 2$:

$$T_l(f_l) = \begin{cases} \frac{1}{1-f_l} & \text{if } f_l < 1 \\ \infty & \text{if } f_l \geq 1. \end{cases}$$

Since $\alpha^2 = 0$, the cost of user 2 is equal to $J^2(\mathbf{f}) = -H^1(\mathbf{f})$. By sending its entire demand r^2 on the link with $\max\{f_1^1, f_2^1\}$, user 2 is able to minimize its cost to $-\infty$ for any routing strategy of user 1. In that case $J^1(\mathbf{f}) = H^1(\mathbf{f}) = \infty$, thus user 1 will respond by sending part of its flow on the link with $\min\{f_1^1, f_2^1\}$. Nevertheless, for any routing strategy $\mathbf{f^1}$, $\max\{f_1^1, f_2^1\} \geq 1/2$ and user 2 will again bring the cost of user 1 to ∞. This process goes on forever and it is clear that, for both bottleneck routing games and additive routing games, there does not exist any Nash equilibrium.

9.5 Conclusions

In this study we investigated routing games where the cost of each agent is represented as a combination of its own performance and that of its *rival*. We established the existence of Strong Nash equilibria in games with bottleneck performance measures and we established the existence of Nash equilibria in games with additive performance measures. For bottleneck routing games and agents with $\alpha^i \in [1/2, 1]$, namely, games where agents care more about their own performance than that of their rivals', we established that any system-optimal flow profile is also a Strong Nash equilibrium, i.e., the Price of Stability and the Strong Price of Stability are equal to 1. We further established that the Price of Anarchy is unbounded and that for fully selfish users (i.e., for all $i, \alpha^i = 1$) the Strong Price of Anarchy is equal to 1. We conclude that a system-optimal flow profile is not only desirable from a system's perspective, but also easy to implement due to its stability. Due to the high PoA and potentially high SPoA for users with a malicious component, a design guideline would be to have a mediator, e.g., a network administrator, propose to all agents in the network to route their flow according to system-optimal operating points. More importantly, this gives a credible guarantee to all agents that the agreed-upon solution will be implemented and no subcoalition will want to deviate.

Moreover, for fully selfish users there is a considerable gap between the SPoA and the PoA. This indicates that the degradation of the network performance at the worst Nash equilibrium is a direct cause of the lack of cooperation between the users, and not due to their selfishness. Indeed, if the users were would cooperate and jointly change their strategy they would bring the system to its optimum, while strictly benefiting themselves.

For additive routing games, we focused on the fundamental load balancing game of routing over parallel links and on agents with $\alpha^i = 1/2$, namely, games where agents view their own performance and that of their rivals with equal importance. We proved that the proportional Wardrop equilibrium (which exists and is unique) is also a Nash equilibrium. Moreover, for a two-player system, we established the uniqueness of the Nash equilibrium. This allows for the PoA and PoS to be bounded by well known current bounds on the Wardrop equilibrium [191]. We also provided an example of the non-uniqueness of the Nash equilibrium for a system with N-players. Finally, we established that, in the general case of agents with $\alpha^i \in [0, 1]$, a Nash equilibrium does not necessarily exist for either bottleneck or additive routing games.

In future research, for bottleneck routing games it will be interesting to consider networks with multiple sources and destination pairs and see how this affects the bounds on the (S)PoS and (S)PoA. Moreover, it remains an open question to bound the SPoA for users with a malicious component in their cost function. For additive routing games, we established the existence of a Nash equilibrium for $\alpha^i = 1/2$ and in [202] the existence is established for $\alpha^i = 1$. It remains an open problem to establish the existence of a Nash equilibrium for the more general case of agents with $\alpha^i \in [1/2, 1]$.

Chapter 10
Stochastic Coalitional Better-Response Dynamics for Finite Games with Application to Network Formation Games

Konstantin Avrachenkov and Vikas Vikram Singh

10.1 Introduction

In a repeated play of a strategic game over infinite horizon, a Nash equilibrium that is played in the long run depends on an initial action profile as well as the way all the players choose their actions at each time. Young [297] considered an n-player strategic game where at each time all the players make a simultaneous move and each player chooses an action that is the best response to k previous games among the m, $k \leq m$, most recent games in the past. In general, this dynamics need not converge to a Nash equilibrium, it may stuck into a closed cycle. Young [297] also considered the case where at each time with small probability each player makes mistake and chooses some non-optimal action. These mistakes add mutations into the dynamics. In general the mutations can be sufficiently small. This leads to the definition of a stochastically stable Nash equilibrium which is selected by the stochastic dynamics as mutations vanish. Young proposed an algorithm to compute the stochastically stable Nash equilibria. He showed that the risk dominant Nash equilibrium of a 2×2 coordination game is stochastically stable. Kandori et al. [154] considered a different dynamic model where at each time each player plays with every other player in a pairwise contest. The pairwise contest is given by 2×2 symmetric matrix game and each player chooses an action which has higher expected average payoff. The mutations are present into dynamics due to wrong actions taken by the players.

Parts of this chapter have been published in RESEARCH REPORT N° 8716, INRIA, HAL Id: hal-01143912.

K. Avrachenkov (✉)
Inria Sophia Antipolis, 2004 Route des Lucioles, 06902 Sophia Antipolis, France
e-mail: K.Avrachenkov@inria.fr

V. V. Singh
Department of Mathematics, Indian Institute of Technology Delhi,Hauz Khas, New Delhi 110016, India
e-mail: vikassingh@iitd.ac.in

E. Altman et al. (eds.), *Multilevel Strategic Interaction Game Models for Complex Networks*, https://doi.org/10.1007/978-3-030-24455-2_10

They showed that the risk dominant Nash equilibrium of a 2×2 coordination game is stochastically stable. That is, for 2×2 coordination games the dynamics given by [154, 297] select the same Nash equilibrium. Fudenberg et al. [102] proposed a dynamics where at each time only one player is selected to choose actions. The mutations with small probability also occur at each time. The risk dominant Nash equilibrium of a 2×2 coordination game need not be stochastically stable under this dynamics.

The equilibrium concepts which are stable against the coalitional deviations are more suitable for the situations where players can a priori communicate, being in a position to form a coalition and jointly deviate in a coordinated way. Sawa [230] introduced such an equilibrium concept which is called (strict) $\mathcal{K}$-stable equilibrium, where $\mathcal{K}$ is a set of all feasible coalitions. A (strict) $\mathcal{K}$-stable equilibrium corresponds to a (strict) strong Nash equilibrium [23] if there is no restriction on coalition formation. As motivated from the application of $\mathcal{K}$-stable equilibria in network formation games considered in [85, 144, 146], we restrict ourselves to only pure actions. A $\mathcal{K}$-stable equilibrium need not always exist and in such case there exist some set of action profiles forming a closed cycle. Recently, some stochastic dynamics due to coalitional deviations have been proposed [26, 187, 230]. Sawa [230] studied the stochastic stability in general finite games where the mutations are present through a logit choice rule. Newton [187] considered the situation where profitable coalitional deviations are given greater importance than unprofitable single player deviations. Avrachenkov et al. [26] studied the stochastic stability for network formation games with teams. In general, the stochastic stability results depend on the way actions being chosen during the infinite play. Some other famous works on stochastic stability in different settings include [97, 101, 102, 104, 156, 188–190, 221].

In this chapter, we consider the coalition formation in a strategic game where at each time players are allowed to form a coalition and make a joint deviation from the current action profile if it is strictly beneficial for all the members of the coalition. Such deviations define a coalitional better-response (CBR) dynamics. We assume that the coalition formation is random and at each time only one coalition can be formed among all feasible coalitions. We also consider the possibility of making wrong decision at each time by the selected coalition. These wrong decisions are made with small probability. These mistakes work as mutations and add perturbations into CBR dynamics. We prove that the perturbed CBR dynamics selects $\mathcal{K}$-stable equilibria or closed cycles, that have minimum stochastic potential among all action profiles, in the long run as mutations vanish. If there is no restriction on coalition formation, CBR dynamics selects all the strong Nash equilibria and closed cycles, i.e., all the strong Nash equilibria and closed cycles are stochastically stable. The similar CBR dynamics can be given for the case where each time a coalition deviate from a current action profile such that all the players from the coalition are at least as well off at new action profile and at least one player is strictly better off. In this case the similar results hold for strict $\mathcal{K}$-stable equilibrium and strict strong Nash equilibrium (SSNE). We apply CBR dynamics corresponding to SSNE to network formation games where nodes (players) of a network form a coalition and make a move to a new network if it offers each player at least as much as it is in the current network and at least one player gets strictly better payoff. We prove that all strongly stable networks and closed cycles are stochastically stable. The CBR dynamics generalizes the stochastic dynamics considered in [26] by considering the general finite games. The stochastic dynamics for pairwise stable networks considered in [147] can also be viewed as a special case of stochastic CBR dynamics.

The chapter is organized as follows. Section 10.2 contains the model and few definitions. We describe the CBR dynamics in Sect. 10.3. Section 10.4 contains the application of CBR dynamics to network formation games. We conclude our paper in Sect. 10.5.

10.2 The Model

We consider an n-player strategic game. Let $N = \{1, 2, \ldots, n\}$ be a finite set of players. For each $i \in N$, let A_i be a finite action set of player i whose generic element is denoted by a_i. Then, $A = \prod_{i=1}^{n} A_i$ is defined as a set of all action profiles whose generic element is denoted by $a = (a_1, a_2, \ldots, a_n)$. The payoff function of player i is defined as $u_i : A \to \mathbb{R}$. We consider the situation where players are allowed to communicate with each other. As a consequence they can form a coalition and revise their strategies jointly. In many situations it may not be feasible to form all types of coalitions. Let $\mathcal{K} \subseteq \mathcal{P}(N) \setminus \phi$ be the set of all feasible coalitions, where $\mathcal{P}(N)$ denotes the power set of N and ϕ an empty set. For a coalition $S \in \mathcal{K}$, define $A_S = \prod_{i \in S} A_i$ whose element is denoted by a_S and a_{-S} denotes an action profile of the players outside S. A coalition of players jointly deviate from a current action profile if new action profile is strictly beneficial for all the players from the coalition. Such deviations are called improving deviations and it leads to the definition of a $\mathcal{K}$-stable equilibrium. In some cases, a coalition of players jointly deviate from a current action profile if at new action profile each player is at least as well off and one player is strictly better off. Such deviations leads to the definition of a strict $\mathcal{K}$-stable equilibrium.

Definition 10.1 An action profile a^* is said to be a $\mathcal{K}$-stable equilibrium if there is no $S \in \mathcal{K}$ and $a \in A$ such that

1. $a_i = a_i^*, \ \forall\, i \notin S$.
2. $u_i(a) > u_i(a^*), \ \forall\, i \in S$.

If $\mathcal{K} = \mathcal{P}(N) \setminus \phi$, a $\mathcal{K}$-stable equilibrium is a strong Nash equilibrium (SNE) [23]. Let $A(S, a)$ be the set of all action profiles reachable from a via deviation of coalition S. It is defined as

$$A(S, a) = \{a' | a'_i = a_i, \ \forall\, i \notin S \text{ and } a'_i \in A_i, \ \forall\, i \in S\}.$$

A coalition always has option to do nothing, so $a \in A(S, a)$. Let $\mathcal{I}_1(S, a)$ be a set of improved action profiles reachable from an action profile a via improving deviations of coalition S, i.e.,

$$\mathcal{I}_1(S, a) = \{a'|a'_i = a_i, \ \forall\, i \notin S \text{ and } u_i(a') > u_i(a), \ \forall\, i \in S\}. \tag{10.1}$$

For an improved action profile $a' \in \mathcal{I}_1(S, a)$, an action profile a'_S of all the players from S is called a *better-response* of coalition S against a fixed action profile a_{-S} of the players outside S. Define, $\overline{\mathcal{I}}_1(S, a) = A(S, a) \setminus \mathcal{I}_1(S, a)$ as a set of all action profiles due to the erroneous decisions of coalition S. The set $\overline{\mathcal{I}}_1(S, a)$ is always nonempty for all S and a because $a \in \overline{\mathcal{I}}_1(S, a)$. A $\mathcal{K}$-stable equilibrium need not always exist. In such a case there exist a set of action profiles lying on a closed cycle and all such action profiles can be reached from each other via an improving path. The definitions of closed cycle and improving path are as follows:

Definition 10.2 (*Improving Path*) An improving path from a to a' is a sequence of action profiles and coalitions $a^1, S_1, a^2, \ldots, a^{m-1}, S_{m-1}, a^m$ such that $a^1 = a$, $a^m = a'$ and $a^{k+1} \in \mathcal{I}_1(S_k, a^k)$ for all $k = 1, 2, \ldots, m-1$.

Definition 10.3 (*Cycles*) A set of action profiles C form a cycle if for any $a \in C$ and $a' \in C$ there exists an improving path connecting a and a'. A cycle is said to be a closed cycle if no action profile in C lies on an improving path leading to an action profile that is not in C.

Theorem 10.1 *There always exists a $\mathcal{K}$-stable equilibrium or a closed cycle of action profiles, or both $\mathcal{K}$-stable equilibrium and closed cycle.*

Proof An action profile is a $\mathcal{K}$-stable equilibrium if and only if it is not possible for any feasible coalition from set $\mathcal{K}$ to make an improving deviation from it to another action profile. So, start at an action profile. Either it is $\mathcal{K}$-stable equilibrium or there exists a coalition that can make an improving deviation to another action profile. In the first case result is established. For the second case the same thing holds, i.e., either this new action profile is a $\mathcal{K}$-stable equilibrium or there exists a coalition that can make an improving deviation to another action profile. Given the finite number of action profiles, the above process either finds an action profile which is a $\mathcal{K}$-stable equilibrium or it reaches to one of previous profiles, i.e., there exists a cycle. Thus, we have proved that there always exists either a $\mathcal{K}$-stable equilibrium or a cycle. Suppose there are no $\mathcal{K}$-stable equilibria. Given the finite number of action profiles and non-existence of $\mathcal{K}$-stable equilibria there must exists a maximal set C of action profiles such that for any $a \in C$ and $a' \in C$ there exists an improving path connecting a and a', and no action profile in C lies on an improving path leading to an action profile that is not in C. Such a set C is a closed cycle. □

An strict $\mathcal{K}$-stable equilibrium can be defined similarly. An action profile a^* in Definition 10.1 is said to be strict $\mathcal{K}$-stable equilibrium if the condition 1 is same and the condition 2 is $u_i(a) \geq u_i(a^*)$ for all $i \in S$ with at least one strict inequality. Further if $\mathcal{K} = \mathcal{P}(N) \setminus \phi$, a^* is a strict strong Nash equilibrium (SSNE). In this case, for a given action profile a and a coalition $S \in \mathcal{K}$ the set of improved action profiles $\mathcal{I}_2(S, a)$ is defined as,

Fig. 10.1 Nash equilibrium and Improving deviations

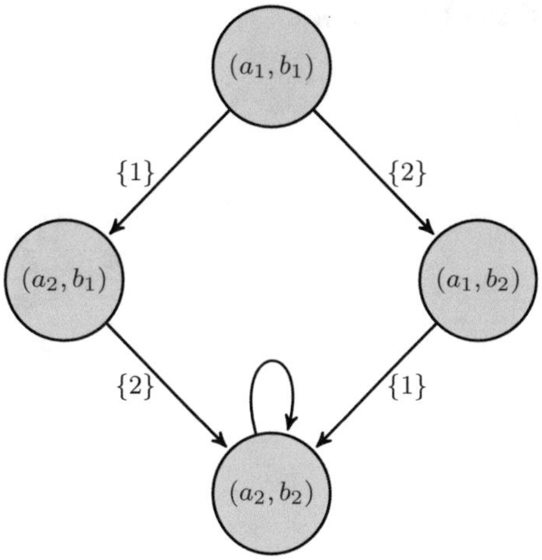

$$\mathcal{I}_2(S, a) = \{a' | a'_i = a_i, \ \forall\, i \notin S, \ \text{and}\ u_i(a') \geq u_i(a), \ \forall\, i \in S, u_j(a') > u_j(a), \quad (10.2)$$
$$\text{for some } j \in S\}.$$

and $\overline{\mathcal{I}}_2(S, a) = A(S, a) \setminus \mathcal{I}_2(S, a)$. The definitions of improving path and cycles can be defined analogously to previous case. A result similar to Theorem 10.1 holds, i.e., there always exists at least a strict $\mathcal{K}$-stable equilibrium or a closed cycle of action profiles or both. A strict $\mathcal{K}$-stable equilibrium is always a $\mathcal{K}$-stable equilibrium, i.e., the set of strict $\mathcal{K}$-stable equilibrium is a subset of the set of $\mathcal{K}$-stable equilibrium.

Now, we give few examples of two player game illustrating the presence of $\mathcal{K}$-stable equilibrium and closed cycle. In particular, we allow only the coalitions of size 1, and hence a $\mathcal{K}$-stable equilibrium is a Nash equilibrium.

Example 10.1 Consider a two player game

$$\begin{array}{c} \\ a_1 \\ a_2 \end{array} \begin{array}{c} \quad b_1 \qquad\quad b_2 \\ \begin{pmatrix} (4, 3) & (2, 5) \\ (6, 1) & (3, 2) \end{pmatrix} \end{array}.$$

Here (a_2, b_2) is the only Nash equilibrium that can be reached from all other action profiles via improving deviations. The situation is described in Fig. 10.1.

A directed edge $(a_1, b_1) \xrightarrow{\{1\}} (a_2, b_1)$ of Fig. 10.1 represents a deviation by player 1. The other directed edges are similarly defined. The self loop at (a_2, b_2) shows that a unilateral deviation from (a_2, b_2) is not possible.

Example 10.2 Consider a two player game

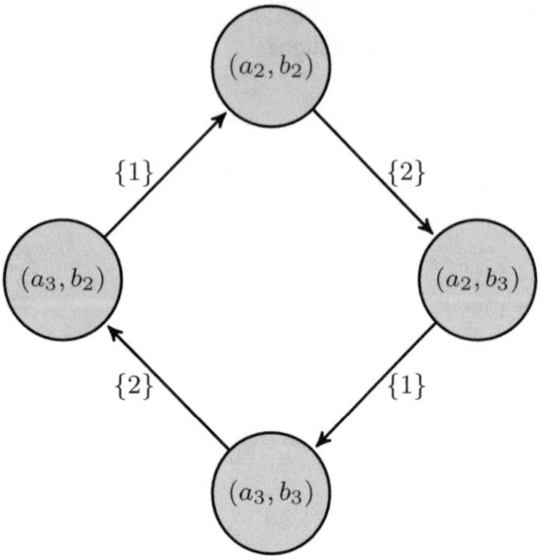

Fig. 10.2 Closed cycle

$$\begin{array}{c} \\ a_1 \\ a_2 \\ a_3 \end{array} \begin{array}{c} \begin{array}{ccc} b_1 & b_2 & b_3 \end{array} \\ \begin{pmatrix} (4,4) & (0,0) & (0,0) \\ (0,0) & (4,5) & (1,6) \\ (0,0) & (2,5) & (6,1) \end{pmatrix} \end{array}.$$

The Example 10.2 has both Nash equilibrium and closed cycle. The action profile (a_1, b_1) is a Nash equilibrium and the closed cycle is given by Fig. 10.2.

10.3 Dynamic Play

We consider the n player strategic game defined in Sect. 10.2 where players can a priori communicate with each other and form a coalition. They jointly deviate from the current action profile to a new action profile if new action profile is strictly beneficial for all the members of the coalition. We consider the coalition formation over infinite horizon. At each time a coalition is randomly formed and it makes an improving deviation from a current action profile to a new action profile according to the improved action profile sets defined by (10.1). That is, at new action profile the actions of the players outside the coalition remain same as before and each player of the coalition is strictly benefited. If there are no such improved action profiles for a coalition then it does not deviate. The same thing repeats at next stage and it continues for infinite horizon. Such deviations define a CBR dynamics. We assume that the coalition formation is random and at each time only one coalition can be formed. If there are more than one improved action profiles for a coalition then each improved action profile can be chosen with positive probability. That is, the CBR dynamics is stochastic in nature. The CBR dynamics defines a Markov chain over a finite set of action profiles A. We also assume that at each time a selected coalition makes mistake and jointly deviate to an action profile where all the members of selected coalition are not strictly benefited. This happens with very small probability. Such mistakes work as mutations and it adds another level of stochasticity in the CBR dynamics. As a consequence we have perturbed Markov chain, see e.g., [24, 25]. We are interested in the action profiles which are going to be selected by the CBR dynamics as mutations vanish. We next describe the stochastic CBR dynamics as discussed above.

10.3.1 A Stochastic CBR Dynamics without Mistakes

At each time $t = 0, 1, 2, \ldots$ a coalition S_t is selected randomly with probability $p_{S_t} > 0$. We assume that at each time selected coalition makes an improving deviation from current action profile a^t, i.e., at time $t+1$, the new action profile is $a^{t+1} \in \mathcal{I}_1(S_t, a^t)$ with probability $p_{\mathcal{I}_1}(a^{t+1}|S_t, a^t)$ where $p_{\mathcal{I}_1}(\cdot|S_t, a^t)$ is a probability distribution over finite set $\mathcal{I}_1(S_t, a^t)$. If there is no improving deviation for coalition S_t, $a^{t+1} = a^t$. Let X_t^0 denotes the action profile at time t, then $\{X_t^0\}_{t=0}^{\infty}$ is a finite Markov chain on set A. The transition law P^0 of the Markov chain is defined as follows:

$$P^0(X_{t+1}^0 = a'|X_t^0 = a) = \sum_{S \in \mathcal{K}; \mathcal{I}_1(S,a) \neq \phi} p_S \, p_{\mathcal{I}_1}(a'|S, a) 1_{\mathcal{I}_1(S,a)}(a') + \sum_{S \in \mathcal{K}; \mathcal{I}_1(S,a) = \phi} p_S 1_{\{a'=a\}}(a'), \tag{10.3}$$

where 1_B is an indicator function for a given set B. It is clear that the $\mathcal{K}$-stable equilibria and closed cycles are the recurrent classes of P^0. A $\mathcal{K}$-stable equilibrium corresponds to an absorbing state of P^0 and a closed cycle corresponds to a recurrent class of P^0 having more than one action profiles.

From Example 10.2 it is clear that in general the closed cycles together with $\mathcal{K}$-stable equilibria can be present in a game. In that case, the CBR dynamics need not converge. In Example 10.2 the CBR dynamics need not converge to Nash equilibrium (a_1, b_1) because once CBR dynamics enter into closed cycle given in Fig. 10.2 then it will never come out of it. The closed cycle $C = \{(a_2, b_2), (a_2, b_3), (a_3, b_3), (a_3, b_2)\}$ is a recurrent class and (a_1, b_1) is an absorbing state of Markov chain P^0 corresponding to the game given in Example 10.2.

We call a game acyclic if it has no closed cycles. The acyclic games include coordination games. There exists at least one $\mathcal{K}$-stable equilibrium for acyclic games from Theorem 10.1. For acyclic games the Markov chain defined by (10.3) is absorbing. Hence, from the theory of Markov chain the CBR dynamics given in Sect. 10.3.1 will be at $\mathcal{K}$-stable equilibrium in the long run no matter from where it starts (see [155]).

10.3.2 A Stochastic CBR Dynamics with Mistakes

We assume that at each time t a selected coalition S_t makes error and deviate from a^t to an action profile where at least one player from the coalition S_t is not strictly better off. We assume that at action profile a^t, coalition S_t makes error with probability $\varepsilon f(S_t, a^t) \in (0, 1)$, where $f(S_t, a^t)$ takes into account the fact that some coalitions can be more prone to make errors than others and that some action profiles may lead to wrong choices more often than others. The parameter ε allows us to tune the frequency of errors. Therefore, at time $t + 1$ the coalition S_t selects an improving deviation with probability $(1 - \varepsilon f(S_t, a^t))$. It selects $a^{t+1} \in \mathcal{I}_1(S_t, a^t)$ according to distribution $p_{\mathcal{I}_1}(\cdot)$ defined in Sect. 10.3.1. By combining the probabilities we obtain that a coalition S_t selects $a^{t+1} \in \mathcal{I}_1(S_t, a^t)$ with probability $(1 - \varepsilon f(S_t, a^t)) p_{\mathcal{I}_1}(a^{t+1}|S_t, a^t)$. The coalition S_t selects a non-improving deviation with probability $\varepsilon f(S_t, a^t)$. Let $p_{\overline{\mathcal{I}}_1}(\cdot|S_t, a^t)$ be a probability distribution over finite set $\overline{\mathcal{I}}_1(S_t, a^t)$. Then, the coalition chooses $a^{t+1} \in \overline{\mathcal{I}}_1(S_t, a^t)$ with probability $\varepsilon f(S_t, a^t) p_{\overline{\mathcal{I}}_1}(\cdot|S_t, a^t)$. If there is no improving deviation, then with probability $\left(1 - \varepsilon f(S_t, a^t)\right)$ the coalition does not modify the action profile, i.e., $a^{t+1} = a^t$, and with the complementary probability selects an action profile in $\overline{\mathcal{I}}_1(S_t, a^t)$ according to the distribution $p_{\overline{\mathcal{I}}_1}(\cdot|S_t, a^t)$. The transition law P^ε of perturbed Markov chain $\{X_t^\varepsilon\}_{t=0}^\infty$ is defined as below:

$$
\begin{aligned}
P^\varepsilon(X_{t+1}^\varepsilon = a'|X_t^\varepsilon = a) = & \sum_{S \in \mathcal{K}; \mathcal{I}_1(S,a) \neq \phi} p_S\big((1 - \varepsilon f(S, a)) p_{\mathcal{I}_1}(a'|S, a) 1_{\mathcal{I}_1(S,a)}(a') \\
& \qquad + \varepsilon f(S, a) p_{\overline{\mathcal{I}}_1}(a'|S, a) 1_{\overline{\mathcal{I}}_1(S,a)}(a')\big) \\
& + \sum_{S \in \mathcal{K}; \mathcal{I}_1(S,a) = \phi} p_S\big((1 - \varepsilon f(S, a)) 1_{\{a'=a\}}(a') \\
& \qquad + \varepsilon f(S, a) p_{\overline{\mathcal{I}}_1}(a'|S, a) 1_{\overline{\mathcal{I}}_1(S,a)}(a')\big),
\end{aligned} \tag{10.4}
$$

for all $a, a' \in A$.

The perturbed Markov chain $\{X_t^\varepsilon\}_{t=0}^\infty$ is irreducible because given nonzero errors it is possible to reach all the action profiles starting from any action profile in a finite number of steps. It is also aperiodic because with positive probability the state does not change. Hence, there exists a unique stationary distribution μ^ε for perturbed Markov chain. However, when $\varepsilon = 0$, there can be several stationary distributions corresponding to different $\mathcal{K}$-stable equilibria or closed cycles. Such Markov chains are called singularly perturbed Markov chains [24, 25]. We are interested in the action profiles to which stationary distribution μ^ε assigns positive probability as $\varepsilon \to 0$. This leads to the definition of a stochastically stable action profile.

Definition 10.4 An action profile a is stochastically stable relative to process P^ε if $\lim_{\varepsilon \to 0} \mu_a^\varepsilon > 0$.

We recall few definitions from Young [297]. If $P^{\varepsilon}(a'|a) > 0$, $a, a' \in A$, the *one step resistance* from an action profile a to an action profile $a' \neq a$ is defined as the minimum number of mistakes (mutations) that are required for the transition from a to $a' \neq a$ and it is denoted by $r(a, a')$. From (10.4) it is clear that the transition from a to a' has the probability of order ε if $a' \notin \mathcal{I}_1(S, a)$ for all S and thus has resistance 1 and is of order 1 otherwise, so has resistance 0. So, in our setting $r(a, a') \in \{0, 1\}$ for all $a, a' \in A$. A zero resistance between two action profiles corresponds to a transition with positive probability under P^0. One can view the action profiles as the nodes of a directed graph that has no self loops and the weight of a directed edge between two different nodes is represented by one step resistance between them. Since P^{ε} is an irreducible Markov chain then there must exist at least one directed path between any two recurrent classes H_i and H_j of P^0 which starts from H_i and ends at H_j. The resistance of any path is defined as the sum of the weights of the corresponding edges. The resistance of a path which is minimum among all paths from H_i to H_j is called as *resistance* from H_i to H_j and it is denoted by r_{ij}. The resistance from any action profile $a^i \in H_i$ to any action profile $a^j \in H_j$ is r_{ij} because inside H_i and H_j action profiles are connected with a path of zero resistance. Now we recall the definition of stochastic potential of a recurrent class H_i of P^0 from [297]. It can be computed by restricting to a reduced graph. Construct a graph $\mathcal{G}$ where total number of nodes are the number of recurrent classes of P^0(one action profile from each recurrent class) and a directed edge from a^i to a^j is weighted by r_{ij}. Take a node $a^i \in \mathcal{G}$ and consider all the spanning trees such that from every node $a^j \in \mathcal{G}$, $a^j \neq a^i$, there is a unique path directed from a^j to a^i. Such spanning trees are called a^i-trees. The resistance of an a^i-tree is the sum of the resistances of its edges. The stochastic potential of a^i is the resistance of an a^i-tree having minimum resistance among all a^i-trees. The stochastic potential of each node in H_i is same and it is a stochastic potential of H_i [297].

Theorem 10.2 *For the stochastic CBR dynamics defined in Sect. 10.3.2, all $\mathcal{K}$-stable equilibria and all the action profiles from closed cycles, that have minimum stochastic potential, are stochastically stable. Furthermore, if one action profile in a closed cycle is stochastically stable then all the action profiles in the closed cycle are stochastically stable.*

Proof We know that the Markov chain P^{ε} is aperiodic and irreducible. From (10.3) and (10.4) it is easy to see that

$$\lim_{\varepsilon \to 0} P^{\varepsilon}(a'|a) = P^0(a'|a), \ \forall\, a, a' \in A.$$

From (10.4) it is clear that, if $P^{\varepsilon}(a'|a) > 0$ for some $\varepsilon \in (0, \varepsilon_0]$, then we have

$$0 < \varepsilon^{-r(a,a')} P^{\varepsilon}(a'|a) < \infty.$$

Markov chain P^{ε} satisfies all three required conditions of Theorem 4 in [297] from which it follows that as $\varepsilon \to 0$, μ^{ε} converges to a stationary distribution μ^0 of P^0 and an action profile a is stochastically stable, i.e., $\mu_a^0 > 0$, if and only if a is contained in a recurrent class of P^0 having minimum stochastic potential. We know that the recurrent classes of Markov chain P^0 are $\mathcal{K}$-stable equilibria or closed cycles. Therefore, all the $\mathcal{K}$-stable equilibria and the action profiles from closed cycles having minimum stochastic potential are stochastically stable. The proof of last part follows from the fact that the stochastic potential of each action profile in a closed cycle is the same. □

Remark 10.1 The stochastic stability results do not depend on the function $f(\cdot)$ or the distributions of $p_{\mathcal{I}_1}(\cdot)$, $p_{\overline{\mathcal{I}}_1}(\cdot)$ and $p = (p_S)_{S \in \mathcal{K}}$.

Corollary 10.1 *If $\mathcal{K} = \mathcal{P}(N) \setminus \phi$, all strong Nash equilibria and all action profiles from closed cycles are stochastically stable for the stochastic CBR dynamics defined in Sect. 10.3.2.*

Proof If $\mathcal{K} = \mathcal{P}(N) \setminus \phi$, all the strong Nash equilibria and closed cycles are the recurrent classes of P^0. Due to the formation of grand coalition it is always possible to reach one action profile from another action profile by at most one error. Then, the resistance r_{ij} between any two distinct recurrent classes H_i and H_j is always 1. Hence, the stochastic potential of each recurrent class of P^0 is $J - 1$, where J is the number of recurrent classes of P^0. In fact, a spanning tree in graph $\mathcal{G}$ includes only $J - 1$ links and each of them has resistance 1. The proof then follows from Theorem 10.2. □

We can have a similar CBR dynamics without mistakes and with mistakes as given in Sects. 10.3.1 and 10.3.2 respectively, if for all $S \in \mathcal{K}$ and $a \in A$ the set of improved action profiles is $\mathcal{I}_2(S, a)$ given by (10.2). We have the following results.

Theorem 10.3 *For a stochastic CBR dynamics corresponding to improved action profile sets defined by* (10.2)*, all strict $\mathcal{K}$-stable equilibria and all the action profiles from closed cycles, that have minimum stochastic potential, are stochastically stable. Furthermore, if one action profile in a closed cycle is stochastically stable then all the action profiles in the closed cycle are stochastically stable.*

Proof The proof follows from the similar arguments given in Theorem 10.2. □

Corollary 10.2 *If $\mathcal{K} = \mathcal{P}(N) \setminus \phi$, all strict strong Nash equilibria and all action profiles from closed cycles are stochastically stable for the stochastic CBR dynamics corresponding to improved action profile sets defined by* (10.2).

Proof The proof follows from the similar arguments given in Corollary 10.1. □

Equilibrium selection in coordination games

We discuss the Nash equilibrium selection, in a 2×2 coordination game, by stochastic CBR dynamics defined in Sect. 10.3.2. Our equilibrium selection results are different from the results given in [154, 297]. We first consider the case where only the coalitions of size 1 are formed. In this case, a $\mathcal{K}$-stable equilibrium is a Nash equilibrium. Consider a 2×2 coordination game,

$$\begin{array}{cc} & \begin{array}{cc} s_1 & s_2 \end{array} \\ \begin{array}{c} s_1 \\ s_2 \end{array} & \begin{pmatrix} (a_{11}, b_{11}) & (a_{12}, b_{12}) \\ (a_{21}, b_{21}) & (a_{22}, b_{22}) \end{pmatrix}, \end{array}$$

where $a_{jk}, b_{jk} \in \mathbb{R}$, $j, k \in \{1, 2\}$ and $a_{11} > a_{21}, b_{11} > b_{12}, a_{22} > a_{12}, b_{22} > b_{21}$. The action set of player $i, i = 1, 2$, is $A_i = \{s_1, s_2\}$. Here (s_1, s_1) and (s_2, s_2) are two Nash equilibria. In this game there are two types of Nash equilibria. One is payoff dominant and another is risk dominant. If $a_{11} > a_{22}, b_{11} > b_{22}$, then (s_1, s_1) is payoff dominant and if $a_{11} < a_{22}$, $b_{11} < b_{22}$, then (s_2, s_2) is payoff dominant. In other cases payoff dominant Nash equilibrium does not exist. From [297], define,

$$R_1 = \min\left\{\frac{a_{11} - a_{21}}{a_{11} - a_{12} - a_{21} + a_{22}}, \frac{b_{11} - b_{12}}{b_{11} - b_{12} - b_{21} + b_{22}}\right\},$$

$$R_2 = \min\left\{\frac{a_{22} - a_{12}}{a_{11} - a_{12} - a_{21} + a_{22}}, \frac{b_{22} - b_{21}}{b_{11} - b_{12} - b_{21} + b_{22}}\right\}.$$

If $R_1 > R_2$, then (s_1, s_1) is a risk dominant Nash equilibrium and if $R_2 > R_1$, then (s_2, s_2) is a risk dominant Nash equilibrium.

The state space of Markov chain is $\{(s_1, s_1), (s_1, s_2), (s_2, s_1), (s_2, s_2)\}$, where (s_1, s_1) and (s_2, s_2) are the absorbing states of Markov chain P^0. From Remark 10.1, the stochastic stability results do not depend on the distributions of $p_{\mathcal{I}_1}(\cdot)$, $p_{\overline{\mathcal{I}}_1}(\cdot)$ and $p = (p_S)_{S \in \mathcal{K}}$ and function $f(\cdot)$. Therefore, we assume that all the distributions are uniform and function $f(\cdot)$ has constant value 1. Under this assumption, the transition probability matrix of perturbed Markov chain is

$$P^{\varepsilon} = \begin{pmatrix} 1 - \frac{\varepsilon}{2} & \frac{\varepsilon}{4} & \frac{\varepsilon}{4} & 0 \\ \frac{1-\varepsilon}{2} & \varepsilon & 0 & \frac{1-\varepsilon}{2} \\ \frac{1-\varepsilon}{2} & 0 & \varepsilon & \frac{1-\varepsilon}{2} \\ 0 & \frac{\varepsilon}{4} & \frac{\varepsilon}{4} & 1 - \frac{\varepsilon}{2} \end{pmatrix}.$$

The unique stationary distribution of P^{ε} is $\mu^{\varepsilon} = \left(\frac{1-\varepsilon}{2-\varepsilon}, \frac{\varepsilon}{2(2-\varepsilon)}, \frac{\varepsilon}{2(2-\varepsilon)}, \frac{1-\varepsilon}{2-\varepsilon}\right)$. As $\varepsilon \to 0$, $\mu^{\varepsilon} \to (\frac{1}{2}, 0, 0, \frac{1}{2})$. That is, both the Nash equilibria are stochastically stable. This happens because we require only 1 mutation to reach from one Nash equilibrium to another Nash equilibrium. Therefore, the resistance from one Nash equilibrium to another Nash equilibrium is 1. Then, the stochastic potential of both the Nash equilibria will be 1. For the case when all types of coalitions can be formed the CBR dynamics always selects a payoff dominant Nash equilibrium whenever it exists because it is an SNE. If a payoff dominant Nash equilibrium does not exist, both the

Fig. 10.3 1-trees

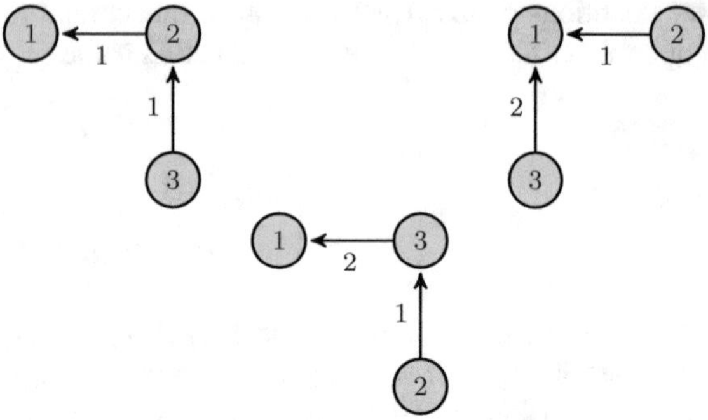

Nash equilibria are strong Nash equilibria and in that case CBR dynamics selects both the Nash equilibria. The stochastic dynamics by Young [297] and Kandori et al. [154] always select a risk dominant Nash equilibrium.

Among symmetric coordination games if we go beyond 2×2 matrix games the result by Young [297] cannot be generalized, i.e., it need not select a risk dominant Nash equilibrium. Consider an example of 3×3 matrix game from [297],

$$\begin{array}{c} \\ s_1 \\ s_2 \\ s_3 \end{array} \begin{array}{c} \begin{array}{ccc} s_1 & s_2 & s_3 \end{array} \\ \begin{pmatrix} (6,6) & (0,5) & (0,0) \\ (5,0) & (7,7) & (5,5) \\ (0,0) & (5,5) & (8,8) \end{pmatrix} \end{array}.$$

In this game, (s_1, s_1), (s_2, s_2) and (s_3, s_3) are three Nash equilibria. The stochastic dynamics by Young [297] selects (s_2, s_2) that is not a risk dominant Nash equilibrium. A Nash equilibrium of an $m \times m$ symmetric coordination game is risk dominant if it is risk dominant in all pairwise contest (see [130]). We now discuss the equilibrium selection by CBR dynamics in above 3×3 coordination game. We first consider the case where only the coalitions of size 1 are formed. The state space of Markov chain is

$$\{(s_1, s_1), (s_1, s_2), (s_1, s_3), (s_2, s_1), (s_2, s_2), (s_2, s_3), (s_3, s_1), (s_3, s_2), (s_3, s_3)\},$$

where (s_1, s_1), (s_2, s_2), and (s_3, s_3) are the absorbing states of Markov chain P^0. We label the states (s_1, s_1) as 1, (s_2, s_2) as 2 and (s_3, s_3) as 3. Then, the resistance from (s_1, s_1) to (s_2, s_2) is denoted by r_{12}, where $r_{12} = 1$. Similarly, $r_{13} = 2$, $r_{31} = 2$, $r_{21} = 1$, $r_{23} = 1$, $r_{32} = 1$. There are three 1-trees as given (Fig. 10.3).

The minimum resistance of a 1-tree among all 1-trees is 2. Hence, the stochastic potential of (s_1, s_1) is 2. Similarly, by constructing 2-trees and 3-trees we can calculate the stochastic potential of (s_2, s_2) and (s_3, s_3). The stochastic potential of (s_2, s_2) and (s_3, s_3) is also 2. Hence, all the Nash equilibria are stochastically stable from Theorem

10.2. For the case where there is no restriction in coalition formation, CBR dynamics selects (s_3, s_3) because it is a strong Nash equilibrium.

10.4 Application to Network Formation Games

In this section we consider the network formation games, see e.g., some recent books [78, 84, 144]. In general, the networks which are stable against the deviation of all the coalitions are called strongly stable networks. In the literature, there are two definitions of strongly stable networks. The first definition is, corresponding to SNE, due to [85]. The second definition is, corresponding to SSNE, due to [146]. A strongly stable network according to the definition of [146] is also strongly stable network according to the definition of [85]. The definition of a strongly stable network according to [146] are more often considered in the literature. We also consider the strong stability of networks according to [146]. We discuss the dynamic formation of networks over infinite horizon. We apply the stochastic CBR dynamics corresponding to SSNE to network formation games to discuss the stochastic stability of networks.

10.4.1 The Model

Let $N = \{1, 2, \ldots, n\}$ be a finite set of players also called as nodes. The players are connected through undirected edges. An edge can be defined as a subset of N of size 2, e.g., $\{ij\} \subset N$ defines an edge between player i and player j. The collection of edges define a network. Let G denotes a set of all networks on N. For each $i \in N$, let $u_i : G \to \mathbb{R}$ be a payoff function of player i, where $u_i(g)$ is a payoff which player i receives at network g.

To reach from one network to another requires the addition of new links or the destruction of existing links. It is always assumed in the literature that forming a new link requires the consent of both the players while a player can delete a link unilaterally. The coalition formation in network formation games has also been considered in the literature. Some players in a network can form a coalition and make a joint move to another network by adding or severing some links, if new network is at least as beneficial as the previous network for all the players of coalition and at least one player is strictly benefited (see [146]). We recall few definitions from [146] describing the coalitional moves in network formation games and the stability of networks against all possible coalitional deviations.

Definition 10.5 A network g' is reachable from g via deviation by a coalition S as denoted by $g \to_S g'$, if

1. $ij \in g'$ and $ij \notin g$ then $\{i, j\} \subset S$.
2. $ij \in g$ and $ij \notin g'$ then $\{i, j\} \cap S \neq \phi$.

The first condition of the above definition requires that a new link can be added only between the nodes which are the part of a coalition S and the second condition requires that at least one node of any deleted link has to be a part of a coalition S. We denote $G(S, g)$ as a set of all networks which are reachable from g via deviation by S, i.e., $G(S, g) = \{g' | g \rightarrow_S g'\}$.

Definition 10.6 A deviation by a coalition S from a network g to a network g' is said to be improving if

1. $g \rightarrow_S g'$,
2. $u_i(g') \geq u_i(g), \ \forall\, i \in S$ (with at least one strict inequality).

We denote $\mathcal{I}_2(S, g)$ as a set of all networks g' which are reachable from g by an improving deviation of S, i.e.,

$$\mathcal{I}_2(S, g) = \left\{g' | g \rightarrow_S g', u_i(g') \geq u_i(g), \ \forall\, i \in S, u_j(g') > u_j(g) \text{ for some } j \in S\right\}.$$

It is clear that $g \notin \mathcal{I}_2(S, g)$ for all S. We denote $\overline{\mathcal{I}}_2(S, g) = G(S, g) \setminus \mathcal{I}_2(S, g)$ as a set of all networks which are reachable from g due to erroneous decisions of S. This set is always nonempty as $g \in \overline{\mathcal{I}}_2(S, g)$ for all S.

Definition 10.7 A network g is said to be strongly stable if it is not possible for any coalition S to make an improving deviation from network g to some other network g'.

A strongly stable network need not always exist and in that case there exists some set of networks lying on a closed cycle and all the networks in a closed cycle can be reached from each other via an improving path. An improving path and a closed cycle in network formation games can be defined similarly to Definitions 10.2 and 10.3, respectively.

Theorem 10.4 *There exists at least a strongly stable network or a closed cycle of networks.*

Proof The proof follows from the similar arguments used in Theorem 10.1. □

10.4.2 Dynamic Network Formation

The paper by Jackson and Watts [147] is the first one to consider the dynamic formation of networks. They considered the case where at each time only a pair of players form a coalition and only a link between them can be altered. We consider the situation where at each time a subset of players form a coalition and deviate from a current network to a new network if at new network the payoff of each player of the coalition is at least as much as at current network and at least one player has strictly better payoff. This process continues over infinite horizon. A coalition can make all possible changes in the network and as a result more than one link can be created or severed at each time. So, we consider the following network formation rules by [146] given below:

- The creation of a link between two nodes requires the agreement of the whole coalition.
- A coalition can create/severe simultaneously multiple links among its members.

The CBR dynamics corresponding to SSNE can be applied to dynamic network formation. That is, at time t a network is g_t and a coalition S_t is selected with probability $p_{S_t} > 0$ and it makes an improving deviation to a new network that is at least as beneficial as g_t for all players of coalition S_t and at least one player of S_t is strictly benefited. So, at time $t+1$ network is $g_{t+1} \in \mathcal{I}_2(S_t, g_t)$ with probability $p_{\mathcal{I}_2}(g_{t+1}|S_t, g_t)$. If an improving deviation is not possible for selected coalition S_t, then $g_{t+1} = g_t$. The above process defines a Markov chain over state space G and its transition probabilities can be defined similarly to (10.3). In general this Markov chain is multichain. We can also assume that at each time selected coalition S_t makes error with small probability $f(S_t, g_t)\varepsilon$. That is, $g_{t+1} \in \mathcal{I}_2(S_t, g_t)$ with probability $(1 - f(S_t, g_t)\varepsilon)p_{\mathcal{I}_2}(g_{t+1}|S_t, g_t)$ and $g_{t+1} \in \overline{\mathcal{I}}_2(S_t, g_t)$ with probability $f(S_t, g_t)\varepsilon p_{\overline{\mathcal{I}}_2}(g_{t+1}|S_t, g_t)$. The transition probabilities of the perturbed Markov chain can be defined similar to (10.4). The presence of mutations makes the Markov chain ergodic for which there exists a unique stationary distribution. We are interested in the stochastically stable networks, i.e., the networks to which positive probabilities are assigned by the stationary distribution as $\varepsilon \to 0$. The stochastic stability analysis similar to the one given in Sect. 10.3.2 holds here. Thus, we have the following result.

Theorem 10.5 *All the strongly stable networks and closed cycles of a network formation game are stochastically stable.*

Proof The proof follows directly from Corollary 10.2. □

10.5 Conclusions

We introduce coalition formation among players in an n-player strategic game over infinite horizon and propose a stochastic CBR dynamics. The mutations are present in the dynamics due to erroneous decisions taken by the coalitions. We show that all $\mathcal{K}$-stable equilibria and all action profiles from closed cycles, that have minimum stochastic potential, are stochastically stable. Similar development holds for strict $\mathcal{K}$-stable equilibrium. When there is no restriction on coalition formation, all SNE and closed cycles are stochastically stable. Similar development holds for SSNE. We apply CBR dynamics to network formation games and prove that all strongly stable networks and closed cycles of networks are stochastically stable.

Chapter 11
Peering Versus Transit: A Game Theoretical Model for Autonomous Systems Connectivity

Giovanni Accongiagioco, Eitan Altman, Enrico Gregori and Luciano Lenzini

We propose a model for network optimization in a non-cooperative game setting with specific reference to the Internet connectivity. The model describes the decisions taken by an Autonomous System (AS) when joining the Internet. We first define a realistic model for the interconnection costs incurred; then we use this cost model to perform a game theoretic analysis of the decisions related to link creation and traffic routing, keeping into account the peering/transit dichotomy. The proposed model doesn't fall into the standard category of routing games, hence we devise new tools to solve it by exploiting specific properties of our game. We prove analytically the existence of multiple equilibria for several scenarios, and provide an algorithm to compute the stable ones. Thanks to the use of simulations we covered those cases for which analytic results could not be obtained, thus analyzing a broad variety of general scenarios, both ad-hoc and realistic. The analysis of the model's outcome highlights the existence of a Price of Anarchy (PoA) and a Price of Stability (PoS), originated by the non-cooperative behavior of the ASes, which optimize their cost function in a selfish and decentralized manner. We further observe the presence of competition between the facilities providing either transit or peering connectivity, caused by the cost differences between these two interconnection strategies.

G. Accongiagioco · E. Gregori (✉) · L. Lenzini
IIT-CNR, Via Moruzzi 1, Pisa, Italy
e-mail: giovanni.accongiagioco@iit.cnr.it

E. Gregori
e-mail: enrico.gregori@iit.cnr.it

L. Lenzini
e-mail: luciano.lenzini@iit.cnr.it

E. Altman
INRIA Sophia Antipolis, 2004 Route des Lucioles, 06902 Valbonne, France
e-mail: eitan.altman@inria.fr

E. Altman et al. (eds.), *Multilevel Strategic Interaction Game Models for Complex Networks*, https://doi.org/10.1007/978-3-030-24455-2_11

11.1 Introduction

The Internet ecosystem is made of tens of thousands Autonomous Systems, interconnected together in a complex and dynamic manner. Roughly speaking, Autonomous Systems (AS) are independently administered networks that dynamically connect together to provide end-to-end reachability. ASes can be grouped in different tiers and categories, depending on the service they offer and the organization they belong to: content providers, access providers, transit providers and so on [291].

The late twentieth-century Internet ecosystem was largely dominated by transit links, where the relationship between the connecting ASes was of "customer to provider" type. This kind of relationship produces a hierarchical pricing scheme, where the customer AS pays its provider for the traffic flowing on the link, both incoming and outgoing; in return, the latter provides a default gateway to reach all Internet's routes. For example, an access provider AS wishing to grant Internet access to the eyeballs (i.e. its end users), needs to establish a link with a transit provider, and pay for the traffic flowing on this link. Transit providers are also known as Network Service Providers (NSP) [211] and their pricing strategy is typically volume-based, metered using the 95th percentile traffic sampling technique (this allows customer ASes to burst, for a limited time period, beyond their committed base rate) [194].

The beginning of twenty-first century brought a new paradigm into the environment, since more and more ASes found it beneficial to establish peering links between them [80]. This kind of relationship is "settlement-free", meaning that the two ASes mutually agree to exchange traffic for free between them, and the only cost they incur is that of laying out the physical interconnection. Peers must agree to each other's policy, which is used to avoid abuse of the peering relationship. Typical clauses include prohibition of using the peer as default gateway (therefore peers cannot be used to reach other Internet's routes) and traffic ratio balancing, meaning that the ratio between incoming and outgoing traffic over the link must not exceed some value (e.g. 2:1) [193].

The exponential growth of peering links was made possible mainly thanks to the deployment of Internet Exchange Points (IXPs) [115]. These interconnection points are facilities through which ASes can exchange traffic, typically by settlement-free (i.e. peering) relationships. The growth of IXPs, in number and in size, made it easy to establish more and more public peering relationships. In fact, by joining an IXP, an AS can potentially peer with all (usually a subset) of the other ASes connected to the same IXP. The pricing strategy of an IXP, with respect to its customers, is typically flat. Each one of them pays a monthly-based fee, depending on the size (speed) of the port they are using and the cost of maintaining the equipment. Thanks to this mechanism, the IXP can share maintenance costs among all its participants [195]. It is worth noting that this pricing strategy doesn't allow standard cost function modeling (like in [203]), since the addition of new participants potentially brings down the costs of an IXP customer.

When an Autonomous System joins the Internet, it needs to decide the connections to lay out with other ASes. While in the last century, as shown above, the decision space for ASes was substantially small, today they have many alternatives: transit or peering, joining one or more IXPs, dealing with distance and geographic issues. In principle, the best strategy for an AS is the one yielding the lowest cost. However, the outcome of its strategy also depends on that of other ASes dealing with the same problem, thus we find it straightforward to analyze the problem in a game-theoretical framework. We propose a model keeping into account the above factors, which can be used to compute the outcome of this problem and the strategy of the players. Realistic modeling of the whole decision space of an AS is an extremely difficult task, therefore in our work we restrict our analysis to the problem of peering versus transit. Nevertheless, a proper understanding of this problem is fundamental to get insight on the behavior of ASes in the Internet environment, as shown by the results of the analysis.

This work brings contributions both from a game theoretic perspective and an engineering perspective. First of all this is, to the best of our knowledge, a novel model to analyze the strategic choices of ASes living in an Internet environment with both technological and economic constraints. The modeling takes into account many realistic elements, which do not fall into standard game frameworks, yet tries to keep the problem mathematically manageable. From a game-theoretic perspective, we prove that our game falls in a specific category for which we both demonstrate the existence of equilibria and provide an algorithm for computing stable solutions. From an engineering perspective, the outcome of the analysis is highly insightful as it shows both the suboptimality of the decentralized solution and the emerging competition, first observed in [6], between the two facilities enabling either transit or peering connectivity: Network Service Providers and Internet Exchange Points.

The remainder of this paper is organized as follows: in Sect. 11.2 we describe the related work. Section 11.3 defines the general model, while Sect. 11.4 gives analytical results for the general model, derives the existence of equilibria and the algorithm to compute them. Section 11.5 analyzes a subcase of the general model, so as to deepen the obtained results and show the inefficiency of the decentralized solution. In Sect. 11.6 we use simulations to study the behavior of the system for several configurations, and in particular to analyze a realistic case-study. Finally, we conclude in Sect. 11.7.

11.2 Related Work

This work relates to the characterization and modeling of the Internet AS-level topology. In this field the majority of graph theoretic models try to reproduce observed Internet topological properties, such as its power-law degree distribution [93], the small-world property [287] and other structural properties (communities, cliques, etc..) [113, 114]. The graph can be reconstructed [119] by either defining some attachment criteria, as in [4, 282, 287, 296] or solving constrained optimization problems for the different nodes, like in [63, 64, 92].

In our paper we rather try to understand network formation as the result of a game between ASes. In this context, it relates to game-theoretic network formation models, which populate both computer science and economics literature (see books [112, 145]). This research branch focuses on proving the existence of equilibria in networks with a fixed number of agents, where links are formed taking into account their preferences in the form of utility functions. The need for mathematical tractability requires simplifications that make these models unrealistic and unable to study real life networks such as the Internet. Other models simulating the dynamics of network formation are agent-based computational models such as GENESIS [176]. In this case authors can include more realistic considerations by skipping the analytical part and simulating the behavior of each agent, hoping to find one of the possibly many equilibria.

In order to keep the problem analytically tractable we do not aim at modeling the whole network formation process. We rather study the interaction between ASes which connect to an existing network in order to serve some demands. A possible modelization of a network where access providers need to select a subset of content providers and fetch traffic from them in a cost-efficient manner is given in [151]. However, the aim of this work is quite different from ours, as it concentrates on the economic analysis of neutral/non-neutral network features, without taking into account the difference between traffic and peering agreements. The models in [74, 87] focus on the competition between backbone providers and on the conditions under which they agree or refuse to establish bilateral peering connections with each other. In [238], authors perform an interesting analysis on network pricing and analyze the economics of private internet exchanges. This kind of peering, known as private peering, has different rules and costs compared with public peering. As explained in the introduction, nowadays Internet is largely dominated by public peering, occurring at IXPs, therefore in our work we concentrate on this last phenomenon, which allows us to give different insights on the present difference between transit and peering.

A partial presentation of the basic components of this work is given in [1, 2]. Specifically, in [2] we analyze a small-scale model, while in [1] some results are extended to the general model. Here we provide an exhaustive framework which: (i) gives the analytic results for the general model and for the small scale model; (ii) presents an organic set of simulations, integrated by a specific case-study used to infer the behavior of the model in the real Internet.

11.3 Scenario

In the following we describe the general scenario under investigation and derive the cost function.

11.3.1 Description

As said earlier, ASes can be grouped in different categories. Throughout the paper we will use the following categorization, closely resembling the one provided by PeeringDB [211]:

Internet Service Provider (ISP) this node gives to eyeballs (i.e. the end users) and lower tiers, access to the Internet and its contents. Each service provider has a traffic demand, hereafter demand, which represents the amount of traffic (uplink+downlink) that it handles.

Content Provider (CP) this node has physical access to the contents users are looking after, therefore an ISP with a demand for his specific content, has to connect to it in order to serve this demand.

Internet Exchange Point (IXP) this is the facility that provides peering connection to all its participants. This means that all the nodes connected to a given IXP can potentially communicate with each other.

Network Service Provider (NSP)[1] this node is located at the highest hierarchical level of the network, meaning that each CP can be reached through it. ISPs can reach CPs by establishing a transit connection with an NSP. In particular, here NSPs can be either Tier-1 ASes, or high-level backbone service providers, whose main interest is to sell transit to lower tiers.

We consider a network with $i \in \{1, ..., I\}$ ISPs, $n \in \{1, ..., N\}$ CPs and $l \in \{1, ..., L\}$ transmit facilities (TF), that can be either the NSPs or the IXPs. Without loss of generality, we impose that TFs $j_1 \in \{1, ..., l_1\}$ are NSPs, while TFs $j_2 \in \{l_1 + 1, ..., L\}$ are IXPs.

The main difference between transmit facilities is that while links to NSPs are established through transit connections, links to IXPs are established through peering connections.

In a *transit connection*, or customer-to-provider (C2P) connection, the cost to the customer is a function of the amount of traffic that crosses the link (typically expressed as $/Mbps).

In a *peering connection*, the price is generally flat and depends on the size of the port the customer buys. Moreover, when peering connections are maintained by an IXP, the costs are shared among all the participants.

Each ISP i has a demand for a CP n, which we indicate as ϕ_i^n. The players of our game are the ISPs, which need to decide how to split their demand among all possible transmit facilities. Figure 11.1 depicts our network scenario.

We indicate with $x_{i,l}^n$ the flow from ISP i to CP n via TF l. The strategy of ISP i is given by the vector $\mathbf{x_i} = (x_{i,1}^1, ..., x_{i,l}^n, ..., x_{i,L}^N) \in \mathbb{R}^{L \times N}$, while the strategy of all the other players is expressed as $\mathbf{x_{-i}} \in \mathbb{R}^{(I-1) \times L \times N}$. The goal of each player is to serve, at the minimum possible cost, his demand $\phi_\mathbf{i} = (\phi_i^1, ..., \phi_i^N) \in \mathbb{R}^N$ by splitting it

[1] ISPs, NSPs and CPs are typically ASes. IXPs are not ASes, even if their infrastructure is under a single administrative control.

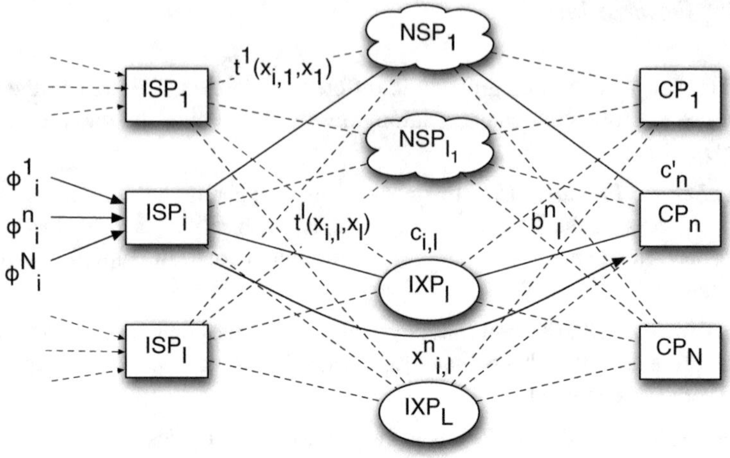

Fig. 11.1 General model

into several flows $x^n_{i,l}$. Please note that we are not dealing with the issue of complete connectivity for the ISPs (in which case, it is sufficient to deploy a single transit connection): our aim is just to enable them to serve their specific demands.

We also indicate as:

- $x_{i,l} = \sum_{n=1}^{N} x^n_{i,l}$ the total flow from ISP i to TF l;
- $x_l = \sum_{i=1}^{I} \sum_{n=1}^{N} x^n_{i,l}$ the total flow at TF l.

Each player, say i, for each transmit facility (NSP or IXP) it connects to, incurs some costs:

TF usage cost this cost depends on the transmit facility used. If it is an NSP, then it is a function of $x_{i,l}$, the flow from the player to the NSP. Otherwise the TF is an IXP and the cost is shared among all the participants, therefore it also depends on the other players, in the form of x_l, the total flow at the IXP. Consequently, this cost can be written as a function:

$$t^l(x_{i,l}, x_l) \tag{11.1}$$

TF capacity cost each link between an ISP and a TF has a fixed capacity $c_{i,l}$; this means that we have a constraint of the form $x_{i,l} \leq c_{i,l}$. While we may introduce it in the problem "as is", this would make the model less manageable. Moreover, due to performance and congestion issues, network operators typically avoid reaching the capacity limit and keep a margin for traffic fluctuations. We can think of this performance degradation as a "virtual cost" for the ISP, and therefore model the constraint as a cost, that increases as the flow over the link approaches the capacity limit (as is typically done in the literature for M/G/1 Processor Sharing queues [203]):

$$\frac{1}{c_{i,l} - x_{i,l}} x_{i,l} \tag{11.2}$$

We are aware that, in reality, network operators adjust this capacity when there is more demand for it, and the interconnection cost grows accordingly. However, this situation can be avoided as long as our working region is sufficiently far away from the saturation point. We will always assume that capacities are symmetric w.r.t. the players, therefore $c_{i,l} = c_l \ \forall i$. Typically the capacity of the NSP can be assumed to be much larger than that of IXPs: $c_{NSP} \gg c_{IXP}$ (see [194, 195]).

CP reachability cost let's indicate with b_l^n the cost of transporting one unit of flow from TF l to CP n. This cost is not relevant from the player's perspective (it is paid by the CP), however it can be used to express the reachability of a given CP. In fact, while all the CPs are connected to the NSPs, an IXP can be connected only to a subset of CPs. This phenomenon can be expressed by putting:

$$b_l^n = \begin{cases} 0 & if \ (l \leq l_1) \ or \ (IXP_l \text{ onnected to } CP_n) \\ \infty & otherwise \end{cases} \tag{11.3}$$

Thanks to all these considerations, the cost function for player i can be expressed as the sum of (11.1), (11.2) and (11.3):

$$C_i(\mathbf{x_i}, \mathbf{x_{-i}}) = \sum_{l=1}^{L} \left(t^l(x_{i,l}, x_l) + \frac{x_{i,l}}{c_l - x_{i,l}} \right) + \sum_{l=1}^{L} \sum_{n=1}^{N} x_{i,l}^n b_l^n \tag{11.4}$$

In order to serve all the demands, each player i has to satisfy the flow constraint: for every CP, the total flow has to be equal to the demand ϕ_i^n. Therefore player i's best response $BR_i(\mathbf{x_{-i}})$ is obtained by minimizing cost function (11.4), subject to the flow constraints (11.5):

$$\begin{cases} BR_i(\mathbf{x_{-i}}) & = \arg\min_{\mathbf{x_i}} C^i(\mathbf{x_i}, \mathbf{x_{-i}}) \\ s.t. & \sum_l x_{i,l}^n = \phi_i^n \quad \forall n \end{cases} \tag{11.5}$$

The vector $\mathbf{x}^* = (\mathbf{x_1^*}, ..., \mathbf{x_I^*}) \in \mathbb{R}^{I \times L \times N}$ is an equilibrium of the game if and only if $\mathbf{x_i^*} \in BR_i(\mathbf{x_{-i}^*}) \ \forall i$, that is, if the strategy of each player is a best response to the strategies of other players.

Throughout the paper we will always refer to the description of Fig. 11.1, however, mutatis mutandis, the results are still valid for scenarios where players are CPs or a mix of CPs and ISPs, as long as the demands are changed accordingly.

11.3.2 Transmit Facility Usage Cost

The TF usage cost is different between the NSPs and the IXPs. More specifically the NSP usage cost is linear in the amount of flow that each player sends to it [194]. Therefore we can write:

$$t^l(x_{i,l}, x_l) = a_l x_{i,l} \qquad l \le l_1 \tag{11.6}$$

where $a_l,\ l \le l_1$ represents the transit price of NSP l per unit of flow. We are aware that, due to economies of scale in the traffic delivery, transit costs are subadditive in reality. However, introducing this aspect would overcomplicate the model, hiding the truly interesting differences between transit and peering. Nevertheless, we are able to show that some of our results still hold for more generic transit cost functions (see below, Theorem 11.5).

For the IXP usage [291], each player has to pay a share of the total cost of IXP maintenance. This share can be expressed as the ratio between the flow sent by player i on IXP l and the total flow crossing that IXP: $\frac{x_{i,l}}{x_l}$. Assume we can write the cost of maintenance of IXP l as a function h_l of the total flow through the IXP, therefore the usage cost is:

$$t^l(x_{i,l}, x_l) = \frac{x_{i,l}}{x_l} h_l(x_l) \qquad l > l_1 \tag{11.7}$$

The cost of maintaining the equipment of an IXP is, in general, a non-linear function of several parameters. In order to keep the problem manageable, we will approximate this cost with that of a single port which handles x_l, the entire flow over the IXP. The cost of a port is a step-wise increasing function, as shown in Fig. 11.2 for the MIX[2] [182], an Italian IXP. This type of cost functions can be modeled (see [183]) by using a function like x^α with $\alpha \in [0.4; 0.7]$. For simplicity, we take $\alpha = 0.5$ as this value provides a fairly accurate fit (shown in Fig. 11.2). Therefore, we express the maintenance cost as:

$$h_l(x_l) = a_l \sqrt{x_l} \tag{11.8}$$

where $a_l,\ l > l_1$ is a constant relating the total flow through IXP l with its maintenance cost. By putting together definitions (11.6), (11.7) and (11.8), the cost function (11.4) can be rewritten as:

$$C^i(\mathbf{x_i}, \mathbf{x_{-i}}) = \sum_{l=1}^{l_1} a_l x_{i,l} + \sum_{l=l_1+1}^{L} \left(\frac{a_l}{\sqrt{x_l}} x_{i,l} \right) + \\ + \sum_{l=1}^{L} \left(\frac{1}{c_l - x_{i,l}} x_{i,l} \right) + \sum_{l=1}^{L} \sum_{n=1}^{N} x_{i,l}^n b_l^n \tag{11.9}$$

[2]Milan IXP—public peering costs available online: http://www.mix-it.net.

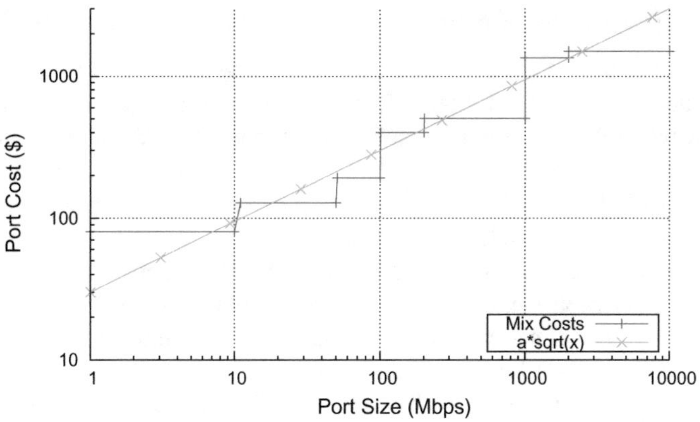

Fig. 11.2 IXP port costs for MIX (log-log scale)

Now, we define these new functions:

$$\begin{cases} f^l\left(\sum_i \sum_n x_{i,l}^n\right) = \begin{cases} a_1 & l \le l_1 \\ \frac{a_l}{\sqrt{\sum_i \sum_n x_{i,l}^n}} & l > l_1 \end{cases} \\ g^l\left(\sum_n x_{i,l}^n\right) = \frac{1}{c_l - \sum_n x_{i,l}^n} \end{cases} \tag{11.10}$$

By using (11.10) and recalling that $x_{i,l} = \sum_n x_{i,l}^n$ and $x_l = \sum_i \sum_n x_{i,l}^n$, we can rewrite (11.9) as:

$$C^i(\mathbf{x_i}, \mathbf{x_{-i}}) = \sum_{l,n} x_{i,l}^n \left[f^l\left(\sum_{i,n} x_{i,l}^n\right) + g^l\left(\sum_n x_{i,l}^n\right) + b_l^n \right] \tag{11.11}$$

Equation (11.11) is the most general expression of the cost function for each player. Please note that (11.11) is in general a non-convex function of $x_{i,l}$, and therefore we cannot directly establish existence of pure equilibria. In particular it does not comply with the general assumptions used for link cost functions in the framework described in [203]. Nevertheless, we cannot avoid dealing with functions of this shape if we want to properly grasp the difference between transit and peering strategies offered, respectively, by NSPs and IXPs.

11.4 General Model

In this section we give analytical results for the general model presented above. Thanks to the peculiar properties of our game, we are able to prove the existence of equilibria under specific conditions, and provide an algorithm to compute them.

11.4.1 The Cost Function

The cost function of Sect. 11.3 takes into account the presence of many NSPs and IXPs. However, while having multiple IXPs is fundamental for understanding how players aggregate around exchange points, especially in presence of reachability constraints (IXPs may be connected to only a subset of CPs), this is not the case for NSPs, due to the fact that their cost is independent from other players' choice, and they are necessarily connected to all possible CPs. Therefore, without loss of generality for our problem, it is safe to consider only a single NSP for $l = 1$, and $L - 1$, for $l = 2..L$, IXPs as we do here. In the following we will always refer to this system, which is a single NSP version of Fig. 11.1.

If we consider a system with I ISPs, N CPs, and L TFs, with $l_1 = 1$, in a possibly disconnected topology, the cost function (11.11) can be rewritten separating the NSP component from the IXPs. Therefore we have:

$$C^i(\mathbf{x_i}, \mathbf{x_{-i}}) = \sum_n x^n_{i,1}\left(a_1 + \frac{1}{c_1 - \sum_n x^n_{i,1}}\right) + \\ + \sum_{l \neq 1} \sum_n x^n_{i,l}\left(\frac{a_l}{\sqrt{\sum_i \sum_n x^n_{i,l}}} + \frac{1}{c_l - \sum_n x^n_{i,l}} + b^n_l\right) \tag{11.12}$$

where the CP reachability cost for the NSP has been removed since we know from (11.3) that $b^n_1 = 0 \ \forall n$. Before making further considerations on cost function (11.12), in the following we review and extend the theory on supermodularity. The obtained results will be of great help when dealing with our game.

11.4.2 Supermodularity

Definition 11.1 (*Supermodular games* [293])

Consider a generic game G, where user's payoffs are given by a utility function $u : \mathbb{R}^k \rightarrow \mathbb{R}$. The utility is supermodular if the following condition holds:

$$u(x \vee y) + u(x \wedge y) \geq u(x) + u(y) \ \forall x, y \in \mathbb{R}^k$$

where $x \vee y$ denotes the componentwise maximum and $x \wedge y$ the componentwise minimum of x and y. The game is supermodular if the utility function of each player is supermodular. If u is twice continuously differentiable, this property is given by the following condition:

$$\frac{\partial^2 u}{\partial z_i \partial z_j} \geq 0 \qquad \forall i \neq j$$

In our case we consider costs rather then utilities and minimization instead of maximization, therefore a game like ours is supermodular iff:

$$\frac{\partial^2 C(\mathbf{x})}{\partial \mathbf{x}_i \partial \mathbf{x}_j} \leq 0 \qquad \forall i \neq j \tag{11.13}$$

Theorem 1 of [11] proves the existence of equilibria for supermodular games, moreover it provides a way of computing them. The proof is based on showing that **best response sequences are monotone** and therefore converge to a limit which is then shown to be a Nash Equilibrium Point (NEP). The monotonicity is a consequence of the "strategic complementarity" of the players: if one of them chooses a strategy x_i that decreases its own cost, this decision is beneficial for the other players too.

Here we relax the results on the existence of equilibria and convergence of best response sequences in supermodular games.

Definition 11.2 (*Symmetric supermodularity*) We define as symmetric supermodular games, those for which (11.13) holds for all strategies $\mathbf{x}_i = \mathbf{x}_j$, meaning that the property holds along the symmetric axis.

Definition 11.3 (*Symmetric best response sequence*) We call symmetric best response sequence a path $\left(\mathbf{x}_i^{(0)}, \mathbf{x}_j^{(0)}, ...\right)$, $\left(\mathbf{x}_i^{(1)}, \mathbf{x}_j^{(1)}, ...\right)$, $\left(\mathbf{x}_i^{(2)}, \mathbf{x}_j^{(2)}, ...\right)$, ..., where $\mathbf{x}_i^{(0)} = \mathbf{x}_j^{(0)} = ...$ and $\forall k$, $\left(\mathbf{x}_i^{(k)}, \mathbf{x}_j^{(k)}, ...\right)$ satisfies $\mathbf{x}_i^{(k)} = \mathbf{x}_j^{(k)} = ...$

Theorem 11.4 *In symmetric supermodular games, pure equilibria exist and are given as the limit of symmetric best response sequences.*

Theorem 11.4, whose proof is reported in Appendix 11.8.1, not only proves the existence of equilibria for symmetric supermodular games, but also gives an algorithm for computing them. Please note that, for this theorem to hold, the game does not need to satisfy (11.13) for all possible strategies, but just along the symmetric path. This result can be applied to our game thanks to Theorem 11.5 and Corollary 11.6.

11.4.3 Analytic Results

Theorem 11.5 and Corollary 11.6 demonstrate existence of equilibria and convergence of the symmetric best response algorithm for the general model. The proofs are reported in Appendix 11.8.1.

Theorem 11.5 *The game defined in (11.12) is symmetric supermodular.*

Sketch of the Proof. The proof is based on showing that (11.13) holds along the symmetric axis for any possible combination of indexes:

$$\left.\frac{\partial^2 C^i(\mathbf{x_i}, \mathbf{x_{-i}})}{\partial x^{\bar{n}}_{j,\bar{l}} \partial x^{n}_{i,l}}\right|_{x^{n}_{i,l} = x^{\bar{n}}_{j,\bar{l}}} \leq 0 \qquad \forall i \neq j,\ \forall l, \bar{l}, n, \bar{n} \tag{11.14}$$

□

It is interesting to note that, as long as the transit cost function t^l of one ISP does not depend on the other ISP, the mixed second derivative (11.14) does not change. Therefore, symmetric supermodularity can be applied to game (11.12) even for more general transit cost functions (as outlined in Sect. 11.3.2). Please note that without the symmetric assumption, the game is neither supermodular, nor submodular, because we cannot say anything about the sign of the mixed derivatives.

Corollary 11.6 *The game defined in (11.12) has at least one pure equilibrium for symmetric demands, given as the limit of a symmetric best response sequence.*

11.4.4 The Simulator

Thanks to the supermodularity property, we have not only proven the existence of equilibria for the symmetric version of our game, but also gave a mean for computing these equilibria, as the the limit of symmetric best response sequences. Therefore, it is natural to deploy a simulator, that exploits this property to compute the stable equilibria of our game.

We implemented in MATLAB [179] the general model (11.11) described in Sect. 11.3. Iteratively, each player performs its best response to the set of other players' strategies as shown in Algorithm 11.1. If the simulation converges, the output **newx** is the NEP for the given input parameters, which are:

- The number of ISPs, TFs and CPs, respectively I, L, N.
- The cost function parameters a_l, c_l, b_l^n and demands ϕ_i^n.
- The $tolerance$ and the **startingpoint**.

Please note that, while we have demonstrated convergence of the best repines dynamic (and therefore of our algorithm) in symmetric cases, we do not have convergence guarantees for asymmetric scenarios. However, we know that if the best response sequence algorithm converges, then it converges to an equilibrium [11]. Therefore we can use our simulator both to study our system in general cases, and to assess convergence for specific cases.

11.5 Minimal Complexity Model (MCM)

After proving existence of equilibria for the general case, here we analyze some subcases in order to understand what kind of equilibria we should expect for specific scenarios. We especially concentrate on the Minimal Complexity Model (MCM), for which we perform an in-depth analysis of: (i) the cost function; (ii) the best response behavior; (iii) the prices of anarchy, stability and fairness.

Algorithm 11.1 Best Response Sequence

startingpoint = ... ▷ Initial strategies
tolerance = ... ▷ NEP stationariety tolerance
newx = **startingpoint**
repeat
  oldx = **newx** ▷ current step game strategies
  for $i = 1; i < I; i++$ **do**
    $\mathbf{x_{-i}} = \mathbf{oldx_{-i}}$ ▷ other players strategy
    $\mathbf{x_i} = \arg\min_{\mathbf{x_i}} C^i(\mathbf{x_i}, \mathbf{x_{-i}})$ ▷ i strategy
      $s.t. \sum_l x^n_{i,l} = \phi^n_i \ \forall n$
    $\mathbf{newx_i} = \mathbf{x_i}$ ▷ next step game strategies
  end for
until $||\mathbf{newx} - \mathbf{oldx}|| < tolerance$

11.5.1 Subcases Analysis

Starting from the general model of the previous section, we have two main specializations:

Fully Connected Topologies Suppose that we have a fully connected topology, meaning that $b^n_l = 0 \ \forall l, n$. In such a case, we can take the summation over n and consider cumulative flows and demands:

$$\begin{cases} x_{i,l} = \sum_n x^n_{i,l} & \textit{cum. flow } ISP_i \rightarrow TF_l \\ \phi_i = \sum_n \phi^n_i & \textit{cum. demand } ISP_i \end{cases}$$

We can now substitute these two variables inside cost function (11.11), thus obtaining an equivalent problem where the strategy of each player is a vector $\mathbf{x_i} = (x_{i,1}, ..., x_{i,l}, ..., x_{i,L}) \in \mathbb{R}^L$. This means that, in fully connected topologies, our system is equivalent to another one where we only have a single CP, and each player has to serve a cumulative demand ϕ_i for that CP. This happens because there are no reachability constraints, therefore from a player's perspective the specific CP from which he has to fetch data is not relevant.

Symmetric IXPs Suppose that all the IXPs have the same costs, capacities and reachability matrix: $a_l = a_{IXP}, c_l = c_{IXP}, b^n_l = b^n_{IXP} \ \forall l \neq 1$. Due to their symmetry, there is an equilibrium where traffic is split equally among them [95], therefore we might think of transforming this problem in an equivalent one having a single IXP with the same reachability matrix and transformed costs and capacities. Unfortunately, we were unable to perform this conversion due to the form of the cost function for the IXPs. In fact, as we see from (11.11) and (11.10), the non linear port cost h_l makes it quite different for players to share small traffic quantities rather then large ones.

The analysis of the two categories highlights that scenarios with multiple CPs can be highly simplified with fully connected topologies, while in the case of multiple IXPs, even if symmetric, the analysis can be quite difficult. Therefore, an easy setting

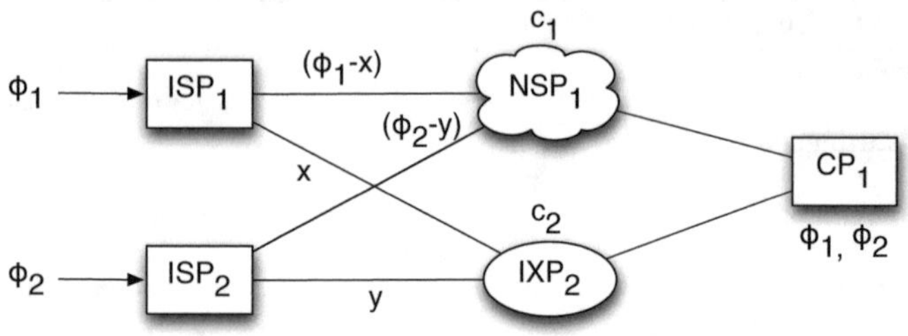

Fig. 11.3 Minimal complexity model

to analyze is the one where we have fully connected topology and just one IXP, due to the fact that we can handle the multiple CPs as if there was just a single one.

Out of all the possible settings to analyze, the easiest is the one where we have just two players. Thus, it is straightforward to define as "**Minimal Complexity Model**" (MCM) the scenario with fully connected topology, $I = 2$ ISP players, $N = 1$ CP (possibly representing an aggregate of all CPs), and $L = 2$ transmit facilities, either the NSP ($l = 1$) or the IXP ($l = 2$). The MCM is depicted in Fig. 11.3.

Thanks to the simplifications of this scenario, the demand of player i can be simply expressed as ϕ_i. With some algebraic manipulations, explicitly shown in Appendix 11.8.2, we can rewrite cost function (11.11) for both players as:

$$\begin{cases} C^1(x,y) & = (\phi_1 - x)\left(a_1 + \frac{1}{c_1-(\phi_1-x)}\right) + \\ & +x\left(\frac{a_2}{\sqrt{x+y}} + \frac{1}{c_2-x}\right) \\ C^2(x,y) & = (\phi_2 - y)\left(a_1 + \frac{1}{c_1-(\phi_2-y)}\right) + \\ & +y\left(\frac{a_2}{\sqrt{x+y}} + \frac{1}{c_2-y}\right) \end{cases} \tag{11.15}$$

where x is the flow sent by player 1 through the IXP and $\phi_1 - x$ is, by constraint, the flow sent through the NSP. The same applies to y for player 2. The best response of player i is thus obtained by minimizing $C^i(x, y)$ defined in (11.15).

While simple, the MCM is interesting on its own as it provides a clear way to study the fundamental difference between transit and peering agreements, shedding light on the emerging competition between NSPs and large IXPs, first observed in [6].

11.5.2 Cost Function Analysis

In order to gain insights on the outcome of the behavior of the best response, here we analyze the cost function. Consider the cost function of player 1 and suppose that the strategy y of player 2 is fixed, so:

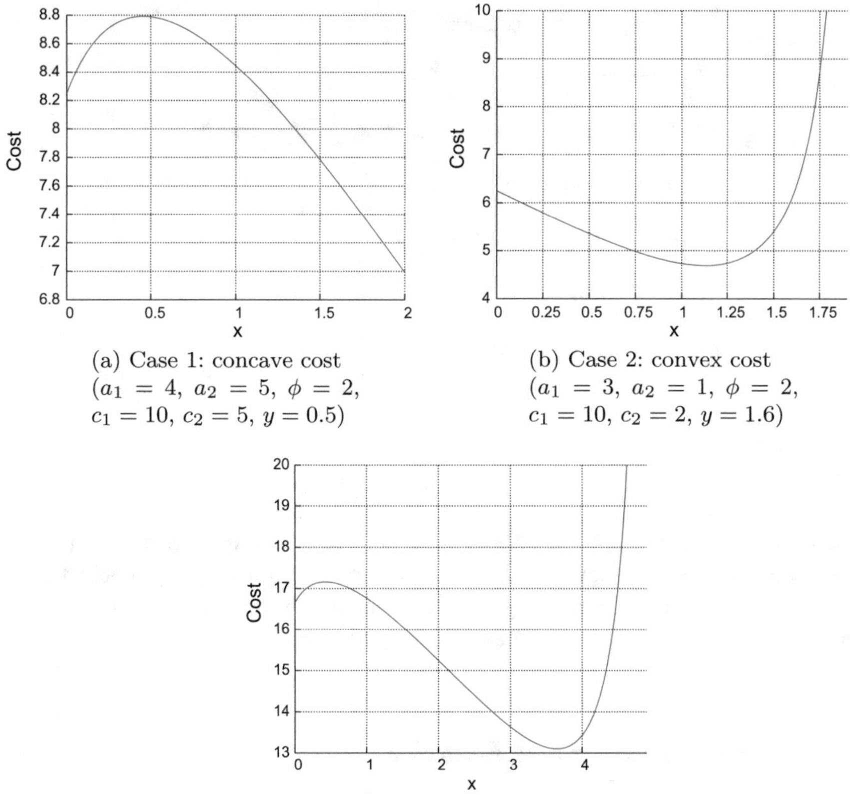

(a) Case 1: concave cost ($a_1 = 4$, $a_2 = 5$, $\phi = 2$, $c_1 = 10$, $c_2 = 5$, $y = 0.5$)

(b) Case 2: convex cost ($a_1 = 3$, $a_2 = 1$, $\phi = 2$, $c_1 = 10$, $c_2 = 2$, $y = 1.6$)

(c) Case 3 : concave/convex cost ($a_1 = 4$, $a_2 = 5$, $\phi = 4$, $c_1 = 10$, $c_2 = 5$, $y = 0.5$)

Fig. 11.4 Cost function

$$C^1(x) = (\phi_1 - x)\left(a_1 + \frac{1}{c_1 - (\phi_1 - x)}\right) + \\ + x\left(\frac{a_2}{\sqrt{x+y}} + \frac{1}{c_2 - x}\right) \tag{11.16}$$

Lemma 11.7 and Theorem 11.8, whose proofs can be found in Appendix 11.8.1, tell us the shape of the cost function.

Lemma 11.7 *The second derivative of the cost function (11.16) is a monotonically increasing function.*

Theorem 11.8 *The cost function (11.16) can be either: always concave, always convex, or first concave and then convex.*

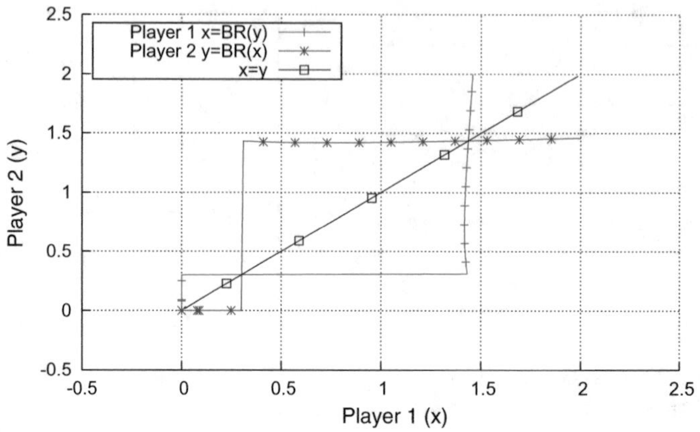

Fig. 11.5 BRI—Case 1: $a_1 = 2, a_2 = 2, \phi = 2, c_1 = 10, c_2 = 3$

Figure 11.4 shows the possible cases. Please note that the shape depends both on the parameters and the strategy y of the other player: while for specific values of y the function might be convex as in Fig. 11.4b, it can also be concave (Fig. 11.4a) and in general is neither convex nor concave, as shown in Fig. 11.4c.

11.5.3 The Best Response Behavior

The analysis of the cost function performed in the previous section, suggests that in our game (11.15), even if the best response procedure converges to a NEP, there might be multiple equilibria, because of the presence of multiple local minima. This assumption can be verified via simulation. First of all, let's show a specific case in which the NEP reached can change, depending on the starting point of the algorithm. We can use the implemented Algorithm 11.1 on the symmetric MCM, which has fully connected topology and symmetric demands, by putting $I = 2, L = 2, N = 1$, $b_l^n = 0 \; \forall l, n$ and $\phi_i^n = \phi$. Given the selected scenario and the best response sequence algorithm, Theorem 11.6 ensures the convergence of the simulation for whatever cost function coefficients. Moreover, thanks to the symmetric property, we can just investigate the strategy of player 1 (x), because player 2 will show exactly the same behavior.

We simulate the following parameters: $a_1 = 4, a_2 = 5, \phi = 4, c_1 = 10, c_2 = 5$. In this case, as previously shown in Fig. 11.4c, the cost function could present multiple local minima, depending on the players' strategies. The simulation has multiple outcomes: if we start from the mean point ($x = 2$, $\phi - x = 2$) we end up in an equilibrium where traffic is split between the IXP and the NSP: $x^* = 3.64$; $\phi - x^* = 0.36$. The IXP is preferred because the usage cost is shared among the two players, however it is not used exclusively due to its smaller capacity not being able to serve

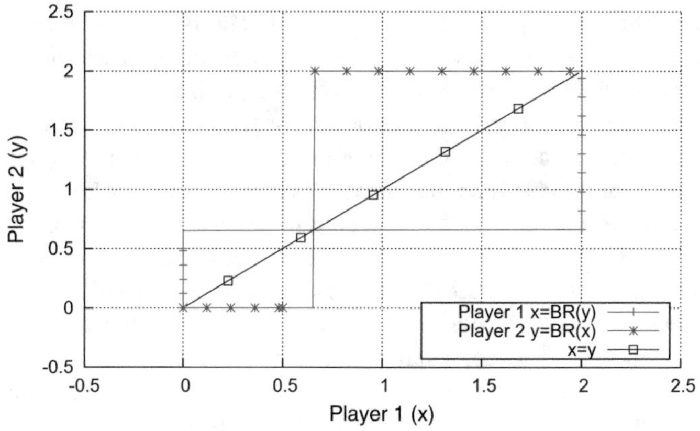

Fig. 11.6 BRI—Case 2: $a_1 = 2, a_2 = 3, \phi = 2, c_1 = 10, c_2 = 4$

all the traffic. With a sufficiently larger capacity, all the traffic would have been routed through the IXP. Otherwise, if we start from a strategy where the majority of traffic is routed through the NSP ($x = 0.4, \phi - x = 3.6$), we end up in an equilibrium where all the traffic flows through the NSP: $x^* = 0$; $\phi - x^* = 4$. This happens because when the IXP is routing a small amount of traffic, the flat port cost is too high to justify its use, therefore the players prefer the NSP. Once the NSP is serving all the traffic no player has an incentive to deviate, because he would pay the whole IXP cost by himself.

As we see, the outcome of the game is highly dependent on the starting point: the IXP is preferred only if it already has, at the beginning, a good amount of flow passing through it, otherwise all players will stick to the NSP. This result is consistent with reality, in fact, the necessary condition for an IXP to emerge is that it has a critical mass (represented by a fraction of the traffic/users in the Internet) which makes the value perceived by a potential participant greater than the cost he would incur in by joining the facility [195].

With the purpose of understanding the number and position of NEPs, we draw the Best Response Intersection (BRI) picture. In this graph, shown in Fig. 11.5, the line with tick marks represents the best response x of player 1 as a function of player 2's strategy y, while the line with cross marks does the exact opposite. The intersection points on the graph mean that both players are playing their best responses, therefore they are Nash Equilibrium Points. As we can see, there are three NEPs and, as expected due to the symmetric property, they are all on the symmetric axis [95]:

Left Equilibrium is in $x = x_L^* = 0$ and corresponds to the scenario where all the traffic is routed through the NSP.

Right Equilibrium is for $x = x_M^* = 1.43$ and is the one where traffic is split between the IXP and the NSP.

Middle Equilibrium happens for $x = x_M^* = 0.31$. This is however a repulsive equilibrium, in fact, as soon as one of the two players deviate, they will never return to this point and reach instead one of the two others equilibria.

These three equilibria can be understood by observing Fig. 11.4c: x_L^* and x_R^* are attractive, and correspond to the minima of the cost function, while x_M^* corresponds to the maximum of the cost function, and is thus repulsive. Of course, the last picture corresponds to the cost function for a specific strategy, therefore it cannot assert the position or the existence of equilibria, however it gives an insight on their meaning.

As we change the game parameters we observe that the shape of the best response is always the same, while the position of x_M^* and x_R^* changes. In particular, as shown in Fig. 11.6, If the ratio $\frac{a_2}{a_1}$ increases (meaning that IXP cost w.r.t. NSP cost increases) then x_M^* gets nearer to x_R^*, making the left equilibrium is easier to reach. Moreover, we observe that if the capacity c_2 of the IXP is large enough, than in the right equilibrium all the traffic will flow through him.

To conclude this analysis, we verify the behavior of the best response in a slightly more complex case, where the number of players is $I = 3$ (it would be difficult to represent more dimensions). The Best Response Intersection (BRI) graph is shown in Fig. 11.7. Just like in the MCM, the picture shows three equilibrium points, obtained by the intersection of the three surfaces representing the players' best responses. As we see from the straight line crossing all such points, the three equilibria are symmetric, with the leftmost (traffic split between NSP and IXP) and the rightmost (all flows through the NSP) being the stable ones.

11.5.4 Price of Anarchy, Stability and Fairness

Social Optimum

We now exploit the MCM to compare the performance of the distributed system, where each Service Provider acts on its own, with that of an ideal centralized system where decisions are took by some external entity. In this case the objective is to minimize the total cost of the two players, given by the summation of the two costs in (11.15):

$$\begin{aligned} C(x, y) = \textstyle\sum_i C^i(x, y) = \\ = (\phi_1 + \phi_2 - x - y)\, a_1 + \tfrac{\phi_1 - x}{c_1 - (\phi_1 - x)} + \\ + \tfrac{\phi_2 - y}{c_1 - (\phi_2 - y)} + \tfrac{x+y}{\sqrt{x+y}} a_2 + \tfrac{x}{c_2 - x} + \tfrac{y}{c_2 - y} \end{aligned} \tag{11.17}$$

Theorem 11.9 and Corollary 11.10, whose proofs can be found in Appendix 11.8.1, explain how to optimize this cost function.

Theorem 11.9 *The cost function (11.17) has a global minimum point. For symmetric demands this minimum point is attained at symmetric strategies, and it is either the*

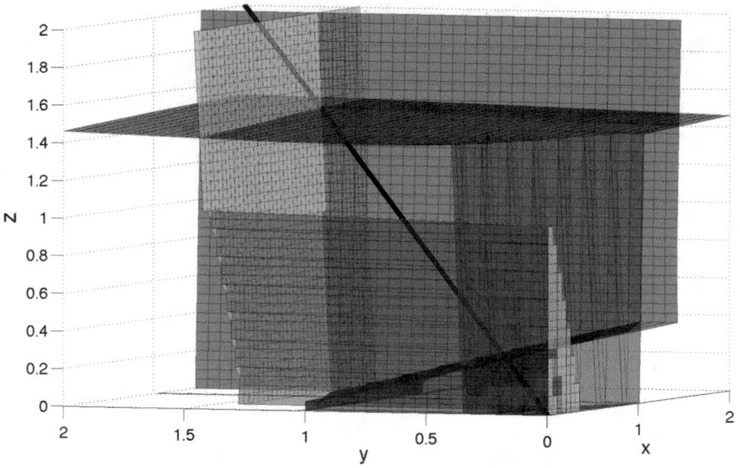

Fig. 11.7 BRI for 3 players: $a_1 = 2, a_2 = 2, \phi = 2, c_1 = 10, c_2 = 3$

left endpoint of the strategy space or the unique local minimum point of its convex part.

Corollary 11.10 *The global minimum point of (11.17) is, for symmetric demands, either the left endpoint of the strategy space or the output of a standard algorithm for convex function optimization that starts from the right endpoint.*

The globally optimal solution to problem (11.17) can thus be computed by comparing the two candidate points highlighted in Corollary 11.10.

Alpha-Fair solution

Another metric for comparison comes from the theory of fairness. A unifying mathematical formulation, known as α-fairness [164], says that given a set of users and utility functions $U_i(x)$, the α-fair solution to the problem of maximizing their utilities is given by:

$$\max_x \left(\sum_i \frac{U^i(x)^{(1-\alpha)} - 1}{1-\alpha} \right)$$

For $\alpha = 0$, this is the same as maximizing the sum of the utilities, thus it gives the social optimum for the problem. The case $\alpha \to 1$ yields the proportional fair share assignment, however this solution is not feasible when we have to deal with cost function rather then utilities, and for $\alpha \to \infty$ it is equivalent to the max-min fairness. For $\alpha = 2$, the formula gives us the "harmonic mean fair" solution, which we investigate here:

$$\max_x \left(\sum_i \frac{U^i(x)^{(1-\alpha)} - 1}{1-\alpha} \right) = \min_x \left(\sum_i -1 - \frac{1}{C^i(x)} \right)$$

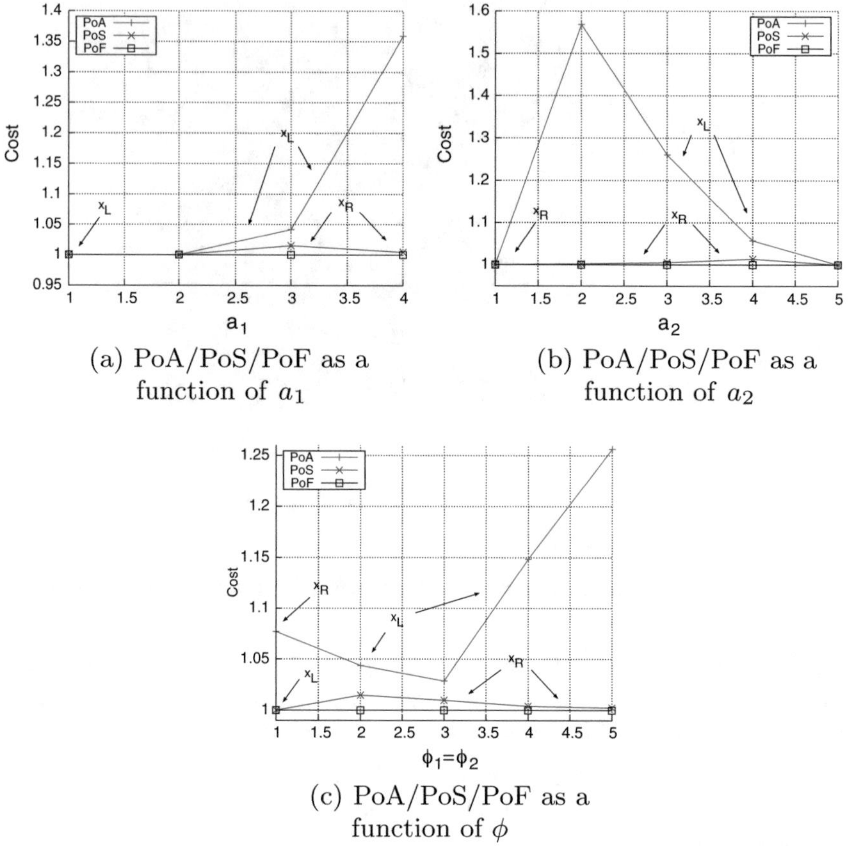

(a) PoA/PoS/PoF as a function of a_1

(b) PoA/PoS/PoF as a function of a_2

(c) PoA/PoS/PoF as a function of ϕ

Fig. 11.8 Price of anarchy, stability and fairness

The solution is computed numerically in the next subsection.

PoA/PoS/PoF Comparison

As is usually done in the literature [151, 164], we define the Price of Anarchy (PoA) as the ratio between the worst decentralized solution (equilibrium) and the social optimum. Similarly, the Price of Stability (PoS) is defined as the ratio between the best equilibrium and the social optimum:

$$PoA = \frac{C\left(x^*_{worst}\right)}{C\left(x_{opt}\right)} \geq 1 \qquad PoS = \frac{C\left(x^*_{best}\right)}{C\left(x_{opt}\right)} \geq 1$$

In our case we have just two attractive equilibria, therefore the best and worst equilibria are either x^*_L or x^*_R. Following the same path, we define the price of fairness as the ratio between the fair and the optimal solution:

$$PoF = \frac{C\left(x_{fair}\right)}{C\left(x_{opt}\right)} \geq 1$$

Algorithm 11.1 has been extended to include numerical computation of the above-defined Prices of Anarchy, Stability and Fairness. We use as general configuration: $a_1 = 3, a_2 = 4, \phi = 2, c_1 = 10, c_2 = 3$ (except for Fig. 11.8c, where c_2 increases as ϕ increases), and show PoA, PoS and PoF as parameters a_1, a_2 and ϕ change. Results are reported in Fig. 11.8. As we see, it is always the case that $PoF = 1$, meaning that the harmonic mean fair solution is equal to the social optimum.

The PoA almost always corresponds to the left equilibrium. An exception to this is the case where there is a small amount of total traffic, shown in Fig. 11.8c for $\phi = 1$: in this case the left equilibrium outperforms the right one, meaning that for small amounts of flow it is not convenient to share costs at the IXP. As ϕ increases, the advantages of sharing become obvious. Figure 11.8a, b show that the PoA increases as a_1 increases and decreases as a_2 increases. An exception to this is the case $a_2 = 1$ of Fig. 11.8b: with these parameters the cost function resembles that of Fig. 11.4b, therefore we have only one equilibrium. The PoS is almost always very low, and it is always caused by the fact that the competition between ISPs reduces the amount of traffic through the IXP, thus reducing their opportunities to share costs.

11.6 Simulations

While the analytic results obtained are interesting on their own, as they shed light both on the inefficiency of the decentralized solution and on the competition between an Internet Exchange Point and a Network Service Provider, driven by the clear differences between transit and peering, one might argue that this topology is a bit simplified to represent the Internet, especially due to the symmetric assumption. Here we explicitly tackle this problem by means of simulation, using our MATLAB implementation to test the behavior of the system. Simulations have been performed using the Best Response Algorithm 11.1: iteratively, each player performs its best response to the set of other players' strategies. If the simulation converges, the output is the equilibrium for the given input parameters, which are:

- the number of ISPs, TFs and CPs, respectively I, L, N;
- the cost function parameters a_l, c_l, b_l^n and demands ϕ_i^n.

Moreover, it is important to set a **startingpoint**, because, as we saw, on startup IXPs need a critical mass, represented by a share of the total traffic in the system, in order to be able to attract players.

11.6.1 Simulating Ad-Hoc Configurations

In the first part of this section, we focus on the study of the system using ad-hoc configurations, in the sense that we use parameters not necessarily connected with the actual Internet. This kind of analysis can be used to study the behavior of the system for generic parameter combinations. This will prove to be especially useful to infer a set of conditions under which the system behaves "better", in the sense that the players obtain a higher utility.

The second part of the section will focus on a regional case-study, where parameters are selected according to the real Internet environment.

Growing Number of ISPs/IXPs

We start with showing the behavior of the system for symmetric cases, for which convergence has been proven, and checking what happens as the number of agents in the system grows. The base configuration used is $I = 2$, $L = 2$, $N = 2$, $b_l^n = 0\ \forall l, n$ and $\phi_i^n = \phi^n = 2$, and all tests have been performed with fully connected topology and symmetric demands. The cost coefficients used are: $a_1 = 1, a_2 = 1.5$, $c_1 = 10, c_2 = 6$. As long as flows and capacities are properly balanced, the existence of multiple CPs does not seem to affect the results of the simulations, therefore here we check what happens when we have either more ISPs or IXPs.

Generic Number of ISPs When the number I of players increase, the benefits of joining an IXP increases as well, due to the fact that costs are shared among multiple participants: in fact, as shown in Fig. 11.9 on the y1 axis, the fraction of traffic flowing through the IXP at the equilibrium increases with I. We recall from Sect. 11.4.4 that IXPs need a critical mass to be used, which in our case corresponds to a fraction of the total traffic in the system. Very interestingly, the y2 axis of Fig. 11.9 shows that this fraction decreases as the number of player grows.

Generic Number of IXPs In order to have an interesting case study as L grows, we test a scenario where the global IXP capacity does not change, therefore $c_l = c_2/(L-1)\ \forall l \neq 1$. This means that instead of having one "large" IXP with a high capacity, we have multiple IXPs with less capacity. In order to have meaningful capacities for the small IXPs, we increased global flows and capacities to: $c_1 = 50$, $c_2 = 25$, $\phi_i^n = \phi^n = 10$. As shown in Fig. 11.10 the fact that IXPs only offer small ports is detrimental for the players, and after a certain point they will all stick to the NSP.

Simulations show that while the IXP critical mass decreases with a larger player base, this effect is counterbalanced by the fact that the critical mass increases with the number of IXPs. Therefore, the results found in the two player case still hold in more realistic scenarios: IXPs need a critical mass to emerge even in scenarios with more ISPs and IXPs, otherwise we still end up in an equilibrium with dominant NSP connectivity.

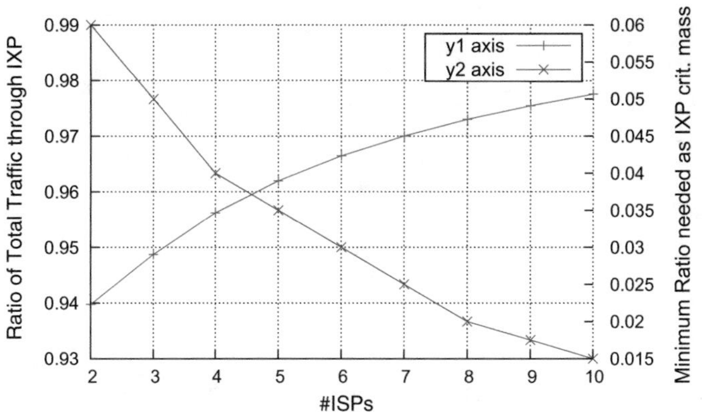

Fig. 11.9 Traffic ratios and equilibrium breakpoint as I grows

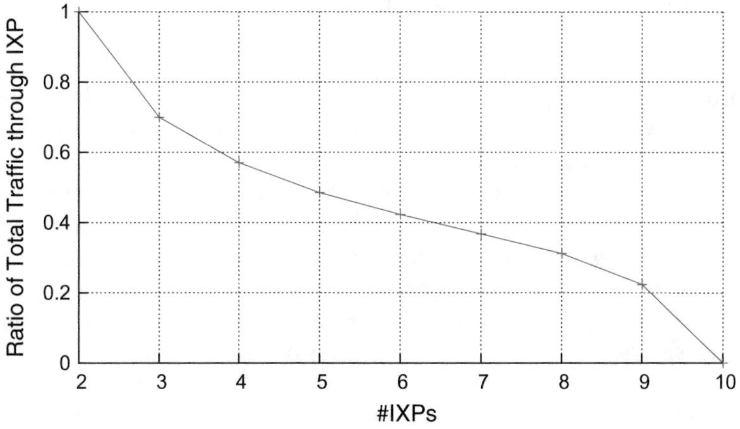

Fig. 11.10 Traffic ratio as L grows

Flow Path Analysis

Here we show results of simulations regarding the path followed by traffic flows. The simulations were performed both for symmetric scenarios, for which convergence has been proven in the general case, and asymmetric scenarios, for which we have no proofs. In fact, as we'll see later on, in this last case it is possible for players to never reach an equilibrium.

Symmetric Case We simulate a scenario with $I = 10$ ISPs, $L = 4$ TFs ($L - 1$ symmetric IXPs) and $N = 4$ CPs. The connectivity matrix is:

$$b_l^n = \begin{cases} \infty & (l, n) = (2, 2) \vee (l, n) = (4, 1) \vee (l, n) = (4, 4) \\ 0 & otherwise \end{cases}$$

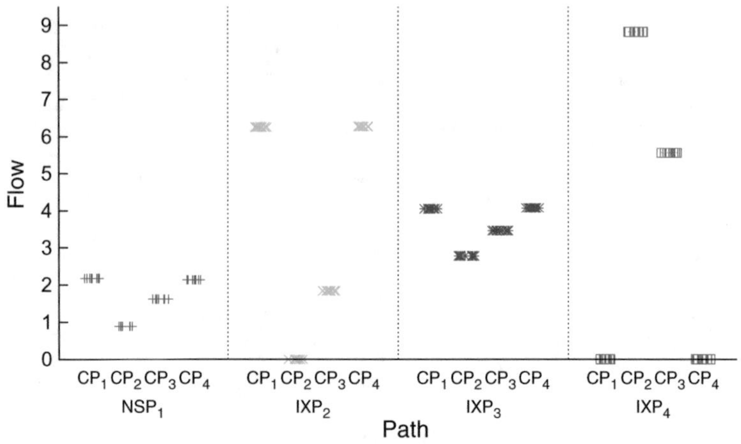

Fig. 11.11 Symmetric case flows scatterplot: $\phi_i^n = 12.5$

The cost parameter a_l has been chosen in order to be similar to present reality [194], therefore we choose $a_1 = a_{NSP} = 3$ and $\forall l \neq 1$, $a_l = a_{IXP} = 30$ as seen in the fit performed for the MIX (Fig. 11.2). All users have symmetric flows $\phi_i^n = 12.5\ \forall i, n$ and their capacities to the facilities are $c_1 = c_{NSP} = 100$ and $\forall l \neq 1$, $c_l = c_{IXP} = 20$, so that $c_{NSP} \gg c_{IXP}$. As already happened in the MCM, depending on the **startingpoint** we notice the existence of multiple equilibria. In fact, if the initial condition is such that one or more IXPs are underutilized, than at equilibrium those IXPs will not be used. This phenomenon corroborates the outcomes of the MCM, showing that indeed even in general scenarios the competition between NSPs and IXPs, and even between IXP themselves, strongly emerges. Differently from the MCM, in this case we observe more than two stable equilibria, since any combination with one or many unused IXPs can be an equilibrium. Suppose now that the **startingpoint** is such that flows are split equally among the facilities, so that all IXPs have the critical mass to attract players. Figure 11.11 shows the scatterplot at equilibrium. In this plot, each dot represents the flow quantity that each user sends on a given path (that is, to a fixed CP through a given IXP). Due to symmetry, we observe that all users will behave symmetrically on the same path, and this is exactly what happens in Fig. 11.11. There is generally a low utilization of the NSP, which rises a little bit for those CPs with a worse reachability matrix (CP_1, CP_4).

Asymmetric Case We now show the impact of asymmetric players' demands. In this case, convergence of the best response sequence is not guaranteed by Corollary 11.6, however, we know that if the simulation converges we certainly reach an equilibrium. We simulate a scenario with exactly the same parameters as in the symmetric case, except that now the demands grow linearly from $\phi_1^n = 10\ \forall n$ to $\phi_I^n = 15\ \forall n$. The average demand is still 12.5, but now the demand of the last player is 1.5 times that of the first one. The scatterplot at equilibrium is shown in Fig. 11.12.Very interestingly, even if demands are asymmetric, paths of flow tend to be almost symmetric for the

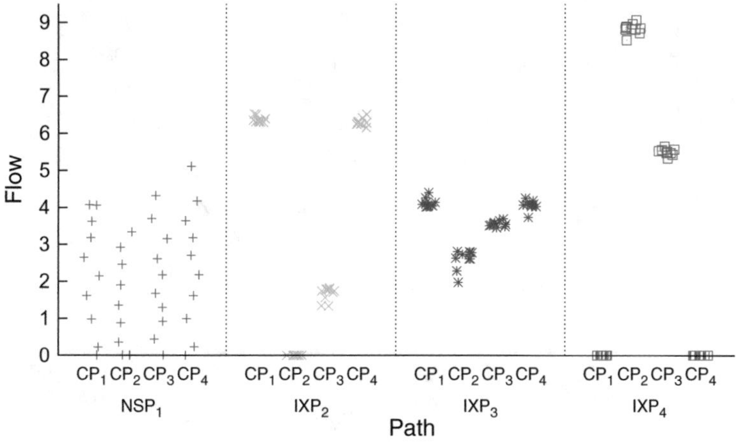

Fig. 11.12 Asymmetric case flows scatterplot: $\phi_i^n = 10 \rightarrow 15$

IXPs, while they spread apart for the NSP. This happens because the benefits of sharing costs at the exchange points is bigger when the traffic ratio is approximately the same between participants, therefore players tend to "symmetrize" around the IXPs. Due to the fact that flows around the IXPs are more or less symmetric, players will send the traffic residual through the NSP, which will see highly asymmetric patterns. In previous case, the asymmetry in players' demands was not very pronounced. Let's now see what happens when the demands go from $\phi_1^n = 6.5\ \forall n$ to $\phi_I^n = 18.5\ \forall n$, meaning that last player demand is nearly three times that of player one. Again, Fig. 11.13 shows the scatterplot at equilibrium. Due to the heavily unbalanced demands, the symmetric patterns around the IXPs are still present, but much less pronounced. While in the previous case equilibrium was driven by the simple rule of "symmetric behavior", in this case the outcome is more difficult to predict. In general, due to the asymmetry, cost benefits of players' for using exchange points decrease, therefore we observe, on average, a higher quantity of flow going through the NSP.

The phenomena emerged through this analysis can provide some preliminary insight on how to devise optimal policies to handle peering traffic at IXPs. More specifically, the "symmetric behavior" rule highlights that it is beneficial to balance traffic as much as possible, therefore IXP owners should create few classes of traffic (namely, few different port sizes), and participants should try to aggregate traffic on these ports, since unbalanced flows must be handled at NSPs and bring to suboptimality. While simulations have been carried out with a limited network size due to computational constraints, the conclusions are fairly general, therefore we expect similar results to hold for larger-scale scenarios.

Non Convergence

Even quite simple scenarios for which we cannot apply Theorem 11.5, might lead to a situation where players' behavior oscillates, never reaching an equilibrium. Consider

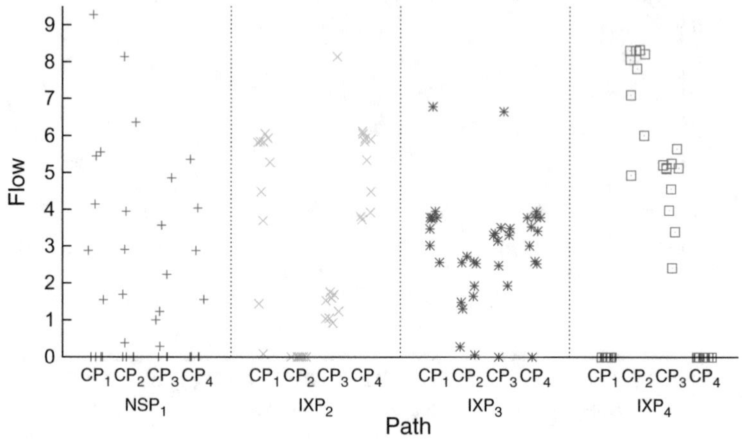

Fig. 11.13 Asymmetric case flows scatterplot: $\phi_i^n = 6.5 \rightarrow 18.5$

a system with two symmetric IXPs and an asymmetric starting point s.t. a group of players send more traffic to one of the IXPs and less to the other, while the other group of players do the opposite. Due to the asymmetric assumption we cannot apply Corollary 11.6, and simulation shows that this scenario might never reach an equilibrium. This happens when players enter a never-ending oscillation between the first and the second group, as detailedly shown in Appendix 11.8.3.

11.6.2 Simulation of a Realistic Scenario

In this part, we instantiate the model on a regional case-study, in order to inspect its behavior for a realistic scenario, and understand which properties of the model apply. We first describe how to set up the scenario, and then the results obtained through simulation.

Simulation Scenario

We focus on a relevant subset of the Internet by focusing on a restricted geographical region: the Italian country. The first thing to do is defining the network topology as in Fig. 11.1, by selecting the ISPs, CPs, IXPs and NSPs. We take as IXPs the Top 3 Italian Internet eXchange Points: Milan Internet eXchange (MIX) [182], Torino Piemonte Internet Exchange (ToP-IX) [259], and Nautilus Mediterranean eXchange (NaMeX) [185]. In order to select the ISPs/CPs, we extracted the list of ASes participating to these IXPs (available on the IXPs websites), and labeled each AS as either ISP or CP, according to the information provided by PeeringDB [211]. Furthermore, we add to the topology a Network Service Provider (NSP), which provides transit services to

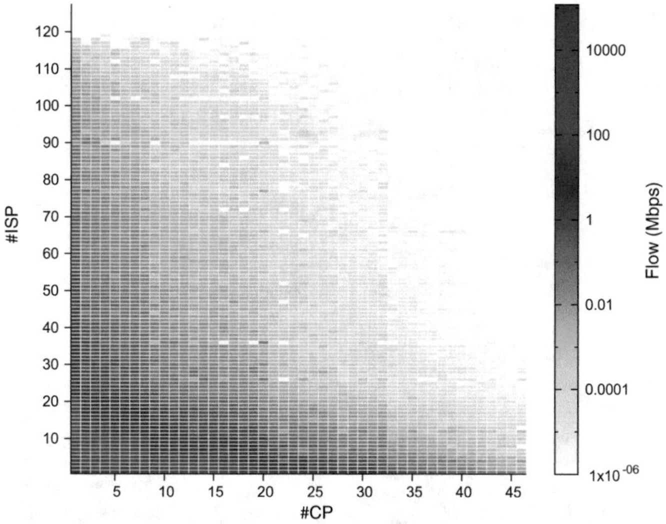

Fig. 11.14 Italian interdomain traffic matrix Φ

the ISPs and CPs, so that they can exchange traffic even if they are not on the same IXP.

The connectivity matrix $\mathbf{B} = \left[b_l^n\right]_{(L \times N)}$ can be directly obtained by considering the IXP participants list: IXP l is connected to CP n only if the Content Provider appears in the IXP participants list. The cost parameter $a_1 = a_{NSP} = 3$ has been chosen in order to be similar to present reality [194], while for IXPs we perform the same fitting procedure as for the MIX (Fig. 11.2) and obtain $\forall l \neq 1, a_l = a_{IXP} \approx 30$. As for the capacities, we assume that the capacity of the NSP is sufficient for routing even the heaviest flow, while each IXP has a capacity equal to $1/6$ of the heaviest flow.

The Interdomain Traffic (flow) Matrix $\Phi = \left[\phi_i^n\right]_{(I \times N)}$ is obtained through the methodology thoroughly described in [3]. Basically, the methodology infers the Interdomain Traffic Matrix (ITM) by exploiting DNS data. To this end, the methodology was applied to the .it DNS traffic recorded at the Institute of Informatics and Telematics of the Italian National Research Council (IIT-CNR). A flavor of the methodology is given in Appendix 11.8.4, while for more information, refer to [3]. The ITM is shown in Fig. 11.14. Please note that the matrix is heavily asymmetric, with a low number of large ISPs and popular CPs carrying almost all the traffic.

Figure 11.15a sums up the parameters of our Italian case-study.

Results

Here we illustrate the results obtained by simulating the Italian case study presented above. The scatterplot at equilibrium is shown in Fig. 11.15. The plot shows results for all the ISPs and the 15 CPs carrying most traffic, due to space limits. Nevertheless,

Parameter	Value
I (#ISPs)	127
N (#CPs)	46
L (#TFs)	1 NSP & 3 IXPs
a_{NSP}	3
a_{IXP}	≈30
$\mathbf{B} = [b_l^n]_{(L \times N)}$	IXP Participants
$\mathbf{\Phi} = [\phi_i^n]_{(I \times N)}$	ITM in Fig. 14.14

(a) Scenario Parameters Table

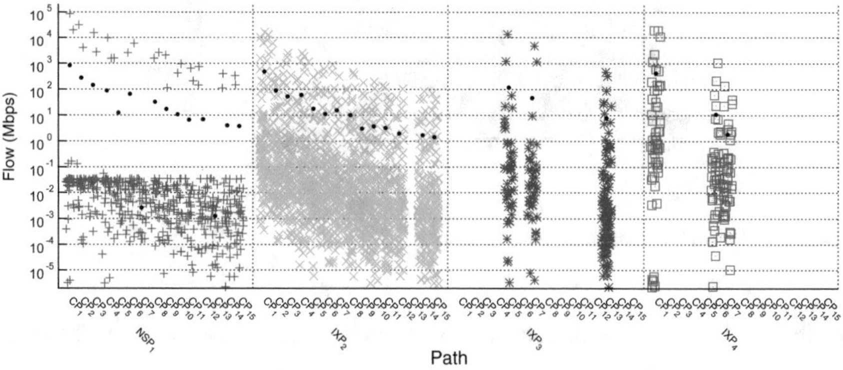

(b) Flows Scatterplot

Fig. 11.15 Realistic Italian scenario

results shown here are still valid when considering the entire CP population. The figure shows a number of interesting results.

First of all, we observe a phenomenon which was missing in the simulations of the previous section, even if it was shown when analyzing the Price of Anarchy as a function of ϕ (Fig. 11.8c: the NSP is preferred over the IXP when the amount of flow to be sent is very small. Very interestingly we notice that IXPs are the to-go facility for flows of average weight (100 kbps to 100 Mbps), while NSPs are only used in the other cases. The high volumes (over 1Gbps) are served by the NSP due to capacity limits of the IXP, while for the low volumes the NSPs are used because they provide an economic advantage over the IXP. However, we still notice low traffic volumes flowing through IXPs. This happens due to an effect of traffic trunking: basically, if an ISP has already bought an IXP port for sending traffic and has not reached the capacity limit, it can take advantage of the IXP also for sending small flows. Instead, when the ISP is only involved in small traffic volumes, there is no need of purchasing IXP ports, therefore the NSP is preferred.

As for the "symmetric behavior" outlined in the previous section, we observe that the effect is mild, due to the fact that the ITM is heavily asymmetric, but still present, as shown by the fact that some zones of the clouds in Fig. 11.15 are more dense than others.

Last, but not least, we observe that the average amount of flow (represented by the black circles in the figure) handled by the largest IXP, is approximately equal to that handled by the NSP, which is confirmed in [6].

11.7 Conclusions

The proposed model gives insight into the economy of different types of Autonomous Systems and the driving forces behind the decisions they make when joining the Internet. The peculiar pricing strategies of players doesn't allow standard modelization, however, by exploiting peculiar properties of the game, we are able to prove analytically the existence of multiple equilibria, and provide an algorithm to compute the stable ones. From a game theoretic perspective, while the theory on supermodularity is well-developed, we relaxed this concept and introduced the new category of Symmetric Supermodular games. Thanks to this we were able to prove existence of equilibria and convergence of best response sequences in our game. This is the first case, to the best of our knowledge, where results on supermodularity are applied even if the property does not hold for the game in general, by showing that it holds along the symmetric path. From an engineering perspective, the outcome of the analysis is highly insightful as it shows different interesting aspects. First of all, we observe the suboptimality of the decentralized solution, originated by the non-cooperative behavior of the ASes, by showing the existence of a Price of Anarchy and Stability. Second, we have shown through a realistic setting that while IXPs should be preferred for medium-size flows, NSPs prevail in the other cases. Moreover, we have shown that also for asymmetric cases the system often reaches an equilibrium. Such equilibrium suggests that players have to "symmetrize" their traffic as much as possible with respect to the peering exchange points, and send their asymmetric traffic quota via the transit service providers. This observation can provide insights on how to devise optimal policies to handle peering traffic at IXPs. Last, but probably most important, we highlight the growing competition between IXPs, providing customers the ability to lay out peering connections, and NSPs, high-level providers selling transit connections, even for large-scale realistic scenarios.

The proposed model was specifically tailored for the Internet environment. This complex system is the result of a human engineering process, which tries to balance social behavior, technological constraints and optimization in the interaction of its building blocks. While all these aspects make game theory a powerful investigation tool, they also require specific design of the models and analysis involved, making it difficult to derive general conclusions for networks which lack these characteristics. Nevertheless, obtained results shed light on the emerging competition, between the two facilities enabling either transit or peering connectivity: Network Service Providers and Internet Exchange Points. This phenomenon, first observed in [6], is a key point driving the evolution of the Internet, thus it is directly related to CONGAS.

11.8 Appendix

11.8.1 Proofs

Proof of Theorem 11.4. Consider a sequence of best responses $\left(\mathbf{x}_i^{(0)}, \mathbf{x}_j^{(0)}, ...\right)$, $\left(\mathbf{x}_i^{(1)}, \mathbf{x}_j^{(1)}, ...\right)$, $\left(\mathbf{x}_i^{(2)}, \mathbf{x}_j^{(2)}, ...\right)$, ... Due to symmetry we can choose this path to be a symmetric best response sequence. From definition 11.2 and by applying the same reasoning as in the original proof [11], we shall get monotone sequences whose limits are equilibria. □

Proof of Theorem 11.5. First of all, we can use constraint (11.5) to reduce the number of variables of our system. In fact if we perform the summation over n on left and right member, and separate the NSP component from the IXPs, we obtain:

$$\sum_n x_{i,1}^n + \sum_{l \neq 1} \sum_n x_{i,l}^n = \sum_n \phi_i^n \tag{11.18}$$

Now we substitute $\sum_n x_{i,1}^n$ taken from (11.18) inside (11.12) and rearrange terms, so as to obtain:

$$\begin{aligned} C^i(\mathbf{x_i}, \mathbf{x_{-i}}) = & \textstyle\sum_{l \neq 1} \sum_n x_{i,l}^n \left(\frac{a_l}{\sqrt{\sum_i \sum_n x_{i,l}^n}} + \frac{1}{c_l - \sum_n x_{i,l}^n} \right) + \\ & + \left(\textstyle\sum_n \phi_i^n - \sum_{l \neq 1} \sum_n x_{i,l}^n\right) \left(a_1 + \frac{1}{c_1 - (\sum_n \phi_i^n - \sum_{l \neq 1} \sum_n x_{i,l}^n)} \right) + \\ & + \textstyle\sum_{l \neq 1} \sum_n x_{i,l}^n b_l^n \end{aligned} \tag{11.19}$$

where all the flow variables have $l \neq 1$. Now we need to compute the mixed second derivatives of Eq. (11.19). We observe that the second term, referring to the NSP, has no mixed components, due to the fact that the cost does not depend on other players' choice, therefore this term becomes zero in the computation. The same happens with the last term, therefore we have:

$$\begin{aligned} \frac{\partial^2 C^i(\mathbf{x_i}, \mathbf{x_{-i}})}{\partial x_{j,\bar{l}}^{\bar{n}} \partial x_{i,l}^n} = & \frac{\partial}{\partial x_{j,\bar{l}}^{\bar{n}}} \left[\left(\frac{a_l}{\sqrt{\sum_i \sum_n x_{i,l}^n}} + \frac{1}{c_l - \sum_n x_{i,l}^n} \right) + \right. \\ & \left. + \textstyle\sum_n x_{i,l}^n \left(-\frac{a_l}{2(\sum_i \sum_n x_{i,l}^n)^{3/2}} + \frac{1}{(c_l - \sum_n x_{i,l}^n)^2} \right) \right] \\ & \forall i \neq j,\ \forall l, \bar{l}, n, \bar{n} \end{aligned} \tag{11.20}$$

Following the same reasoning previously done, we observe that the second and forth term in (11.20) do not depend on $x_{j,\bar{l}}^{\bar{n}}$, therefore their contribution in the final derivative is zero. Moreover, we observe that first and third term only have flow variables with index l, therefore for any $\bar{l} \neq l$ the whole derivative becomes zero:

$$\frac{\partial^2 C^i(\mathbf{x_i}, \mathbf{x_{-i}})}{\partial x^{\bar{n}}_{j,\bar{l}} \partial x^n_{i,l}} = 0 \quad \forall i \neq j,\ \forall l \neq \bar{l},\ \forall n, \bar{n} \tag{11.21}$$

while in the other case we have:

$$\frac{\partial^2 C^i(\mathbf{x_i}, \mathbf{x_{-i}})}{\partial x^{\bar{n}}_{j,l} \partial x^n_{i,l}} = -\frac{a_l}{2\left(\sum_i \sum_n x^n_{i,l}\right)^{3/2}} + \frac{3a_l \sum_n x^n_{i,l}}{\left(\sum_i \sum_n x^n_{i,l}\right)^{5/2}} \qquad \forall i \neq j,\ \forall l, n, \bar{n} \tag{11.22}$$

Please note that, regardless of the chosen $n, \bar{n}$, the derivatives are all the same. In order to prove symmetric supermodularity, we have to show that property (11.13) holds for both (11.21) and (11.22). While in the first case this is trivial, for the second one we multiply (11.22) by the positive quantity $\left(\sum_i \sum_n x^n_{i,l}\right)^{3/2}$, thus obtaining that:

$$\operatorname{sgn}\left(\frac{\partial^2 C^i(\mathbf{x_i}, \mathbf{x_{-i}})}{\partial x^{\bar{n}}_{j,l} \partial x^n_{i,l}}\right) = \operatorname{sgn}\left(\frac{a_l}{4} \cdot \frac{\sum_n x^n_{i,l} - 2\sum_{j \neq i} \sum_n x^n_{j,l}}{\sum_i \sum_n x^n_{i,l}}\right) \tag{11.23}$$

Along the symmetric axis we have that $x^n_{i,l} = x^n_{j,l}\ \forall i \neq j,\ \forall l, n$, meaning that each couple (i, j) of players send, to a fixed CP n through a given IXP l, the same quantity of flow. With this condition, Eq. (11.23) is always negative, therefore (11.12) is symmetric supermodular. □

Proof of Corollary 11.6. By hypothesis the demands satisfy $\phi^n_i = \phi^n_j\ \forall i \neq j,\ \forall n$. Therefore ISPs keep playing along the symmetric axis [95], and we obtain this result by combining Theorems 11.4 and 11.5. □

Proof of Lemma 11.7. The second derivative of function (11.16) w.r.t. x is:

$$\frac{\partial^2 C^1(x)}{\partial x^2} = \frac{2c_1}{(c_1 - (\phi_1 - x))^3} + \frac{2c_2}{(c_2 - x)^3} + \frac{3a_2 x}{4(x+y)^{\frac{5}{2}}} - \frac{a_2}{(x+y)^{\frac{3}{2}}} \tag{11.24}$$

We have that $c_1 \gg c_2, \phi_1$, therefore the first term is negligible when trying to check the sign of this derivative. Given that $x, y > 0$, $c_2 > x$ and $x \leq \phi_1$, the second term is an always positive increasing function. For a fixed value of y, the summation of the third and forth term is an always negative increasing function. Therefore (11.24) is a monotonically increasing function, as it is the summation of two increasing functions. □

Proof of Theorem 11.8. According to Lemma 11.7 the second derivative of (11.16) is monotonically increasing. Therefore we can only have three cases:

$$\begin{cases} \frac{\partial^2 C^1(x)}{\partial x^2} < 0\ \forall x & \Rightarrow C^1 \text{ always concave} \\ \frac{\partial^2 C^1(x)}{\partial x^2} > 0\ \forall x & \Rightarrow C^1 \text{ always convex} \\ \exists! \bar{x}\ s.t.\ \left.\frac{\partial^2 C^1(x)}{\partial x^2}\right|_{x=\bar{x}} = 0 & \Rightarrow C^1 \text{ concave in } [0; \bar{x}] \\ & \quad \text{and convex in } [\bar{x}; c_2] \end{cases}$$

□

Proof of Theorem 11.9. The strategy space is $x \in [0, \min(c_2, \phi_1)[$ and $y \in [0, \min(c_2, \phi_1)[$. Within this set, C is continuous, therefore, by Weierstrass' theorem, it has a global minimum, which might be either a stationary point or one of the interval endpoints. In order to find all the stationary points, we study the function gradient:

$$\begin{cases} \frac{\partial C(x,y)}{\partial x} = -a_1 - \frac{c_1}{(c_1-(\phi_1-x))^2} + \frac{a_2}{2\sqrt{x+y}} + \frac{c_2}{(c_2-x)^2} \\ \frac{\partial C(x,y)}{\partial y} = -a_1 - \frac{c_1}{(c_1-(\phi_2-y))^2} + \frac{a_2}{2\sqrt{x+y}} + \frac{c_2}{(c_2-y)^2} \end{cases} \tag{11.25}$$

By hypothesis we have symmetric demands: $\phi_1 = \phi_2 = \phi$. If we sum the two equations in (11.25) we have that:

$$-\frac{c_1}{(c_1-(\phi-x))^2} + \frac{c_2}{(c_2-x)^2} = -\frac{c_1}{(c_1-(\phi-y))^2} + \frac{c_2}{(c_2-y)^2}$$

which clearly is true only for symmetric strategies, therefore we must have $x = y$.

The capacity of the NSP is typically $c_1 \gg \phi, x$ therefore we can simplify $\frac{c_1}{(c_1-(\phi-x))^2} \approx \frac{1}{c_1}$. In order to find the stationary points we need to find the roots of equation:

$$-a_1 - \frac{1}{c_1} + \frac{a_2}{2\sqrt{2x}} + \frac{c_2}{(c_2-x)^2} \tag{11.26}$$

Unfortunately, this is a fifth-degree polynomial, therefore we don't have an explicit solution. Consider now the derivative of Eq. (11.26):

$$-\frac{a_2\sqrt{2}}{8x\sqrt{x}} + \frac{2c_2}{(c_2-x)^3} \tag{11.27}$$

This is a monotonically increasing function (as it is the sum of two increasing functions), that goes to $-\infty$ for $x \to 0$ and to $+\infty$ for $x \to c_2$, therefore it has one root. As a consequence, we know that Eq. (11.26), representing the first derivative of C, is convex and has limits $+\infty$ for $x \to 0, c_2$. Therefore, it has a unique minimum point corresponding to the root of equation (11.27).

Given the form of its first derivative (11.26), the cost function (11.17) is concave in its first part (where the derivative decreases) and convex in the second part (where the derivative increases). The points of minimum of the concave part are its two endpoints, while the convex part has a unique minimum point. The right endpoint of the concave part is inside the convex part (C is continuous), therefore the convex

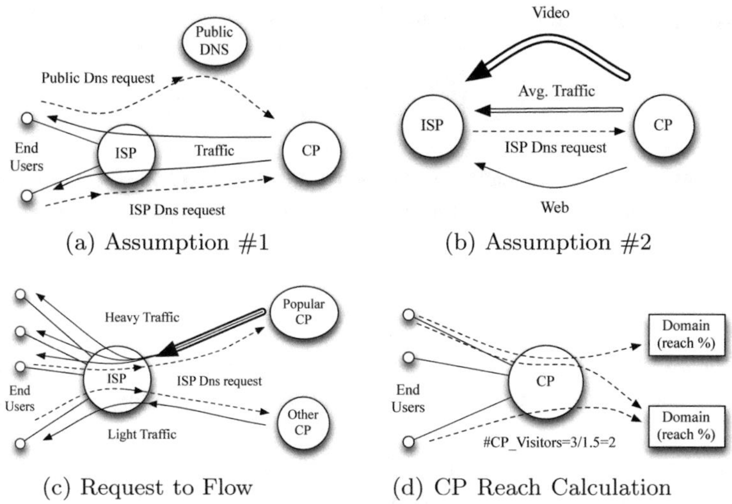

Fig. 11.16 DNS to traffic: assumptions and calculations

minimum is an improvement over it. Henceforth, the global minimum point is either the minimum point of the convex part, or the left endpoint of the concave part, which is also the left endpoint for $C(x, y)$. □

Proof of Corollary 11.10. Given such an algorithm, we execute it on the function giving as initial point the right endpoint of the strategy set, where we know that the function is convex. The output of the algorithm is the local minimum of the convex part, therefore according to Theorem 11.10 the global minimum point is either this point or the left endpoint. □

11.8.2 MCM Cost Function Derivation

Consider the model in Fig. 11.3. First of all, given that the topology is fully connected, meaning that $b_l^n = 0 \; \forall l, n$, we can erase all those terms. Moreover, we have $I = 2,\ N = 1,\ L = 2$, where $l = 1$ is the NSP and $l = 2$ is the NSP, therefore we can explicitly rewrite the general cost function (11.11) for the two players as:

$$\begin{cases} C^1(\mathbf{x_i}, \mathbf{x_{-i}}) & = x_{1,1}^1 \left(a_1 + \frac{1}{c_1 - x_{1,1}^1} \right) + \\ & + x_{1,2}^1 \left(\frac{a_2}{\sqrt{x_{1,2}^1 + x_{2,2}^1}} + \frac{1}{c_2 - x_{1,2}^1} \right) \\ C^2(\mathbf{x_i}, \mathbf{x_{-i}}) & = x_{2,1}^1 \left(a_1 + \frac{1}{c_1 - x_{2,1}^1} \right) + \\ & + x_{2,2}^1 \left(\frac{a_2}{\sqrt{x_{1,2}^1 + x_{2,2}^1}} + \frac{1}{c_2 - x_{2,2}^1} \right) \end{cases} \tag{11.28}$$

Constraints shown in (11.5) can also be rewritten as:

$$\begin{cases} x^1_{1,1} + x^1_{1,2} = \phi_1 \\ x^1_{2,1} + x^1_{2,2} = \phi_2 \end{cases} \tag{11.29}$$

where ϕ_1 and ϕ_2 are the demands from player 1 and 2 to the CP, respectively. For the sake of readability, we apply the following variable renaming to our problem:

$$\begin{cases} x = x^1_{1,2} \\ y = x^1_{2,2} \end{cases} \tag{11.30}$$

Meaning that x is the flow sent by player 1 to CP 1 (the only one present) through the IXP, while y is the analogous for player 2. Due to constraints (11.29) we have that:

$$\begin{cases} x^1_{1,1} = \phi_1 - x \\ x^1_{2,1} = \phi_2 - y \end{cases} \tag{11.31}$$

By substituting in (11.28) the variables defined in (11.30) and the ones obtained in (11.31), we obtain the final form of the cost functions for the two players, shown in (11.15).

11.8.3 Non Convergence Case-Study

Consider a system with fully connected topology, $I = 16$ ISPs, $L = 3$ TFs (one NSP and two symmetric IXPs) and $N = 1$ CP. Once again we use cost parameters $a_{NSP} = 3$ and $a_{IXP} = 30$. Capacities are $c_{NSP} = 200$ and $c_{IXP} = 70$ and we even take symmetric demands $\phi_i^n = 50\ \forall i, n$. Furthermore, we separate players in two groups: players i with $i \leq 8$ belong to the first group, while players for $i > 8$ take part into the second group.

According to Corollary 11.6, equilibrium can be reached by following a sequence of symmetric best responses. Instead, we set an asymmetric starting point, depending on the group:

$$\begin{cases} x_{i,l} = 24.9 & if\ (l = 1 \wedge i \leq 8) \vee (l = 2 \wedge i > 8) \\ x_{i,l} = 25.1 & if\ (l = 2 \wedge i \leq 8) \vee (l = 1 \wedge i > 8) \\ x_{i,l} = 0 & otherwise \end{cases}$$

Simulation shows that players never reach an equilibrium, and go through a never-ending oscillation between two points:

$$\begin{cases} x^-_{i,l} = 0\ x^+_{i,l} = 50 & if\ (l = 1 \wedge i \leq 8) \vee (l = 2 \wedge i > 8) \\ x^-_{i,l} = 50\ x^+_{i,l} = 0 & if\ (l = 2 \wedge i \leq 8) \vee (l = 1 \wedge i > 8) \\ x^-_{i,l} = x^+_{i,l} = 0 & otherwise \end{cases}$$

This happens because, on each iteration, each player of the first group sees the second group of players on a different IXP, and finds it beneficial to deviate on that TF. The same happens for the players of the second group, which in turn deviate altogether to the IXP of the first group. After the deviation, situation is reversed, therefore the two groups keep deviating all the time, never reaching an equilibrium. Please note that using a symmetric starting point would immediately lead to an equilibrium where $x_{i,l} = 25$ if l is an IXP, and zero otherwise.

11.8.4 Inferring ITM from DNS Data

Here we briefly summarize our methodology for inferring the Interdomain Traffic Matrix (ITM) by exploiting DNS data [3]. The DNS data we use, was obtained by exploiting the Passive DNS monitoring system developed at IIT-CNR [79] and used to collect data for all the .it domains. The monitoring system produces logs containing all the information regarding each DNS request performed by a client to the DNS server, including the client AS and the query performed. Therefore, the first step consists of building a DNS Requests Matrix. This matrix contains the number of requests performed by a Source AS to a Destination AS. Please note that since we use DNS traffic for inferring this matrix, we are only capturing traffic flowing in client-server environments, while we are discarding peer-to-peer traffic.

The second step consists of transforming the DNS request matrix into an Interdomain Traffic Matrix. The transformation is carried out through 4 steps, which implement a set of assumptions so as to map each DNS request to an amount of traffic generated by a single client. The 4 steps are depicted in Fig. 11.16b, and are reported below:

Step 1 The DNS request of a client can be performed by either its ISP or a Public DNS Server (e.g. Google DNS). While, in the first case, the endpoints of the request will be the same as that of the traffic flow, in the second case this is not true (see Fig. 11.16a). Therefore, when converting DNS requests to Traffic flows, we have to discard all those Source ASes which are not ISPs.

Step 2 The amount of traffic flowing between two ASes due to a DNS request, depends on the application type (e.g. Video, Web, etc..). Since we are not interested in differentiating between flows, we can easily solve this problem by using a constant quantity obtained by averaging the traffic generated by different sources (see Fig. 11.16b). We indicate as *#avg_request_traffic* the average traffic generated by a request from a single user. This can be estimated by averaging the amount of data downloaded from several webpages, with either text, video, etc.

Step 3 In order to transform a DNS request into a traffic flow, we have to consider that each DNS request performed by a client to the DNS server is cached at the ISP, and potentially used for all the other clients. Moreover, the number of clients interested in the same content, depends on the popularity of the content itself (see Fig. 11.16c).

More precisely, we can represent the relationship between a request between source and destination AS and the number of users that can benefit from it as:

$$\#ISP\_Users * \frac{\#Content\_Visitors}{\#Total\_Visitors}$$

The $\#ISP\_Users$ is calculated, for each ISP, by computing the ratio between the IP address space of each ISP and the total IP space; afterwards we estimate the user population by multiplying this quantity by the total population of the country.

Step 4 The quantity $\frac{\#Content\_Visitors}{\#Total\_Visitors}$ is also known as *Reach* (%), and has been measured for several websites by Alexa [9]. It basically expresses the popularity of a given content, with respect to all other contents. While the metric has been measured for websites, we need to convert it to ASes, since we are interested in interdomain traffic. In order to do so, we: (i) discard all Destination ASes which are not Content Providers hosting websites ranked by Alexa, since these CPs are the only ones for which we are able to estimate the *Reach* metric; (ii) calculate the *Reach* of a CP as the the sum of all *Reaches* of its websites, normalized by the average number of requests of the CP clients. To understand this formula, we shall look at Fig. 11.16d. In the Figure, there is a CP hosting two websites and 2 clients making requests. As we see, there is one visitor for the first domain, and two visitors for the second domain, therefore, if we simply sum the $\#Content\_Visitors$ for the CP, we obtain 3, while the users effectively accessing the CP are only 2. However, if we normalize this quantity by the average number of requests of the CP clients ($(2 + 1)/2 = 1.5$ in the example), we can correct this overestimation.

Algorithm 11.2 Traffic Generator

```
// Input Data
DNS_Req = < Source_AS, Dest_AS, #Requests >
AS_population =< ISP, #Users >
AS_reach = < CP, Reach % >
// Output Data
Traffic = < Source_AS, Dest_AS, Flow >

avg_request_traffic = ...                          ▷ Average Traffic Generated
                                                     by a Request (Single User)
for all Row ∈ DNS_Req do
    src = Row[Source_AS]
    dst = Row[Dest_AS]
    req = Row[#Requests]
    if src ∈ ISP && dst ∈ CP then
        Traffic[src, dst] = req * avg_request_traffic *
            * AS_population[src] * AS_reach[dst]
    end if
end for
```

Once we have computed all the input data, the DNS Request Matrix can be transformed into an Interdomain Traffic Matrix. Basically each request between a CP and an ISP is transformed to a flow quantity equal to:

$$\#avg\_request\_traffic * ISP\_Users[ISP] * Reach[CP]$$

The final conversion procedure is described in Algorithm 11.2. For more information on the methodology used for inferring the ITM by exploiting DNS data, and on the results obtained through it, refer to [3].

Chapter 12
A Net Neutrality Perspective for Content Down-Streaming over the Internet

Alexandre Reiffers-Masson, Yezekael Hayel, Eitan Altman and Tania Jimenez

12.1 Introduction

The *Network Neutrality* issue has been at the center of debate worldwide lately. Some countries have established laws so that principles of network neutrality are respected. Among the questions that have been discussed in these debates there is whether to allow agreements between service and content providers, i.e. to allow some preferential treatment by an operator to traffic from some providers (identity-based discrimination).

Network neutrality is an approach of providing network access without unfair discrimination between applications, nor between content, nor between the specific source of the traffic. Hahn and Wallsten [120] wrote that net neutrality "usually means that broadband service providers charge consumers only once for Internet access, do not favor one content provider over another, and do not charge content providers for sending information over broadband lines to end users." There are two applications or services or providers that require the same network resources, and one is offered a better quality of service (delays, speed, etc.) or is cheaper to access, then there is discrimination. Our study in this paper is related to the latter point but taking into account also the quality of service for the customers. Historically, the neutrality of the access to the Internet has characterized the first steps of the development of the Internet and much of the industrial activity that uses the Internet. Alternative non-neutral approaches have been recently pushed forward by Internet Service Providers (ISPs), content providers (CPs) and by equipment providers (EPs). Deviating from its original neutral character may have far-reaching consequences on the whole e-economy and on the society in which the Internet has become so central.

A. Reiffers-Masson · Y. Hayel · E. Altman (✉) · T. Jimenez
University o Cote d'Azur, INRIA,2004 Route des Lucioles, 06902 Sophia Antipolis, France
e-mail: eitan.altman@inria.fr

E. Altman
Lincs, 23 Avenue d'Italie,Paris, France

E. Altman et al. (eds.), *Multilevel Strategic Interaction Game Models for Complex Networks*, https://doi.org/10.1007/978-3-030-24455-2_12

This has pushed many countries to take regulation actions to determinate the future characteristics of the Internet and some of which have been followed by legislation on the matter. A key step in the policy making has been the launching of public consultations in various countries (USA, UK, France and others) as well as at the EU level. All society sectors have been invited to answer these consultations. The global objectives in these consultations are related to this net neutrality debate and have the following two major goals:

- Contribute to the network neutrality debate in proposing a new economic model to compare the expected outcome of non-neutral approaches with neutral ones in terms of the quality of service offered.
- Study competition aspects related to network neutrality and particularly, the exclusive contracts between CPs with ISPs.

Nowadays, the users have access to several CPs that provide the same content. For example, CPs like Netflix®, Hulu®, M-GO, etc. provide video content and CPs like Deezer®, iTunes® and Spotify® provide music content. Also, the CPs have different pricing policies such as flat-rate or pay-as-you-go, or options like buy a song or a full Long Play (LP). Our goal in this paper is to analyze the impact of non-neutral pricing policies introduced through agreements betweens the ISPs and the CPs on the network economic system. In fact, the decision about from which CP the users request contents is based on two main features: **price** and **quality**.

In the Internet, the access price and the content price are set by the ISPs and CPs, respectively, and access prices are set independently of the content price. In our framework, CPs competitively set content price to maximize their profit. Recently, CPs and ISPs introduced a new pricing policy in order to play a more important role in this techno-economical market. The pricing policy is based on the exclusive contracts between the ISPs and the CPs. In such contracts, the subscribers get exclusive preferential access to the content of a CP that is in agreement with the ISP which provides them last mile connectivity. Such pricing policies induces a non-neutral Internet, as the end users have different access charges for the same content from the same CP; i.e., such pricing schemes lead to identity based discrimination. A particular example of such non-neutral pricing behavior happened recently in Europe. Orange® is a French ISP, and Deezer® is a French music streaming service provider (a CP). According to the Financial Times[1]: "As part of the deal with Deezer®, Orange® will make available a special mobile-only tariff for pay-monthly customers, to avoid the \$9.99 standalone cost of Deezers top package." Therefore, customers with an Internet subscription with Orange will have preferential offer for listening music on the website Deezer®. This type of non-neutral collusion between a CP and an ISP arises in streaming services offered over the Internet, which accounts for half of the global internet traffic (see Fig. 12.2). Another example, in the USA, is related to Apps/softwares where several CPs (if we see an App as a content) are also competing over the Internet. In fact, Google® has an agreement with Verizon®, a wireless

[1] Tim Bradshaw, *Deezer takes on Spotify with Orange deal*, ft.com, September 7, 2011.

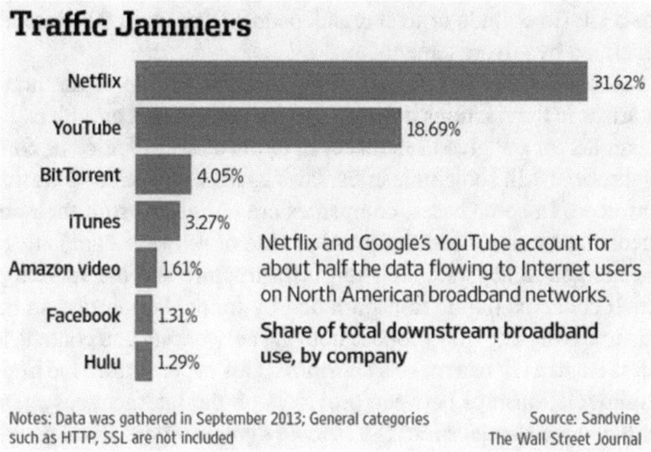

Fig. 12.1 The large share of data flow to internet users from CP streaming service like Netflix and YouTube [2]

ISP. This partnership is expressed in the form of three free user accounts for Google Apps[2] to the Verizon® customers.

Apart from the revenues from selling content, the CPs earn revenue also through advertisements, and the amount of advertisement revenue depends on the level of activity or the demand for their contents. The latter case is related to the quality of the content provided by a CP, which is assumed to be related to the number of subscribers asking/downloading contents. We consider this interaction between subscribers as a routing game framework. Therefore, each Internet user decides the way to split their demand among several content providers. Evaluating the quality of service (or experience) in communication networks is not an easy task. Many papers have focused on this objective, and evaluation of the quality perceived by customers in a network neutrality model is non-trivial. Our approach considers a routing game framework for modeling the interactions between the subscribers downstream flows. Routing games provide a natural framework to model interactions among the users and characterize the quality of service perceived by them at different CPs as a function of the content price.

We model this complex interacting system within a two-sided market framework as a multistage game model composed of a congestion game at the lower level and a noncooperative pricing game at the upper level. It is natural to expect that exclusive contracts between the CPs and the ISPs can have different impacts on the equilibrium of the multistage game and therefore on the costs and profits of the different players. We would like to analyze these aspects of non-neutrality through a rigorous mathematical framework. We would also like to analyze the collusion

[2]Google Apps for Verizon: *Google Apps for Business now available for Verizon customers*, 01/24/2011 posted by Monte Beck, Vice President of Small Business Marketing, Verizon.

decision for a CP (to collude or to stay independent) and also take into account the revenue generated by advertisements.

In this paper, we study the impact of a particular feature of the non-neutrality debate that arises in the relations between service and content providers, i.e. the possibility that an ISP or CP give preference, in terms of access price or content price, to certain subscribers. In some industries, laws against such kind of vertical monopolies are enforced. In some cases, companies are obliged to split their activity into separate specialized companies; this was the case of railways companies in Europe which were obliged to separate their rail infrastructure and the service part of the activity which concerns public transportation by trains. In contrast, in the telecom market, the same company may propose both the networking and content services, or similarly an ISP and a CP can make a *collusion*. This paper studies the implication of such economic relationships between providers on the Internet ecosystem. Specifically, we will address the question: Is a *Pricing agreement/collusion* between an ISP and a CP good for subscribers? We suggest here a novel point of view of a *pricing agreement/collusion*: an issue actively discussed in the ongoing net neutrality debate. Usually in the network neutrality debate the problem of agreement or disagreement between ISP and CP is a vertical foreclosure (Degradation of traffic) [69]. This type of problem has been observed in France between Free (a French ISP) and Google.[3] In our framework, if a CP and an ISP have a *pricing agreement*, then a subscriber of the ISP mentioned above does not have to pay for the access to this CP's content.

Our main contribution in this paper is to model and analyze the effect of new pricing policies proposed by some of the ISPs and the CPs that enter into exclusive contracts. We first consider ISPs as passive players that connect CPs to end users in a two-sided market model, and study the behavior of CPs and the end users in the new pricing regime using a multi-stage game framework. The subscribers interact through their downstream traffic carrying their content. Their behavior is influenced by the congestion that occurs at the CP-side in the access link between CPs and ISPs. The CPs decisions are influenced by the "price war" as most of them sell the same kind of contents (multimedia contents as movies, LPs, popular TV shows, etc.), and also to attract more users as this will increase their advertisements revenues. A non-neutral aspect of the access to contents results from the exclusive contracts and therefore we are interested to study the impact of such contracts between CPs and ISPs on the equilibrium performances of the market.

The main contributions of this work are:

1. In the two-sided market framework, we consider a hierarchical game in which the higher level is a normal form non-cooperative game between the CPs (content price), and at the lower level, a non-cooperative (non-atomic) routing game between the subscribers (user set demand for content).
2. In the hierarchical game, we first study the sub-game perfect equilibria (SPE) of the routing game between subscribers. Depending on the pricing policies of the CPs and the perceived congestions levels at each CPs, the subscribers compete to minimize the overall delay (download time) and the total cost for the

[3]"An ad-block shock France versus Google", The Economist, January 12, 2013.

content. Results have been obtained and are described in Sect. 12.3.5. We look for existence of a symmetric SPE by studying best-response functions.
3. Based on the SPE of users demands, we study a non-cooperative game between the CPs and determine their content price at equilibrium. In our multi-stage game framework, the CPs compete through their content prices and aim to maximize their revenues from content and advertisements.
4. Assuming that each CP is in collusion with an ISP (inducing a preferential tariff for his content to some specific subscribers), we study the new equilibrium of the multi-stage game. The analysis process is based on backward induction: we first determine the SPE for the subscriber game, and then the Nash equilibrium for the CPs game.
5. Finally, we study another decision step for the CPs—whether or not to collude with an ISP. CPs decide first whether to collude with an ISP or not, and then set the content price. Such a decision by a CP impacts other CPs' decisions to collude or not and also their content prices.

12.2 Related Works

Several recent papers in the literature deal with game-theoretic models for network neutrality analysis. The survey article [13] provides summary of various issues discussed in the network neutrality debate. The authors describe the models used to analyze various issues in the net neutrality debate and compare the results. In the following section, we discuss the literature relevant to the proposed research.

The two-sided model proposed in [86] investigates the effect of network neutrality regulation in both monopoly and duopoly setting between ISPs. The authors show that neutrality increases social welfare in the duopoly case. However, investment decisions and congestion effects are not taken into account into the model. Our two-sided market model is similar to [184] by considering fixed number of CPs and ISPs. The authors compare the return on investment under one-sided or two-sided pricing, and they show that this amount is comparable. The congestion effect is not considered in this paper. Investment mechanisms in a two-sided market has been proposed in [192]. The model consists of two interconnected ISPs represented as profit maximizing firms that choose quality investment levels and then compete in prices for both CPs and consumers. The authors consider a large number of CPs and consumers. The revenue of the CPs come from advertising only. In our framework, we also consider the revenue generated form content selling, which is important for multimedia streaming CP. The game-theoretic model proposed is a 6-stage game in which several competitions occur at each level between CPs, ISPs and consumers.

The effect of investments on the quality of services (QoS) is studied in [192]. The authors consider a model where the interactions between CPs and also between consumers come from their choice of ISP, that impacts also their quality of service (QoS). The latter depends on the congestion at each ISP, at the CP side, which is a function of the mass of consumers and CPs connected to it. The investment decision of each ISP determines the QoS in a deterministic manner. The ISP receives payment from both sides—CPs and consumers. The authors consider two scenarios: neutral and non-neutral. They show that investments are higher in the non-neutral regime because it is easier to extract revenue through appropriate CP pricing. Interestingly, the congestion effect is taken into account at the CP level. Specifically, the authors compute the value of the content by dividing the quality level associated to each CP by the mass of CPs that connect to the same platform in order to incorporate congestion effects: more CPs in a platform generate more congestion, reducing value. Then, from this point of view, this paper deals also with a congestion feature that impacts the subscriber behavior as in our framework. But, the important difference from our setting is about the interaction model between subscribers which is based on a routing game.

Models from queuing theory have been used to analyze the effect of investments in [51]. The congestion effect is taken into account as the average sojourn time in an M/M/1 queue, and the investment decision of the ISP affects directly the service rate of this queue. In [186], the authors study the discrimination effect at the service level. Particularly, the authors look at the effects of net neutrality regulation on the investment incentives of ISPs and CPs. The QoS based on a congestion model is also expressed by an M/M/1 queue. Their results are ambiguous, but key effects are exposed. The difficulty is that the average sojourn time in an M/M/1 queue is not linear and then closed-form solutions are usually difficult to obtain. Another model, proposed in [160], is based on a queuing congestion model. The model is related to content analysis, and the authors show that strategic quality degradation and non-neutral ISP reduces content variety. Routing game based model is used in [5] to study a non-cooperative game between subscribers. Subscribers play a non atomic selfish routing game and CPs control both flows and prices, while in our proposal CPs control only prices and subscribers determine the source of the traffic they wish to download from. Another preliminary work [150] is related to exploration of the effects of content-specific pricing, including multiple CPs providing different types of content. But the competition between providers is not considered. The authors have analyzed various theoretical aspect of collisions in routing games [12] and in nonneutral networks [16] [?]. In [12], the authors studied the effect of collusions in routing games and extended the performance metric *price of collusion* [132] to evaluate the effect of collusion on both the colluding and non-colluding players.

12.3 Multistage Game for Two-Sided Market Framework

The general mathematical framework is a multistage game composed of several non-cooperative games at each level. The task description provides technical details on the type of solution concept that will be studied and how to get it. We also describe the role of each player in the multi-stage game coupled with their actions and utilities.

12.3.1 Model

We consider a general economic model of content distribution over the Internet as a two-sided market with competition [20]. In fact, the ISPs provides a platform connecting end-users or subscribers to the Internet Content Providers (CPs). We assume that several CPs are able to distribute the same global content over the different ISPs. If we think about music streaming, most of the same artists are on the different CP like iTunes, Deezer, Spotify, etc. The same remark applies to movies streaming CPs.

We consider a fixed number M of CPs. The streaming traffic is carried through the network by high level ISPs which have direct links to all CP. In practice, when a subscriber requests a specific content from a CP, this streaming flow needs to traverse several different networks. This complex transit of traffic between networks is governed by a variety of agreements between different tier providers. We consider the local monopolistic ISP that provides the last mile connection service to CPs. This type of model assumption is usually considered in two-sided market models like [186] and [86].

Finally, those ISPs are connected to subscribers (as a mass). We consider in our system N different ISPs. Each subscriber is connected to a single ISP and cannot access directly any CP. As each subscriber is connected to a unique ISP, we make the abuse of identifying a subscriber with its ISP. Then we talk about "subscriber n" instead of "subscriber connected to ISP n". We consider in our study the download traffic; contents on the Internet transit mostly from servers to clients (video-on-demand, movie broadcast, etc). The source of each content flow is a CP and the corresponding destination is a subscriber.

Requests about contents are generated from each subscriber n (i.e. subscriber connected to ISP n and reached all CP m. At each CP m, any request from a subscriber n induces a traffic rate (music or videos streaming traffic, downloading, etc.) $x_n^m > 0$, which is therefore aggregated and sent to the subscriber mass connected to ISP n. We assume that the total traffic flow from ISP n to its subscriber is $\phi_n = \sum_{m=1}^{M} x_n^m$. This total traffic called also total demand ϕ_n, is an average value of the total amount of traffic downloading to the mass of subscribers connected to ISP n. The subscriber decision about requesting content from one CP to the other depends on the quality of service, which is expressed here in terms of CP-side congestion. Indeed, on the subscriber side, the ISP can dimension correctly the downstream (from ISP to subscribers) network as the total traffic is fixed. But, at the CP side, the total upstream (from CP to ISP) traffic is not fixed, it is equal to $\sum_{n=1}^{N} x_n^m$ and depends on the number of requests to this CP. Then, we model the CP-side congestion effect coming from the interaction of the demand of all the subscribers to this CP as a non-cooperative routing game. In fact, the congestion level in a link depends on the total traffic trough this link. For example on Fig. 12.1, we observed that the link between CP M and ISP M is more congested compared to the access links of other CPs. In fact, all subscribers connected to ISP N request content to CP M and not any other CP. The traffic rate that goes trough this link is important. Whereas the access link between

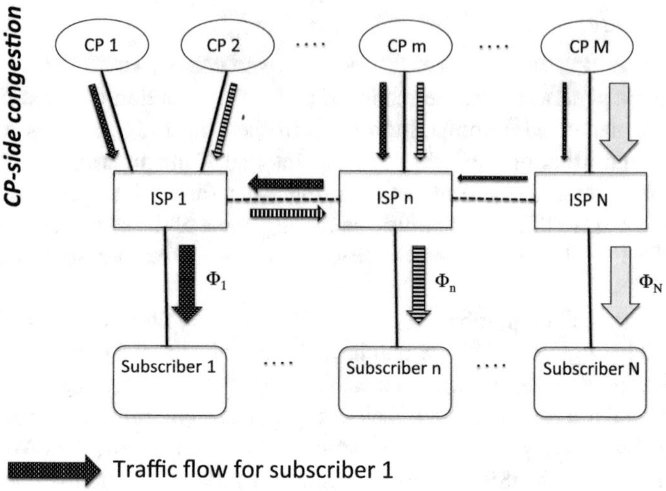

Fig. 12.2 Traffic flows in our two-sided market framework. Congestion occurs at the CP-side level in the last-mile connection service to ISP

CP 1 and ISP 1 is less congested as its traffic is composed only from a small proportion of the subscriber 1 total traffic ϕ_1. Then, the first modeling technique that will be used in this work is ***routing game*** which is related to algorithmic game theory [?] and Wardrop equilibrium concept [285]. Indeed, we consider a noncooperative routing game where the decision of subscriber connected to ISP n, is how to split his download traffic ϕ from all the CP, i.e. the decision variables for subscriber connected to ISP n is the vector $x_n = (x_n^1, x_n^2, \ldots, x_n^M)$.

Subscriber costs: We denote by p^m the charge, per unit of traffic, that a subscriber has to pay in order to download traffic from CP m. Then, for a traffic quantity x_n^m, the subscriber n has to pay $x_n^m p^m$ to the CP m. We consider that a subscriber prefers to download his content from the less crowded CP, due to congestion cost effect at the CP link. This congestion cost depends on the total download traffic generated at each CP m, that is $\sum_n x_n^m$. Let $D^m : \mathbb{R}^+ \mapsto \mathbb{R}^+$ be the congestion cost function at CP m (exactly the congestion is suffered on the access link between CP m and its ISP) which we assume to be convex and increasing. The congestion cost perceived by a subscriber n who downloads a traffic rate x_n^m from CP m is equal to $x_n^m \cdot D^m(\sum_n x_n^m)$.

Then, the total cost (content price + congestion cost) for a subscriber connected to ISP n is given by:

$$C_n(x_n, x_{-n}, p) = \sum_m x_n^m \left[D^m(\sum_n x_n^m) + p^m \right], \tag{12.1}$$

where $x_n = \{x_n^1, \ldots, x_n^M\}$ is the decision vector for the subscriber connected to ISP n, x_{-n} is the decision vectors of all the other subscribers connected to other ISPs and $p = \{p^1, \ldots, p^M\}$ is the content price vector of all the CP. This latter price is expressed per unit rate. Each subscriber n will minimize his cost function under his demand constraint:

$$\min_{x_n} C_n(x_n, x_{-n}, p) \quad \text{such that} \quad \sum_{m=1}^{N} x_n^m = \phi_n.$$

CP Profits: The revenue of CP m is defined by:

$$\Pi(x, p^m, p^{-m}) = p^m \sum_n x_n^m,$$

where x_n^m is the traffic flow from CP m to ISP n, p^m is the price charged by CP m, p^{-m} is the price vector of all the other CP. Each CP determines his content price p^m in order to maximize his profit, taking into account the traffic flow (demand) requested by subscribers. Then, this problem can be solved by considering a ***multistage game***.

Multistage game: The players in our two-sided framework are the CPs and the subscribers. The multistage game we study consists of the following stages:

1. Content prices decisions: All CP determine simultaneously their content price p_m.
2. Demand splitting decisions: All subscriber determine simultaneously their downloading streaming rate x_n^m from each CP m.

We can see that for the moment the ISP plays the role of a platform between the CPs and the subscribers, and that they are inactive in the economic market studied. We plan to introduce active ISP as decision makers in the next step of this proposal. We solve this multistage game by considering sub game perfect Nash equilibrium and we use a backward induction technique. We assume that the CPs update/change their prices at a slower timescale compared to the decision request of subscribers. Then, the routing game between the subscribers is solved by considering the content prices as fixed. That is why we consider the previous stages ordering. Moreover, the multistage game is closely related to an Equilibrium Problem with Equilibrium Constraints (EPEC) [251] and can be written as follows:

$$\forall m \in \{1, \ldots, M\}, \quad \max_{p^m} p^m \sum_n \underline{x}_n^m(\mathbf{p}), \quad \text{(Leader)}$$

such that

$$\forall n \in \{1, \ldots, N\}, \quad \underline{x}_n^m(\mathbf{p}) \in \arg\min_{x_n} C_n(x_n, x_{-n}, p). \quad \text{(Follower)}$$

Generally speaking, this type of multistage game is not trivial to study but closed-form solutions can be obtained when the game is symmetric. We thus assume that the decision variable of each CP m is in an interval, i.e., $p^m \in [0, p_{\max}]$. The system is totally symmetric, in the sense that the quantity of traffic ϕ is the same for all subscribers n, and congestion cost functions do not depend on m. In all our mathematical analysis we consider a linear congestion cost function $D(x) = ax$ as in [5]. Based on this symmetry property of the game, we can use results from [149] and assume the existence of a symmetric equilibrium for our multistage game. First, this symmetric assumption can be justified by the fact that in a large network, we can approximate the behavior of many subscribers with only one subscriber which has the average characteristics of all the subscribers. Secondly, this assumption allows us to obtain explicit formulations of the equilibrium of our multistage game with a routing game as constraint.

12.3.2 Neutral Scenario

The neutral scenario denotes the case in which each CP decides independently its content price. In other words, ISPs do not influence the content price decision of the CP. Therefore, there is no price discrimination between subscribers in order for them to access to the content of different CPs. In this case, we have the following preliminary result.

Theorem 12.31 *For all CP m, there exists a unique symmetric equilibrium* $(x_n^m, p^m) = (\underline{x}, \underline{p})$ *of our multistage game, given by* $\forall (n, m) \in \{1, \ldots, N\} \times \{1, \ldots, M\}$*:*

$$\underline{x}_n^m = \frac{\phi}{M} \quad \text{and} \quad \underline{p} = (N-1)\phi a.$$

We observe that due to the competition between the subscribers and also between the CPs, the total downstream rate of each subscriber is equally splitte between all the CPs. Also, based on this result, we are able to determine the cost of each subscriber and the revenue of each CP at the equilibrium situation of our multistage game. In fact, this proposition gives the equilibrium prices and the value of the total traffic generated by each CP at equilibrium. Note that we have uniqueness of this total traffic at equilibrium:

$$\forall m, \ \sum_{n=1}^{N} \underline{x}_n^m = \frac{N\phi}{M}.$$

The cost for a subscriber connected to CP n at the equilibrium is given by:

$$C_n(\underline{x}, \underline{p}) = \phi^2 a(N - 1 + \frac{N}{M}).$$

The revenue for any CP m is:

$$\Pi_m(\underline{x}, \underline{p}) = \frac{(N-1)N}{M}\phi^2 a.$$

12.3.3 Non-neutral Scenario

We consider in this second scenario that each ISP n makes an agreement with a CP. Then, we assume that the number of ISP is equal to the number of CP, i.e. $M = N$. In order to simplify the notations, n is the index of the CP which has an agreement with ISP n. These agreements or collusions, imply that the content price p^n is equal to 0 for the traffic generated from the CP n to the ISP n. Then, the total cost for the subscriber connected to ISP n becomes:

$$C_n^v(x_n, x_{-n}, p) = \sum_{m \neq n} x_n^m \left[D(\sum_n x_n^m) + p^m \right] + x_n^n D(\sum_n x_n^n),$$

where $\mathbf{p}$ is the vector (size $N-1$) of the content prices for all CP except n. The revenue of the CP m is now:

$$\Pi^m(p^m, p^{-m}) = p^m \sum_{n \neq m} \underline{x}_n^m(p).$$

Then, we have now a similar multistage game in which the cost function of the subscribers is slightly modified. The sub game perfect equilibrium solution of the routing game between the subscribers is no more symmetric and we define the following variables. Let y_n^n be the quantity of traffic requested by a subscriber connected to an ISP n from the CP n associated to its ISP, and y_n^m the download traffic requested by a subscriber connected to an ISP n from CP m, with $m \neq n$.

Theorem 12.32 *For all CP m, there exists for all $(i, n, m) \in \{1, \ldots, I\} \times \{1, \ldots, N\}^2$ a symmetric equilibrium $(y_n^m, y_n^n, p^m) = (\underline{z}, \underline{y}, \underline{q})$, which is given by:*

$$\underline{q} = a\phi \frac{(N+1)}{3N-1}, \quad \underline{z} = \frac{\phi}{N}(\frac{2N-2}{3N-1}) \quad \text{and}$$

$$\underline{y} = \frac{\phi}{N}(1 + \frac{(N-1)(N+1)}{3N-1}).$$

We observe that a subscriber does not download all its content from the CP which has an agreement with its ISP. In fact, part of its downstream traffic will be from other CPs. We are now able to express the cost for the subscriber connected to ISP n, at the equilibrium, is given by:

$$C_n^v(\underline{y}, \underline{z}, \underline{q}) = \phi^2 a + 2a\phi^2(\frac{N-1}{3N-1})^2(\frac{N+1}{N}).$$

The reward for CP m at the equilibrium is:

$$\Pi_m^v(\underline{q}, \underline{y}, \underline{z}) = 2a\phi^2(\frac{N-1}{3N-1})^2(\frac{N+1}{N}).$$

The first remark is that in our context of interaction between subscribers, their optimal decision implies that each subscriber downloads part of his demand from other CPs than the privileged one, even if they have to pay for. Another remark is that the download traffic from the privileged CP, $\underline{y}$, has a bounded limit of $\frac{\phi}{3}$ when the number of provider N tends to infinite. In the neutral context, all the download rates converge to 0. Thus, it means that by making agreement or collusion with an ISP, each CP has a minimum quantity guarantee of traffic to send. It is an important result for dimensioning CP network infrastructure and also when considering the advertising incomes. This part will be considered in the future works section.

12.3.4 Comparison Between Neutral and Non-neutral

The previous mathematical results help us to study the effects of the agreement between CPs and ISPs. We are able to determine which scenario (the neutral or non-neutral one) induces lower costs for subscribers. The next theorem proves that even if the neutral scenario implies symmetry and free will for the subscribers about the CP, as they pay the same price for accessing the content, the non-neutral scenario is even better for the subscribers.

Theorem 12.33 *Let us assume that $M = N$. At equilibrium, the agreement between service and content providers is costless for the subscribers, i.e., for all subscribers connected to ISP n we have:*

$$C_n^v(\underline{y}, \underline{z}, \underline{q}) < C_n(\underline{x}, \underline{p}). \tag{12.2}$$

This result seems counterintuitive, in fact, agreements between companies are usually not allowed by government of several countries in order to protect consumers. However, according to our theorem, in our setting, such agreements between CPs and ISPs are not harmful for the subscribers and are even better in terms of costs.

12.3.5 A Competition over Agreement Between CPs

In the past section we have imposed economic agreements between ISP and CP. The scenario was the following, CPs set prices that subscribers have to pay, but not the economic topology of the network (in other words if there are pricing agreements or not between CPs and ISPs). We propose now to let each CP decides by himself to make an agreement (strategy A) with an ISP or not (strategy NA). We prove that the strategy vector $(NA, \ldots, NA)$ is a pure Nash equilibrium. In order to prove this result, we first assume that all the subscribers except one, called n', play NA. We will prove that n' has no interest in not playing NA. However to do so, we need to compute the utility of this CP n' if he plays A. The next proposition gives his utility when he plays A and the others play NA.

Proposition 12.27 *The utility of the CP n' when he makes an agreement and all the other CP do not, is given by:*

$$\Pi'_n(\underline{p}_{n'}, \underline{p}_{-n'}, \underline{x}_{m'}, \underline{x}_m) = \frac{a\phi(N+1)(2N^2+N-1)}{(4N^3-5N^2+2N-1)}$$
$$+\frac{a\phi(N-1)(2N^2+N-1)}{(N+1)(4N^3-5N^2+2N-1)} \times \left(\frac{(4N^2+5N+1)}{4N^2-N+1} - \frac{2(2N^3+N^2-N)}{4N^3-5N^2+2N-1}\right)$$

With the help of the previous proposition, we have the next theorem:

Theorem 12.34 *If $\phi > 1$ the utility of the player m is lower when he plays NA than when he plays A, i.e.*

$$\Pi'_m(\underline{p}_{m'}, \underline{p}_m, \underline{x}_{m'}, \underline{p}_{m'}) < \Pi_m(\underline{p}, \underline{x}).$$

This first result implies that $(NA, \ldots, NA)$ is a nash equilibrium.

This last theorem reduces the importance of the agreement. Indeed, we recall that according to Theorem 12.33, agreements are beneficial for subscribers. However, the last Theorem 12.35 proves that CPs are not interested in making such agreements in a selfish situation. Therefore, policy makers may want to force agreements by, for example, designing new rules and then reduce the cost for the subscribers.

12.4 A Competition over Agreement Between CPs

In the past section we have imposed economic agreements between ISP and CP. The scenario was the following, CPs set prices that subscribers have to pay, but not the economic topology of the network (in other words if there are pricing agreements or not between CPs and ISPs). We propose now to let each CP decides by himself to make an agreement (strategy A) with an ISP or not (strategy NA). We prove that the strategy vector $(NA, \ldots, NA)$ is a pure Nash equilibrium. In order to prove this result, we first assume that all the subscribers except one, called n', play NA. We will prove that n' has no interest in not playing NA. However to do so, we need to compute the utility of this CP n' if he plays A. The next proposition gives his utility when he plays A and the others play NA.

Proposition 12.28 *The utility of the CP n' when he makes an agreement and all the other CP do not, is given by:*

$$\Pi'_n(\underline{p}_{n'}, \underline{p}_{-n'}, \underline{x}_{m'}, \underline{x}_m) = \frac{a\phi(N+1)(2N^2+N-1)}{(4N^3-5N^2+2N-1)}$$
$$+\frac{a\phi(N-1)(2N^2+N-1)}{(N+1)(4N^3-5N^2+2N-1)} \times \left(\frac{(4N^2+5N+1)}{4N^2-N+1} - \frac{2(2N^3+N^2-N)}{4N^3-5N^2+2N-1}\right)$$

With the help of the previous proposition, we have the next theorem:

Theorem 12.35 *If $\phi > 1$ the utility of the player m is lower when he plays NA than when he plays A.*

$$\Pi'_m(\underline{p}_{m'}, \underline{p}_m, \underline{x}_{m'}, \underline{p}_{m'}) < \Pi_m(\underline{p}, \underline{x})$$

This first result implies that $(NA, \ldots, NA)$ is a nash equilibrium.

The previous theorem minimize the importance of agreement. Indeed, we recall that according to Theorem 3, agreements is more interting for subscribers that a neutral situation. However, Theorem 4 prove that agreements it is not a good scenario for the CP and so finally it emeges that the neutral situation will emerge.

12.5 Conclusions and Perspectives

From the research that has been carried out in this chapter, it is possible to conclude that agreement between CPs and IPSs is not an harmful situation for the subscribers. To reach this conclusion, we propose a theoretical model that captures two scenarios (depending on the parameters), a neutral and a non-neutral one. We were able to find mathematical closed form expression of the Nash equilibrium in each scenario and we prove mathematically that an agreement between an ISP and a CP induces a lower cost for subscribers than a neutral situation when such agreements are not possible. The second outcome of our chapter is that CPs will not make agreements selfishly, and so incentives from the government is necessary in order to improve the welfare of the subscribers. We prove that the situation where each CP does not make an agreement is a pure Nash Equilibrium. The findings of the chapter suggest that our approach and results could be useful for government and policy makers for the design of new regulation rules in the telecom market. More research into the generalization of our model is still necessary before obtaining a definitive answer to whether or not agreements between CPs and IPSs are beneficial for the subscribers. A first generalization could be to consider more general topology and asymmetric scenarios.

Appendix

Proof of Theorem 12.31 First let

$$L_n(\mathbf{x_n}, \mathbf{x_{-n}}, \mathbf{p}, \lambda_n) =$$

$$\sum_m x_n^m \left[D^m(\sum_n x_n^m) + p^m \right] - \lambda_n(\sum_m x_n^m - \phi)$$

the Lagrangian function associated to the cost function $C_n(\cdot)$. We look for a symmetric equilibrium between the CPs, i.e. for the noncooperative pricing game at the upper layer. Then we assume that CPs $m' \in \{1, \ldots, M\} - \{m\}$ play q and one CP, say m, plays p^m. We want to find some q where the best reply of CP m against q is q. First, we have to determine the equilibrium flows between the subscribers, depending on those prices, i.e. $\underline{\mathbf{x}}(p^m, q)$ for all p^m and q. We look for $\underline{\mathbf{x}}(p^m, q)$, a solution of the following system:

$$\begin{cases} \dfrac{\partial L_n}{\partial x_n^m}(\mathbf{x_n}, \mathbf{x_{,-n}}, p^m, q, \lambda_n) = 0 \\ \sum_m x_n^m = \phi, \ \forall (n, m) \end{cases}.$$

This game has a strong symmetric property as all subscribers are interchangeable. Then, we can restrict ourselves to two strategies x and y where x is a request for CP m' and y is for CP m. This induces a great simplification in the analysis of our complex hierarchical game. Thanks to [149], the previous system is equivalent to the following one:

$$\begin{cases} x\dfrac{\partial D}{\partial x}(Nx) + D(Nx) + q - \lambda = 0 \\ \\ y\dfrac{\partial D}{\partial y}(Ny) + D(Ny) + p^m - \lambda = 0 \\ \\ (M-1)x + y = \phi. \end{cases}$$

We denote by $\underline{x}$ and $\underline{y}$ a solution of this system. We consider the linear cost function $D(x) = ax$ and thus we have to solve the following linear system:

$$\begin{cases} ax + a(Nx) + q - \lambda = 0 \\ ay + a(Ny) + p^m - \lambda = 0 \\ (M-1)x + y = \phi. \end{cases}$$

If $\underline{x} < 0$ (or $\underline{y} < 0$) then $\underline{x} = 0$ (or $\underline{y} = 0$). And if $\underline{x} > \phi$ (or $\underline{y} > \phi$) then $\underline{x} = \phi$ (or $\underline{y} = \phi$).

Let's now consider the CP m and how it's going to optimize its revenue, which is the function $R^m(p^m, q) = p^m \underline{y} N$. Its best reply against all other CPs that play q is given by p^m solution of $\frac{\partial R^m}{\partial p^m}(p^m, q) = 0$. We have to find a certain p which is a solution of $\frac{\partial R^m}{\partial p^m}(p, p) = 0$. We denote this equilibrium by $\underline{p}$. Considering the linear cost function, we obtain:

$$\underline{p} = (N-1)\phi a.$$

If $\underline{p} > p_{max}$ then $\underline{p} = p_{max}$. We have a particular interest in the case where $p_{max} > (N-1)\phi a$. Then the equilibrium flow from CP n to ISP m is $\underline{x}_n^m = \frac{\phi}{M}$.

Proof of Theorem 12.32 In order to compute an equilibrium for the game with agreements, we can use the method described previously.

We consider the following Lagrangian function:

$$L_n^v(\mathbf{x_n}, \mathbf{x_{-n}}, \mathbf{p}, \lambda_n) =$$

$$\sum_{m \neq n} x_n^m \left[D^m\left(\sum_n x_n^m\right) + p^m \right] + x_n^n [D(n) x_n^n] - \lambda_n \left(\sum_m x_n^m - \phi\right).$$

As previously, we assume that CPs $m' \in \{1, \ldots, N\} - \{m\}$ play q and the CP m plays p^m. Now again there are several symmetries: we can see that there are two types of subscribers. Each subscriber of each type are interchangeable. Type 1 is subscribers with an agreement with CP m. Type 2 is subscribers without an agreement with CP m. The variables of Type 1 subscribers are:

- x is the flow from CP m',
- y is the flow from CP m.

The variables of Type 2 subscribers are:

- u is the flow from CP m,
- v is the flow from the CP with has an agreement,
- w is the flow from all the other CP except CP m and CP with the agreement.

Due to symmetry, the system is equivalent to the following one with 5 variables (x, y, u, v, w):

$$
\begin{cases}
x\dfrac{\partial D}{\partial x}(x+v+(N-2)w)+D(x+v+(N-2)w)+q-\lambda_1=0 \\[2ex]
v\dfrac{\partial D}{\partial v}(x+v+(N-2)w)+D(x+v+(N-2)w)-\lambda_2=0 \\[2ex]
w\dfrac{\partial D}{\partial w}(x+v+(N-2)w)+D(x+v+(N-2)w)+q-\lambda_2=0 \\[2ex]
y\dfrac{\partial D}{\partial y}(y+(N-1)u)+D(y+(N-1)u)-\lambda_1=0 \\[2ex]
u\dfrac{\partial D}{\partial u}(y+(N-1)u)+D(y+(N-1)u)+p^m-\lambda_2=0 \\[2ex]
(M-1)x+y=\phi. \\[2ex]
(M-2)w+u+v=\phi
\end{cases}
$$

Considering the linear cost function $D(x) = ax$. We have to solve the linear system that are given below:

$$
\begin{cases}
ax+a(x+v+(N-2)w)+q-\lambda_1=0 \\[2ex]
av+a(x+v+(N-2)w)-\lambda_2=0 \\[2ex]
aw+a(x+v+(N-2)w)+q-\lambda_2=0 \\[2ex]
ay+a(y+(N-1)u)-\lambda_1=0 \\[2ex]
au+a(y+(N-1)u)+p^m-\lambda_2=0 \\[2ex]
(M-1)x+y=\phi. \\[2ex]
(M-2)w+u+v=\phi
\end{cases}
$$

We denote $\underline{x}$, $\underline{y}$, $\underline{u}$, $\underline{v}$, $\underline{w}$ the solution of the previous system. The revenue of CP m is $R_v^m(p^m, q) = p^m \times (N-1)\underline{u}$. To compute the equilibrium price we need to find

p which solves $\frac{\partial R_v^m}{\partial p^m}(p, p) = 0$. If $\underline{q} > p_{max}$ then $\underline{q} = p_{max}$. We have a particular interest in the case where $p_{max} > a\phi\frac{(N+1)}{3N-1}$. Equilibrium price is given by:

$$\underline{q} = a\phi\frac{(N+1)}{3N-1}.$$

Then the equilibrium flow from CP n to ISP m, $n \neq m$ is $\underline{z} = \frac{\phi}{N}(\frac{2N-2}{3N-1})$, and the equilibrium flow from CP n to ISP n is

$$\underline{y} = \frac{\phi}{N}(1 + \frac{(N-1)(N+1)}{3N-1}).$$

Proof of Theorem 12.33 We compare the expressions of the subscriber cost in each case. When an agreement exists between service and content providers, the cost for any subscriber connected to CP n is given by:

$$C_n^v(\underline{y}, \underline{z}, \underline{q}) = \phi^2 a + 2a\phi^2(\frac{N-1}{3N-1})^2(\frac{N+1}{N}).$$

We have to compare this expression with the following one, which is the cost for any subscriber also connected to CP n but in the case where no agreements are possible between service and content providers, that is:

$$C_n(\underline{x}, \underline{p}) = \phi^2 a(N - 1 + 1).$$

comparing these expressions, we get:

$$C_n^v(\underline{y}, \underline{z}, \underline{q}) < C_n(\underline{x}, \underline{p}).$$

Proof of Proposition 12.28 We have the same methodology as in the other proofs. We assume that CPs $m'' \in \{1, \ldots, N\} - \{m', m\}$ plays q, CP m, plays p^m, and CP m' plays r. CP m' is the only one with an agreement with the ISP m'.

In order to compute the equilibrium between the subscribers, for all (p^m, q, r), we have to define all the strategy of all subscribers. Let x the request of a subscriber of ISP n for CP m, y the request of a subscriber of ISP n for CP m'' and z the request of a subscriber of ISP n for CP m' for all $n \in \{1, \ldots, N\} - \{m'\}$. And let x' the request of a subscribers of ISP $n' = m'$ for CP m, y' the request for CP m'' and z' the request for CP m' with $n \in \{1, \ldots, N\} - \{m'\}$. We have now to solve:

$$\begin{cases} x\dfrac{\partial D}{\partial x}((N-1)x+x')+D((N-1)x+x')+q-\lambda_1=0 \\ x'\dfrac{\partial D}{\partial x'}((N-1)x+x')+D((N-1)x+x')+q-\lambda_2=0 \\ y\dfrac{\partial D}{\partial y}((N-1)y+y')+D((N-1)y+y')+p^m-\lambda_1=0 \\ y'\dfrac{\partial D}{\partial y'}((N-1)y+y')+D((N-1)y+y')+p^m-\lambda_2=0 \\ z\dfrac{\partial D}{\partial z}((N-1)z+z')+D((N-1)z+z')+r-\lambda_1=0 \\ z'\dfrac{\partial D}{\partial z'}((N-1)z+z')+D((N-1)z+z')+r-\lambda_2=0 \\ (N-2)x+y+z=\phi \\ (N-2)x'+y'+z'=\phi \end{cases}$$

We denote by $\underline{x}$, $\underline{y}$, $\underline{z}$, $\underline{x}'$, $\underline{y}'$, $\underline{z}'$ a solution of this system. We consider the linear cost function $D(x)=ax$ and thus we have to solve the following linear system:

$$\begin{cases} ax+a((N-1)x+x')+q-\lambda_1=0 \\ ax'+a((N-1)x+x')+q-\lambda_2=0 \\ ay+a((N-1)y+y')+p^m-\lambda_1=0 \\ ay'+a((N-1)+y')+p^m-\lambda_2=0 \\ az+a((N-1)z+z')+r-\lambda_1=0 \\ az'+a((N-1)z+z')+r-\lambda_2=0 \\ (N-2)x+y+z=\phi \\ (N-2)x'+y'+z'=\phi \end{cases}$$

The solution of this system is:

$$x=\frac{\phi a(N+1)+p-2q+2r}{Na(N+1)}$$

$$y = \frac{\phi a(N+1) - (N-1)p + (N-2)q + 2r}{Na(N+1)}$$

$$z = \frac{\phi a(N+1) + p + (N-2)\,q - 2(N-1)\,r}{Na(N+1)}$$

$$x' = \frac{\phi a(N+1) + p - 2\,q - (N-1)r}{Na(N+1)}$$

$$y' = \frac{\phi a(N+1) - (N-1)p + (N-2)\,q - (N-1)r}{Na(N+1)}$$

$$z' = \frac{\phi a(N+1) + p + (N-2)q + (N-1)^2 r}{Na(N+1)}$$

And after some computation, in order to compute the price competition between CP we finally obtain the next utility function at equilibrium of CP m':

$$\Pi'_m(\underline{p}_{m'}, \underline{p}_m, \underline{x}_{m'}, \underline{p}_{m'}) = \frac{a\phi(N+1)\left(2N^2+N-1\right)}{\left(4\,N^3 - 5\,N^2 + 2\,N - 1\right)}$$

$$+\frac{a\phi(N-1)\left(2\,N^2+N-1\right)\left(\frac{(4\,N^2+5\,N+1)}{4\,N^2-N+1} - \frac{2\,(2\,N^3+N^2-N)}{4\,N^3-5\,N^2+2\,N-1}\right)}{(N+1)\left(4\,N^3-5\,N^2+2\,N-1\right)}$$

Proof of Theorem 12.35 First we have to simplify the expression of Π'_m:

$$\Pi'_m(\underline{p}_{m'}, \underline{p}_m, \underline{x}_{m'}, \underline{p}_{m'}) = \frac{a\phi(N+1)\left(2N^2+N-1\right)}{\left(4\,N^3 - 5\,N^2 + 2\,N - 1\right)}$$

$$+\frac{a\phi(N-1)\left(2\,N^2+N-1\right)\left(\frac{(4\,N^2+5\,N+1)}{4\,N^2-N+1} - \frac{2\,(2\,N^3+N^2-N)}{4\,N^3-5\,N^2+2\,N-1}\right)}{(N+1)\left(4\,N^3-5\,N^2+2\,N-1\right)}$$

$$\Leftrightarrow \Pi'_m(\underline{p}_{m'}, \underline{p}_m, \underline{x}_{m'}, \underline{p}_{m'}) = \frac{a\phi(N+1)(N+1)(N-\frac{1}{2})}{\left(4\,N^3 - 5\,N^2 + 2\,N - 1\right)}$$

$$+\frac{a\phi(N-1)(N+1)(N-\frac{1}{2})\left(\frac{(4\,N^2+5\,N+1)}{4\,N^2-N+1} - \frac{2\,(2\,N^3+N^2-N)}{4\,N^3-5\,N^2+2\,N-1}\right)}{(N+1)\left(4\,N^3-5\,N^2+2\,N-1\right)}$$

$$\Leftrightarrow \Pi'_m(\underline{p}_{m'}, \underline{p}_m, \underline{x}_{m'}, \underline{p}_{m'}) = \frac{a\phi(N+1)^2(N-\frac{1}{2})}{\left(4\,N^3 - 5\,N^2 + 2\,N - 1\right)}$$

$$+\frac{a\phi(N-1)(N-\frac{1}{2})\left(\frac{(4N^2+5N+1)}{4N^2-N+1}-\frac{2(2N^3+N^2-N)}{4N^3-5N^2+2N-1}\right)}{(4N^3-5N^2+2N-1)}$$

$$\Leftrightarrow \Pi'_m(\underline{p}_{m'},\underline{p}_m,\underline{x}_{m'},\underline{p}_{m'}) = \left[\frac{a\phi(N-\frac{1}{2})}{(4N^3-5N^2+2N-1)}\right]$$

$$\times\left[(N+1)^2+(N-1)\left(\frac{(4N^2+5N+1)}{4N^2-N+1}-\frac{2(2N^3+N^2-N)}{4N^3-5N^2+2N-1}\right)\right].$$

Now we can compare Pi'_m and Pi_m. We can see that if we have $P(N) < (N-1)$ with:

$$P(N) = \frac{\Pi'_m(\underline{p}_{m'},\underline{p}_m,\underline{x}_{m'},\underline{p}_{m'})}{a\phi}$$

$$\times\left[(N+1)^2+(N-1)\left(\frac{(4N^2+5N+1)}{4N^2-N+1}-\frac{2(2N^3+N^2-N)}{4N^3-5N^2+2N-1}\right)\right]$$

which means that:

$$\Pi'_m(\underline{p}_{m'},\underline{p}_m,\underline{x}_{m'},\underline{p}_{m'}) < \Pi_m(\underline{p},\underline{x}).$$

After some algebras, we obtain that $P(N) < (N-1)$ which proves the proposition. Then, the decision vector where each CP does not make an agreement is a Nash equilibrium.

Part IV
Intermittency in Complex Systems

Duccio Piovani, Jelena Grujić and Henrik J. Jensen

High dimensional complex systems both physical and biological exhibit intermittent dynamical evolution consisting of stretches of relatively little change interrupted by often sudden and dramatic transitions to a new meta-stable configuration [243]. Such transitions can have crucial consequences when they occur in, say, ecosystems or financial markets and it is therefore important to develop methods that are able to identify precursors, warning signals and ideally techniques to forecast the transitions before they take place. We will expect that the mechanisms behind the rapid rearrangement may be different in different systems.

The widespread presence of this type of behaviour has fascinated people for many year now, and several different approaches have bee attempted to explain the emergence of transitions in complex systems. Concepts coming from statistical physics are often used to describe this problem, and their application has been attempted by researcher belonging to domains apparently very far from physics. In the past years people have developed a method pertinent to systems in which the transition takes the form as a bifurcation captured by a robust macroscopic variable, which emerges from the micro dynamics. A precursor of the systemic change can then be identified from the critical slowing down and enhanced fluctuations exhibited by this macroscopic collective degree of freedom [235, 237, 236] as a change in some external parameter drives the system towards the bifurcation point.

Here we consider an alternative scenario suggested recently in [66] in which the transitions are induced by intrinsic fluctuations at the level of the individual components which propagates to the macroscopic systemic level and thereby triggers a change in the overall configuration. Our approach is relevant to systems in which the available configuration space evolves as a consequence of the dynamics. One may think of a new and more virulent virus being created through a mutation of an existing strain (e.g. the SARS virus in 2003), or a new economic agent arriving in the market (e.g. the dot-com bubble in 1997–2000).

We describe below our methodology through applications to two models. First we consider the Tangled Nature (TNM) model of evolutionary ecology [79], which has had considerable success in reproducing both macro-evolutionary aspects such as the intermittent mode of extinctions [129] and ecological aspects such as species abundance distributions [27] and species area laws [172]. We also present results for transitions in a model with a very different type of dynamics, namely a high dimensional replicator with a stochastic element of mutation [244, 260], the Stochastic Replicator model (SRM). We will demonstrate in the following chapters that the replicator system with this element of stochasticity exhibit intermittency. Given the broad relevance of the replicator dynamics (population dynamics, game theory, financial dynamics, social dynamics etc.), success in forecasting transitions in this model may indicate that our method can be useful in many different situations [12].

Despite their different general mechanisms, the two models can be pictured in the same way. Their stochastic dynamics is characterised by a huge number of fixed points, and when the system randomly falls into one of them it enters a quiescent period of little change. Eventually the intrinsic stochastic fluctuations will allow the population of hitherto empty parts of configuration space, which may effectively serve as a random kick that is able to drive the system away from the local minimum and towards the chaotic regime where the system undergoes a high dimensional adaptive walk searching for another (metastable) fixed point.

Indeed both the nature of the fixed points and their stability varies significantly. Some fixed points are controlled by only a few interacting components while others involve many. Some are very stable while others less so leading to a very broad distribution of time spend in the metastable configurations of a given fixed point. The dynamics of the transitions between metastable configurations—the adaptive walk mentioned above—can also differ much. It can happen that the system is "trapped" between two or more attractors and switches between them before being pushed away. The transitions that lead from a fixed point to the other can be both sudden or slow and differ in magnitude. The point to be stressed is that the phenomenon we are trying to predict is highly heterogeneous and one has to bare this well in mind when interpreting the results.

That said, our claim is that we are able, in both models, to understand which kind of intrinsic stochastic fluctuation will be able to push the system out of its stable configuration. Indeed through a mean field description of the stochastic dynamics we can infer the Jacobian, from which by Linear Stability Analysis (LSA) we can identify the unstable eigendirections responsible for the destruction of the current metastable configuration. As will be shown in the following of this part, monitoring the relationship (vectorial overlap) between the existing configuration and the unstable mean field dangerous eigendirections allows to forecast approaching transitions with a high accuracy.

Chapter 13
Models with Adaptive Intermittent Behaviour

Duccio Piovani, Jelena Grujić and Henrik J. Jensen

In this chapter we will present the two models we have used as test cases: the Tangled Nature Model (TNM) and the Stochastic Replicator Model (SRM). As we will see in the following despite substantial differences these two models have a very similar behaviour. At the microscopic level the dynamics unfolds at a constant pace, and nodes stochastically gain or loose weight depending on their interaction with the rest of the system. On the other hand, at the systemic level both models present an intermittent dynamics, and the system switches or jumps from one fixed point to the other. For both models we will also go through the steps that lead to a mean field approximation of the dynamics.

13.1 Tangled Nature Model

The Tangled Nature Model, is a stochastic model of evolutionary ecology formalised at the microscopic level of individuals who can reproduce, mutate and die. As we

D. Piovani
Head of Data Science, nam.R, 4 rue Foucault, 75116 Paris, France

J. Grujić
Department of Computer Science, Sciences and Bioengineering Sciences, Artificial Intelligence Laboratory, Vrije Universiteit Brussel, Pleinlaan 2, 1050 Brussels, Belgium

Computer Science Department, Faculty of Science, Machine Learning Group, Université Libre de Bruxelles, Boulevard du Triomphe, CP 212, 1050 Brussels, Belgium

H. J. Jensen (✉)
Department of Mathematics and Centre for Complexity Science, Imperial College London, South Kensington Campus, London SW7 2AZ, UK
e-mail: h.jensen@imperial.ac.uk

E. Altman et al. (eds.), *Multilevel Strategic Interaction Game Models for Complex Networks*, https://doi.org/10.1007/978-3-030-24455-2_13

will see despite its simplicity the TNM is able to reproduce macro-evolutionary aspects such as the intermittent mode of extinctions [122] and ological aspects such as species abundance distributions [18] and species area laws [165]. Indeed the entire taxonomic hierarchy at the macroscopic level of species emerges from the dynamics at the microscopic level of individuals. Recently, in [38] the authors have shown how the dynamics in the TNM is a spontaneous non-equilibrium physical process, where the entropy increases in time while the free energy decreases.

The model is embedded in a random and constant interaction network, where every node represents a species and every link the pairwise interaction between species. The reproduction probability of individuals does not depend on a predefined fitness function but only depends on the web of interactions. The interactions change over time together with the change in the extant species and the fluctuations in the number of individuals that belong to them.

This implies that same species can be fit or unfit depending on which other individuals populate the system. Rabbits in an environment rich in carrots and poor in foxes would be fit, and their numbers would rapidly increase. On the other hand the same exact rabbits embedded in opposite conditions would find the environment harsh and probably go extinct. Being the number of individuals in constant evolution, so will the fitness of all the individuals, together with their reproduction rate.

Species who's reproduction rate is constantly lower than a constant death rate will eventually go extinct, while in the opposite case they will survive. The system will therefore organise itself around strong *mutualistic* interactions, i.e. interactions that are beneficial to all species involved, which form the core of the configuration, surrounded by cloud of new mutants which appear and quickly disappear.

This means that the stability of any configuration is threatened by the constant appearance of new species. There is always the possibility that a new mutant, may cause a fatal decrease (increase) in the fitness of other previously (un-fit) fit species. When this happens the system quickly abandons the configuration and enters a chaotic phase which eventually ends when new mutualistic configurations are restored.

The emergence of a macroscopic intermittent dynamics is in full agreement with the idea of a punctuated equilibrium formulated by Gould and Eldredge [110] to describe the tempo and mode of macroevolution as inferred from palaeontoligical data. Indeed in the TNM the system is usually found in long periods of relative stability, called qESS (quasi-Stable Evolutionary Stable Strategies), during which the configuration of the system changes only due to stochastic fluctuations. The name was chosen to recall the notion coming from game theory ESS stressing at the same time the stochastic nature of the model with the quasi. These quasi-stable configurations are then interrupted by sudden burst of activity, called quakes or transitions, where the network of extant types is reorganised.

Once again it is important to stress that the appearance of these transitions is not given by a change in the values of a parameter, nor by the an external perturbation. The quakes are an emergent property of the internal dynamics of the model. While at the microscopic level of individuals the evolution is constant, at the macroscopic level of species it switches between two different phases. This type of behaviour is common to many other complex systems and therefore success in forecasting the transitions in the TNM can be seen as a step to the creation of a more general mathematical framework.

13.1.1 Model Description

In the TNM, an agent is represented by a sequence of binary variables with fixed length L, denoted as $\mathbf{S}^\alpha = (S_1^\alpha, S_2^\alpha, \cdots, S_L^\alpha)$, where $S_i^\alpha = \pm 1$. Thus, there are 2^L different sequences, each one represented by a vector in the genotype space: $\mathcal{S} = \{-1, 1\}^L$. In a simplistic picture, each of these sequences represents a genome uniquely determining the phenotype of all individuals of this genotype. We denote by $n(\mathbf{S}^\alpha, t)$ the number of individuals of type $\mathbf{S}^\alpha$ at time t and the total population as $N(t) = \sum_{a=1}^{2^L} n(\mathbf{S}^\alpha, t)$. We define the distance between different genomes $\mathbf{S}^\alpha$ and $\mathbf{S}^\beta$ as the Hamming distance: $d_{\alpha\beta} = \frac{1}{2L}\sum_{i=1}^{L} |S_i^\alpha - S_i^\beta|$.

A time step is defined as the succession of one annihilation and of one reproduction attempt. During the killing attempt, an individual is chosen randomly from the population and killed with probability p^{kill} constant in time and independent of the type. During the reproduction process, a different randomly chosen individual $\mathbf{S}^\alpha$ successfully reproduces with probability:

$$p^{\text{off}}(\mathbf{S}^a, t) = \frac{\exp\left(H(\mathbf{S}^\alpha, t)\right)}{1 + \exp\left(H(\mathbf{S}^\alpha, t)\right)} \tag{13.1}$$

which depends on the occupancy distribution of all the types at time t via the weight function:

$$H(\mathbf{S}^\alpha, t) = \frac{k}{N(t)} \sum_{\beta \neq \alpha} \mathbf{J}(\mathbf{S}^\alpha, \mathbf{S}^\beta) n(\mathbf{S}^\beta, t) - \mu N(t). \tag{13.2}$$

In Eq. (13.2), the first term couples the agent $\mathbf{S}^a$ to one of type $\mathbf{S}^\beta$ by introducing the interaction strength $\mathbf{J}(\mathbf{S}^\alpha, \mathbf{S}^\beta)$, whose values are randomly distributed in the interval $[-1, +1]$. In Appendix A we go though the details of how the single values are generated.

For simplicity and to emphasise interactions we here assume: $\mathbf{J}(\mathbf{S}^\alpha, \mathbf{S}^\alpha) = 0$. The parameter k scales the interactions strength and μ can be thought of as the carrying capacity of the environment. An increase (decrease) in μ corresponds to harsher (more favourable) external conditions.

The reproduction is asexual: the reproducing agent is removed from the population and substituted by two copies $\mathbf{S}^{\alpha_1}$ and $\mathbf{S}^{\alpha_2}$, which are subject to mutations. A single mutation changes the sign of one of the genes: $S_i^\alpha \to -S_i^\alpha$ with probability p^{mut}. Similarly to a Monte Carlo sweep in statistical mechanics, the unit of time of our simulations is a *generation* consisting of $N(t)/p^{\text{kill}}$ time steps, i.e. the average time needed to kill all the individuals at time t.

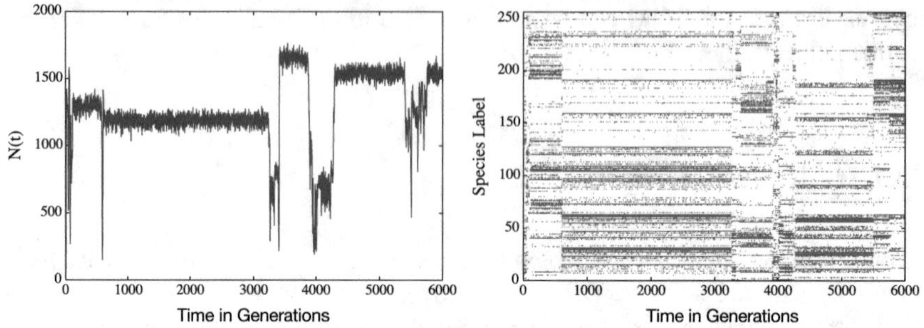

(a) Total Number of Individuals (b) Occupation Plot

Fig. 13.1 **a** Total population as a function of time (in generations) for a single realization of the TNM. The punctuated dynamics is clearly visible: quasi-stable periods alternate with periods of hectic transitions, during which $N(t)$ exhibits large amplitude fluctuations. **b** Occupancy distribution of the types. The genotypes are labelled arbitrarily and a dot indicates a type which is occupied at the time t. These figures are obtained with parameters $L = 8$, $p^{\mathrm{mut}} = 0.2$, $p^{\mathrm{kill}} = 0.4$, $K = 40$ and $\mu = 0.07$

These microscopic rules generate intermittent macro dynamics. The system is persistently switching between two different modes: the meta-stable states and the transitions separating them. The qESS states are characterised by small amplitude fluctuations of N(t) and stable patterns of occupancies of the types (Fig. 13.1, respectively left and right panel). However, these states are not perfectly stable and configurational fluctuations may trigger an abrupt transition to a different qESS state.

The transitions consist of collective adaptive random walks in the configuration space while searching for a new metastable configuration and are related to high amplitude fluctuations of N(t).

13.2 The Stochastic Replicator Model

In this section we will tackle the second test case for our method, the stochastic replicator model (SRM) a stochastic model, based on the replicator dynamics. Since Maynard Smith introduced evolutionary game theory [246], game theory itself has been mainly studied and developed as a mathematical framework to study Darwinian evolution. The deterministic version of the replicator dynamics is used routinely in a large variety of applications, precisely because of its relation to game theory and is therefore expected to be of relevance to the description of high dimensional socio-economic or biological systems [241, 278]. This suggests that if our method works in the SRM the procedure can be of broad relevance.

Despite that fact that the replicator equation is mainly being used to describe biological or social economical ecosystems, both of which are embedded on co-evolving large webs of interaction, it is usually studied in low dimensions, i.e. with few strategies present in the system.

Here our view point is to make proper contact between theory and real systems. It is therefore important to consider large numbers of co-evolving strategies, who can appear, change and leave the system, because this is exactly what happens in the systems we want to describe. Besides working with large matrices to capture the limit of many strategies one simply needs to allow the dimension of the pay-off matrix to vary as the number of strategies changes due to extinction and creation events. This version of the replicator dynamics set-up was studied by Tokita and Yasutomi in [294]. The authors focused on the emerging network properties. Here we continue this study but with an emphasis on the intermittent nature of the macro-dynamics. Despite sharing the same spirit of the Tangled Nature, the details of the dynamics are quite different, and the model is able to reproduce intermittent dynamics at the macroscopic level.

13.2.1 Model Description

In this model the configuration vector $\mathbf{n}(t) = (n_1(t), \ldots, n_d(t))$ tells us the frequency of players, choosing a given strategy. This means that the components $n_i(t) \in [0, 1]$ for all $i = 1, 2, \ldots, d$, and the actual number of individuals is not included in the description of the system but only how they distribute on the different strategies.

We start the simulations by generating the dxd payoff matrix $\mathbf{J}$ of the game that will tell us the payoffs of every pairwise combination of strategies. Each strategy distinguishes itself from the others in its payoffs or interactions with the rest of the strategy space. We have used the same interaction network used in the Tangled Nature model, however we found that matrices with payoffs uniformly distribute on the interval $(-1, 1)$ exhibit the same behaviour as matrix of the form used for the Tangled Nature model. However, if the payoffs are drawn from a power law distribution with no second moment, the dynamics becomes different and the intermittent behaviours is not so clear any more.

In the initial configuration, $n_o < d$ randomly chosen strategies start with the same frequency $n_i = \frac{1}{n_o}$. All the other possible strategies are non active, i.e. the corresponding components $d - n_o$ of the occupation vector $\mathbf{n}(0)$ are $n_i(0) = 0$. The empty strategies can only become populated by one of the *active* strategies mutating into them. Once this happens their frequency will evolve according to the replicator equation. At each time step we calculate the *fitness*, $h_i(t) = \sum_j J_{ij} n_j(t)$ of each active strategy and compare it with the average fitness $\bar{h}(t) = \sum_{ij} J_{ij} n_i(t) n_j(t)$, exactly as expected in a replicator dynamics. Each frequency is then updated according to

$$n_i(t+1) = n_i(t) + \left(\sum_j J_{ij} n_j(t) - \sum_{ij} J_{ij} n_i(t) n_j(t) \right) \cdot n_i(t) \qquad (13.3)$$

The stochastic element, of the otherwise deterministic dynamics, consists in the following updates. With probability p^{mut} each strategy mutates into another one, this is done by transferring a fraction α_{mut} of the frequency from the considered strategy to another strategy. The label of the mutant strategy is chosen in the vicinity of the first by use of a normal distribution $N(i, \sigma)$ centred on label i with variance σ. The closer the labels of two strategies the more likely it is for one to mutate into the other.

It should be noted that as long as the payoff matrix is random and uncorrelated in its indices, no similarity criteria between strategies does really exists (2 similar strategies interact in a completely different way with the environment). The parameter has been introduced only to control the level of disorder in the system. Higher values of σ mean that a single strategy can populate more nodes, and therefore the configuration space is explored more rapidly.

When the frequency of a strategy i goes below a preset extinction threshold $n_i(t) < n^{ext}$, the strategy is considered extinct and its frequency is set to zero $n_i(t+1) = 0$. Right after an extinction event the system is immediately renormalised in order to maintain the condition $\sum_i n_i(t) = 1$. For the simulations unless stated differently we have used the following parameter set $d = 256$, $\text{n}^{\text{ext}} = 0.001$, $\alpha^{mut} = 0.01$, $p^{mut} = 0.2$.

In Fig. 13.2 one can clearly distinguish the intermittent dynamics of the system. Both figure (a) and (b) show that the system is jumping from one configuration to the other, and its switching from a phase of little or no activity to a chaotic phase during which the strategies drastically change their frequencies.

Furthermore we can see that both the stable phases and the transitions can be quite different among them. It is very instructive to take a closer look tat Fig. 13.2, because it gives a good understanding of the high heterogeneity of the phenomenon we are trying to understand and describe. The stable phases characterised by a constant value of the frequencies differ from one another, in length, stability and number of players.

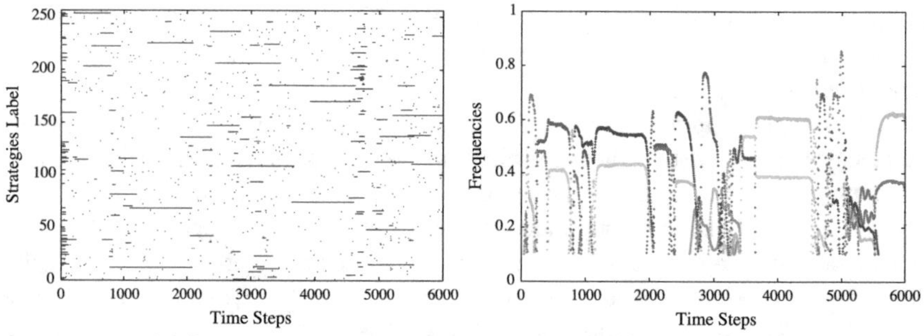

(a) Occupation Plot (b) Active Strategies

Fig. 13.2 **a** Occupancy distribution of the types. The genotypes are labelled arbitrarily and a dot indicates a type which is occupied at the time t. The punctuated dynamics is clearly visible: quasi-stable periods alternate with periods of hectic transitions. **b** We present the frequencies of the strategies. Each colour belongs to a different strategy. Once again the transitions from fixed point to another is clear. The value of the parameters is $d = 256$, $\text{n}^{\text{ext}} = 0.001$, $\alpha^{mut} = 0.01$, $p^{mut} = 0.2$

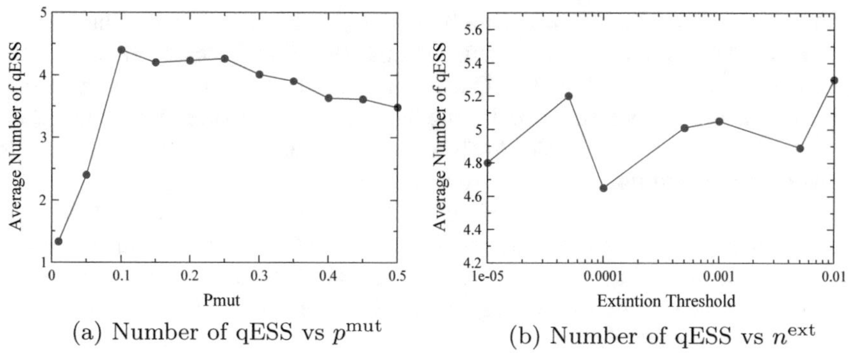

Fig. 13.3 In these plots we analyse the role the two parameters p^{mut} and n^{ext} respectively the mutation probability and the extinction threshold, play in the dynamics. We have measured the average number different qESS tanking 10^3 different simulations of 10^4 time steps. As expected the higher p^{mut} the less fixed points it explores, while the extinction rate doesn't seem to play any particular role. This second result suggests that the dynamics is dominated by the wild types

Fig. 13.4 In this figure we show the distribution of the durations of qESS. A we can see the curve is well fitted by a power law distribution $x^k$with exponent $k = -1.7$

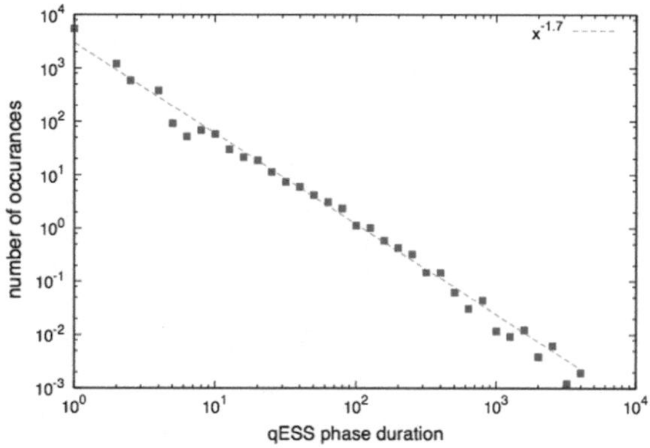

And so do the transitions: some are sharp while some other slowly gain momentum. The great result is that these are properties emerge from the simple dynamics we have just outlined.

In Fig. 13.3a we see that, as one may expect the number of qESS decreases as p^{mut} increases, which confirms that the mutation rate plays a crucial role in the formation of the intermittent behaviour. For very low values there is no intermittency, which means that once the system enters a fixed point is has more difficulty to leave it. While for very high values of the parameter the system fails to ever stabilize and the number of qESS decreases once again. One could argue that by decreasing p^{mut} one is just slowing down the pace of the dynamics, i.e. increasing its characteristic timescale, which would explain the decrease of the number of fixed points explored.

The answer to this observation is of crucial importance. One has to always bare in mind that all the definitions we are giving are completely dependent on a specific timescale as well as on a scale magnitude of change. By increasing the resolution of the observation what seems to be a continuos variation may become intermittent and vice versa. For this reason a change in timescale introduces a fundamental and profound difference in the system. One has to understand which are the interesting scales. This is an easy task when dealing with natural phenomena where the problem itself exists in a given scale. Analysing data coming from models of course its different given their typical qualitative description.

Furthermore in the right panel of Fig. 13.3b, we see that the system's behaviour is quite independent from the extinction threshold. This suggest that the dynamics is strongly dominated by the wild-types and that the presence of strategies with low frequency in the system is absolutely marginal. Finally, in Fig. 13.4 we see that the distribution of the duration of the qESS follows a power law distribution which tells us that there is no characteristic duration for the stable phases in this model.

Chapter 14
Outline of the Forecasting Procedure

Duccio Piovani, Jelena Grujić and Henrik J. Jensen

As said in the introduction of this part, our aim is to develop a general procedure to forecast transitions in systems that present intermittent dynamics at the systemic level. In this chapter we will outline a forecasting procedure based on a Linear Stability Analysis (LSA). Indeed we will go through the mathematics behind the theory, showing and explaining how we have exploited its results to build a stability indicator that can be used as an early warning sign of the arrival of a transition. The aim is to justify the method analytically, and prove it can be useful for a broad range of systems with different properties. In the following sections we go through the main steps necessary to build the indicator, giving a general description that will be completely independent on the details of the system. Of course applications to different systems will require specific adjustments and considerations. We will then apply the results of this chapter to the two test cases in the following chapters.

D. Piovani
Head of Data Science, nam.R, 4 rue Foucault, 75116 Paris, France

J. Grujić
Department of Computer Science, Sciences and Bioengineering Sciences, Artificial Intelligence Laboratory, Vrije Universiteit Brussel, Pleinlaan 2, 1050 Brussels, Belgium

Computer Science Department, Faculty of Science, Machine Learning Group, Université Libre de Bruxelles, Boulevard du Triomphe, CP 212, 1050 Brussels, Belgium

H. J. Jensen (✉)
Department of Mathematics and Centre for Complexity Science, Imperial College London, South Kensington Campus, London SW7 2AZ, UK
e-mail: h.jensen@imperial.ac.uk

E. Altman et al. (eds.), *Multilevel Strategic Interaction Game Models for Complex Networks*, https://doi.org/10.1007/978-3-030-24455-2_14

14.1 Mean-Field Description

Being the types of systems we want to describe intrinsically stochastic, the first step towards building the stability indicator is to establish a mean field approximation of the stochastic dynamics in order to obtain a set of deterministic equations. We define the state vector $\mathbf{n}(t) = (n_1(t), \ldots, n_d(t))$, where d is the dimension of the system and where every component $n_i(t)$ tells us the weight of agents at time t. In different systems the weight can represent the number of individuals belonging to a given species, the amount of capital of a given company or the fraction of players using a given strategy just to name a few examples.

We establish the average flow of the weight variable n_i between different types of individual agents. The mean field time evolution is of the form

$$\mathbf{n}(t+1) - \mathbf{n}(t) = \mathbb{T}(\mathbf{n}(t)) \cdot \mathbf{n}(t) \quad (14.1)$$

where the matrix $\mathbb{T}$ is the mean field evolution matrix, which will contain all the contributions of the processes involved in the dynamics. In a birth death process type of dynamics for example it will describe: death, reproduction and mutation. In the systems we want to describe typically the flow of n_i will be the result of interactions between agents, and their weight. This explains why the matrix $\mathbb{T}$ depends on the configuration of the system $\mathbf{n}(t)$.

As said, the systems to which we want to apply the method will have intermittent macro dynamics which means they will be generally found in a stationary configuration, i.e. in a fixed point, and it is indeed the neighbourhood of the fixed point we want to analyse through our mean field approximation. Hence its not the entire dynamics that needs to be described by the deterministic equations, but a good characterisation of the quasi stable phases would be enough for our method to work. We therefore hope that by numerically applying the mean field equations to a stationary configuration its stability is reflected. Namely we exploit the fact that during a stable phase

$$\mathbf{n}(t+1) \simeq \mathbf{n}(t) \simeq \mathbf{n}^* \qquad \forall t \in \Delta t_{\text{stable}} \quad (14.2)$$

where $\mathbf{n}^*$ is a fixed point for which

$$\mathbb{T}(n^*) \cdot \mathbf{n}^* = 0 \Longrightarrow \mathbb{T}(\mathbf{n}(t)) \cdot \mathbf{n}(t) \simeq 0 \qquad \forall t \in \Delta t_{\text{stable}} \quad (14.3)$$

If this result is confirmed it means that during a metastable period the mean field approximation well describes the system. This would suggest that we can use Eq. (14.1) to study local stability properties of the neighbourhood of the fixed points $\mathbf{n}^*$.

14.2 Linear Stability Theory: Continuum Approximation

We will here proceed using the continuum approximation, thus considering the left hand side of Eq. (14.1) as a derivative, so that now

$$f(\mathbf{n}(t)) = \frac{d\,\mathbf{n}(t)}{d\,t} \simeq \mathbb{T}(\mathbf{n}(t))\mathbf{n}(t) \tag{14.4}$$

This approximation is justified if

$$\frac{|\mathbf{n}(t+1) - \mathbf{n}(t)|}{|\mathbf{n}(t)|} << 1 \tag{14.5}$$

so if the change in one time step is small if compared to the vector.

14.2.1 *Linearization*

In order to understand how the system behaves in the neighbourhood of a fixed point we can linearise Eq. (14.3) about the fixed point $\mathbf{n}^*$. We therefore introduce a perturbation $\mathbf{n}(t) \to \mathbf{n}^* + \delta\mathbf{n}(t)$, where

$$\frac{\|\delta\mathbf{n}\|}{\|\mathbf{n}^*\|} << 1 \Longrightarrow \|\mathbf{n}^*\| \simeq \|\mathbf{n}(t)\| \tag{14.6}$$

We want linearise Eq. (14.4), by introducing the perturbation we obtain

$$f(\mathbf{n}^* + \delta\mathbf{n}) = \mathbb{T}(\mathbf{n}^* + \delta\mathbf{n})(\mathbf{n}^* + \delta\mathbf{n}) \tag{14.7}$$

and by exploiting the result in Eq. (14.3) and neglecting the second order we get

$$f(\mathbf{n}^* + \delta\mathbf{n}) = \big(\mathbb{T}(\mathbf{n}^*) + \partial_{\mathbf{n}}\mathbb{T}(\mathbf{n}^*)\delta\mathbf{n}\big)\big(\mathbf{n}^* + \delta\mathbf{n}\big) + \mathrm{o}(\delta\mathbf{n}^2) \tag{14.8}$$

which means that

$$\frac{\mathrm{d}\,(\delta\mathbf{n}(t))}{\mathrm{d}t} \simeq \mathbb{M}(\mathbf{n}) \cdot \delta\mathbf{n}(t) \tag{14.9}$$

where

$$\mathbb{M} = \big(\mathbb{T}(\mathbf{n}^*) + \partial_{\mathbf{n}}\mathbb{T}(\mathbf{n}^*) \cdot \mathbf{n}^*\big) \tag{14.10}$$

or for every component

$$M_{ij} = \left(T_{ij}(n^*) + \sum_k \frac{\partial T_{ik}}{\partial n_j} n_k^*\right) \tag{14.11}$$

is the *stability matrix* or Jacobian of the system. As we will see its eigenspace contains precious information on the stability of the system.

We will start by solving the case in which $\mathbb{M}$ is diagonizable and will move on to the more general case afterwards. As we will see the two cases will produce the same results with a slight difference in the procedure.

The fact that the matrix is diagonalizable implies that the *algebraic* and *geometric multiplicities* coincide $\forall\lambda$, where λ are the eigenvalues. The eigenvectors in this case form a linear independent set of vectors, which allows us to express the displacement vector, or any other vector as

$$\delta\mathbf{n}(t) = \sum_{k=1}^{n} c_k(t)\mathbf{e}_k \tag{14.12}$$

where $\mathbf{e}_k$ are $\mathbb{M}$'s eigenvectors and $c_k(t)$ the coefficients of the expansion. By substituting Eq. (14.12) in Eq. (14.9) we can solve the equation in the basis formed by $\{\mathbf{e_k}\}$:

$$\frac{d\delta\mathbf{n}(t)}{dt} = \begin{pmatrix} \dot{c}_1(t) \\ \vdots \\ \dot{c}_d(t) \end{pmatrix} \simeq \begin{pmatrix} \lambda_1 & & \\ & \ddots & \\ & & \lambda_d \end{pmatrix} \begin{pmatrix} c_1(t) \\ \vdots \\ c_d(t) \end{pmatrix} \tag{14.13}$$

We can now solve the d first order differential equations in the new coordinates $c_k(t)$:

$$c_k(t) = c_k(0)\mathrm{e}^{\lambda_k t} = c_k(0)\mathrm{e}^{\lambda_k^R t}\mathrm{e}^{\mathrm{i}\lambda_k^I t} \tag{14.14}$$

where λ^R and λ^I are respectively the real and imaginary part of the the eigenvalue. In a symmetric matrix of course $\lambda^I = 0$ but here we are sketching the general case. By looking at Eq. (14.14) it is clear that components with $\lambda_i^R > 0$ will diverge in time no matter what the imaginary part does. On the contrary if $\lambda_i^R < 0$ a perturbation in that direction will exponentially die out. This allows us to say that the directions indicated by eigenvectors relative to eigenvalues with positive real part are unstable. Such eigenvectors form the unstable subspace $\mathbb{S}^+$, while those with $\mathrm{Re}(\lambda) < 0$ form the stable subspace $\mathbb{S}^-$. What this result is telling us is that if the dynamics were completely deterministic, it would be completely dominated by its $\mathbb{S}^+$, because after only a few time steps $c_k(t) \to 0$ if $\mathrm{Re}(\lambda_k) < 0$ and $c_k(t) \to \infty$ if $\mathrm{Re}(\lambda_k) > 0$.

Exploiting this result, if the mean-field approximation is able to at least partially describe the stochastic dynamics and we manage to embed the system in such eigenspace, we would be able to distinguish to some degree between dangerous and harmless perturbations. The closer the actual dynamics to the mean-field the more precise the description will be. Perhaps this will give us prediction power on the next transition that should occur after the system suffers a perturbations parallel to a dangerous direction.

If on the other hand the Jacobian is non-diagonizable it means there is degeneracy in its eigenvalues, and the two multiplicities don't coincide for every λ. This means that the number or linearly independent eigenvectors is lower that the dimensions of the system which implies they will not *cover* the entire space. The best we can do in this case is introduce the *generalised eigenvectors* or *power vectors*. These vectors have the property of being root of a power of the characteristic polynomial, but not of the polynomial itself, and together with the eigenvectors they form a linear independent set of vectors that can be used as basis.

In this case we can write Eq. (14.12) as

$$\delta \mathbf{n} = \sum_{\lambda} \sum_{i=1}^{m_\lambda^a} c_\lambda^i \mathbf{e^i}_\lambda \tag{14.15}$$

where $\mathbf{e}$ are the set of eigenvectors and *power vectors* together and m_λ^a is λ's algebraic multiplicity. In this basis the Jacobian is in what is called its *Jordan normal form*: the matrix is organised in *Jordan blocks* and is said to be block diagonal. A *Jordan block* is a square matrix with zero's everywhere except along the diagonal and super diagonal, with each element of the diagonal consisting in the eigenvalue λ and each element of the super diagonal consisting of a 1. Every block belongs to a different eigenvalue and the dimension of the block is given by $m_{\lambda_i}^a$. Of course a Jordan block of $dim = 1$ is formed only by its λ and includes strictly only one eigenvector and no power vectors

An example of a 6x6 matrix in Jordan normal form, formed by three different blocks is

$$\begin{pmatrix} \lambda_1 & 0 & 0 & 0 & 0 & 0 \\ 0 & \lambda_2 & 1 & 0 & 0 & 0 \\ 0 & 0 & \lambda_2 & 0 & 0 & 0 \\ 0 & 0 & 0 & \lambda_3 & 1 & 0 \\ 0 & 0 & 0 & 0 & \lambda_3 & 1 \\ 0 & 0 & 0 & 0 & 0 & \lambda_3 \end{pmatrix} \tag{14.16}$$

where we have a block of dim = 1 for λ_1 of dim = 2 for λ_2 and dim = 3 for λ_3. In this case solving Eq. (14.9) is slightly more complicated.

Exactly how we have previously solved the equation for every λ_i separately here we have to solve it for every block. We will start by solving a 3 − d block and then we will extend the result to the general case:

$$\frac{d}{dt} \begin{pmatrix} c_1(t) \\ c_2(t) \\ c_3(t) \end{pmatrix} = \begin{pmatrix} \lambda & 1 & 0 \\ 0 & \lambda & 1 \\ 0 & 0 & \lambda \end{pmatrix} \cdot \begin{pmatrix} c_1(t) \\ c_2(t) \\ c_3(t) \end{pmatrix} \tag{14.17}$$

so the system of equation we have to solve looks like

$$\begin{cases} \dot{c_1}(t) = \lambda c_1(t) + c_2(t) \\ \dot{c_2}(t) = \lambda c_2(t) + c_3(t) \\ \dot{c_3}(t) = \lambda c_3(t) \end{cases} \tag{14.18}$$

we can start by solving the homogeneous differential equation for c_3, which trivially becomes $c_3 = c_3(0)e^{\lambda t}$. We substitute the result in the equation of c_2 which becomes

$$\dot{c}_2(t) = \lambda c_2(t) + c_3(0)e^{\lambda t} \tag{14.19}$$

to solve this non homogeneous differential equation we exploit the results of the variation of parameter's method where the solution in the sum of the homogeneous solution c_3^h and the particular solution c_2^p. Explicitly for the homogeneous on obtains $c_2^h = c_2(0)e^{\lambda t}$ while for the particular solution $c_2^p = c_2(t)e^{\lambda t}$ where the coefficient is considered as time dependent. By substituting $c_2^p = c_2(t)e^{\lambda t}$ in the left hand side of Eq. (14.19) we obtain

$$\dot{c}_2(t)e^{\lambda t} + \lambda c_2(t)e^{\lambda t} = \lambda c_2(t)e^{\lambda t} + c_3(0)e^{\lambda t} \tag{14.20}$$

which yields

$$\longrightarrow c_2(t) = c_2(0) + c_3(0)t$$

and the particular solution becomes

$$c_2^p(t) = [c_2(0) + c_3(0)t]\, e^{\lambda t} \tag{14.21}$$

By adding the homogeneous solution we arrive to the complete solution of the equation, which has the form

$$c_2(t) = c_2^p + c_2^h = [2c_2(0) + c_3(0)t]\, e^{\lambda t} \tag{14.22}$$

and by repeating the same procedure in the first equation we obtain

$$c_1(t) = \left[2c_1(0) + 2c_2(0)t + \frac{1}{2}c_3(0)t^2\right] e^{\lambda t} \tag{14.23}$$

The result can be easily generalised for d-dim block where the d-th equation is the homogeneous

$$c_d(t) = c_d(0)e^{\lambda t} \tag{14.24}$$

and all the others, $\forall k \in [1, d-1]$, have the form

$$c_k(t) = \left[2 \cdot \sum_{i=k}^{d-1} \frac{c_i(0)t^{i-k}}{(i-k)!} + \frac{c_d(t)t^{d-k}}{(d-k)!}\right] e^{\lambda t} \tag{14.25}$$

Looking at Eq. (14.25) we see that the coefficients of the components that belong to the blocks with $d > 1$ will grow in time as power of t. But this growth will be killed by the exponential decay in the case $\lambda^R < 0$ and will be neglectable in the opposite case. Finally we can say that the stability and instability conditions stay the same,

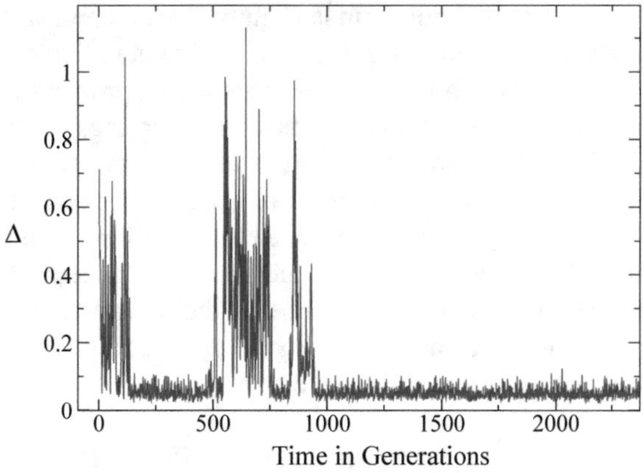

Fig. 14.1 In this figure we show the quantity $\Delta = \frac{\|\mathbf{n}(t+1)-\mathbf{n}(t)\|}{\mathbf{n(t)}}$. We can see how the system during a qESS changes less than 10%

the only difference being the fact that the unstable directions can be seen by a *power vector* and not a normal eigenvector. So by computing the power vectors we are able to apply the method even in systems where the Jacobian is not diagonisable.

14.3 TNM Mean Field Description

In order to apply the procedure just presented, we need to approximate the dynamics just presented using a mean field formalism. In this section we will describe the approximation step by step. In the TNM there are multiple sources of stochasticity, namely reproduction, mutations and deaths. To formulate a mean field equation we have to average out these sources.

From Fig. 14.1 we can see that the quantity in Eq. (14.5) is indeed small

$$\frac{\|\mathbf{n}(t+1)-\mathbf{n}(t)\|}{\mathbf{n(t)}} < 0.1 \tag{14.26}$$

during the stable phase. A continuum approximation seems like a plausible approach. We therefore build a mean field version of the derivative $f = \frac{dn(t)}{dt}$.

As seen previously the killing process is quite simple: at each time step with probability p^{kill} a randomly chosen individual is removed from the system, which implies that the occupation number of the species it belongs to decreases of 1 unit $(\Delta n_i = -1)$. The probability of choosing an individual belonging to the ith species is $\rho_i = \frac{n_i}{N}$, so the killing term becomes

$$\rho_i \cdot p_{\text{kill}} \cdot (-1) \tag{14.27}$$

which is the quantity that species i will loose in average at each time step for the killing term.

The reproduction term is slightly more complicated given the presence of mutations. At each time step a randomly chosen individual is selected for asexual reproduction. The reproduction happens with a probability p^{off} given by Eq. (13.1). We have to take into account the fact that offsprings can both mutate ($\Delta n_i = -1$), there can be only one mutation ($\Delta n_i = 0$), or no mutations ($\Delta n_i = +1$). These results are independent from the number of actual genes that mutate during the reproduction. Here we only want to formalise how mutations influence the occupation number of the ith species, therefore all mutations are treated in the same way, because they all influence the occupancy number in the same way, namely $\Delta n = -1$.

By defining the probability of no mutations $p_o = (1 - p^{\text{mut}})^L$ the mean field reproduction term therefore becomes

$$\rho_i(t) \cdot p_i^{\text{off}}(t) \left[p_o^2 - (1 - p_o)^2\right] = \alpha\, \rho_i(t)\, p_i^{\text{off}}(t) \tag{14.28}$$

where $\alpha = (2p_o - 1)$ is a constant.

The third term we have to consider is the *backflow effect*, which describes the event of $\mathbf{S}^i$ being populated by mutations occurring during the reproduction happening elsewhere. This term will take into account the probabilities of choosing $\mathbf{S}^j$ at time t for the reproduction process, and the probability of $\mathbf{S}^j$ reproducing and mutating into $\mathbf{S}^i$. The term has the form

$$2 \cdot \sum_j \rho_j(t) p_j^{\text{off}} p_{j\to i}^{\text{mut}} \tag{14.29}$$

where the 2 comes from the details of the reproduction process, and the $p_{j\to i}^{\text{mut}}$ comes is the probability of type $\mathbf{S}^j$ to mutating into type $\mathbf{S}^i$. For this to happen, d_{ij} genes will have to mutate, i.e. $1 \to -1$ or vice versa, in order to bridge the hamming distance of the two species. Given the presence of only two mutually exclusive outcomes (a gene can either mutate or not mutate) the probability has the binomial form, namely

$$p_{i\to j}^{\text{mut}} = p_{\text{mut}}^{d_{ij}} \cdot (1 - p_{\text{mut}})^{L - d_{ij}} \tag{14.30}$$

We have excluded the binomial coefficient because the order and position of the mutations does matter, and therefore we have to neglect the possible permutations. To mutate from $\mathbf{S}^i \longrightarrow \mathbf{S}^j$ there is only one possible combination of mutations.

Putting together all these effects we finally find the mean field equation for this model, namely

$$n_i(t+1) - n_i(t) = \frac{1}{N} \sum_{j \in 2^L} \{ \left(p_j^{\text{off}}(t)\,(2p_o - 1) - p^{\text{kill}}\right) \cdot \delta_{ij} + \tag{14.31}$$

$$p_j^{\text{off}} \cdot p_{j\to i}^{\text{mut}} \cdot \left(1 - \delta_{ij}\right)\} n_j(t)$$

where

$$T_{ij} = \left(p_j^{\text{off}}(t)\,(2p_o - 1) - p^{\text{kill}}\right) \cdot \delta_{ij} + p_j^{\text{off}} \cdot p_{j\to i}^{\text{mut}} \cdot \left(1 - \delta_{ij}\right) \tag{14.32}$$

is the mean-field evolution matrix of the system. The term with δ_{ij} takes into account the processes that happen in i, namely reproduction and killing, while the term in $(1 - \delta_{ij})$ formalises the *back flow* effect.

Now proceeding just like previously shown, we can linearise Eq. (14.32) about a fixed point $n^* \to n^* + \delta n$. This is done by substituting Eq. (14.32) into Eq. (14.10) which yields the specific form of the stability matrix for the Tangled Nature Model

$$\mathbb{M}_{ij} = (\alpha p_j^{\text{off}} - p^{\text{kill}})\delta_{ij} + 2(1 - \delta_{ij}) p_j^{\text{off}} p_{j\to i}^{\text{mut}} \tag{14.33}$$

$$+ \sum_k \left[\alpha\delta_{ik} + (1 - \delta_{ik}) \cdot p_{k\to i}^{\text{mut}}\right] \frac{\partial p_{\text{k}}^{\text{off}}}{\partial n_j} n_k^*$$

This is the mean field matrix we use for our linear stability analysis of the stochastic fixed points. To us its a useful tool to gain some insight in the neighbourhood of a given stochastic configuration. We can consider it as a plausible guess on the next move the system will make.

$$\|\mathbf{n}(t+1) - \mathbf{n}(t)\|_{\text{stoch}} \simeq \|\mathbb{T}(\mathbf{n}(t))\mathbf{n}(t)\|_{\text{mf}} \tag{14.34}$$

From the details of the model it follows that the mean-field jacobian in Eq. (14.33) is non symmetric

$$M_{ij} \neq M_{ji} \tag{14.35}$$

This implies that the matrix it is not necessarily diagonalisable and λ, $\mathbf{e}_\lambda \in \mathbb{C}$. It appears that in order to study the stability we will have to express $\mathbb{M}$ in *Jordan* form and find out the structure of its generalised eigenspace.

But by analysing the spectrum of $\mathbb{M}$ we have found that there is a high degeneracy in the stable subspace $\mathbb{S}^-$, but not in the unstable subspace $\mathbb{S}^+$. The eigenvalues with positive real part are always distinct.

$$\lambda_i \neq \lambda_j \qquad \forall \lambda_i, \lambda_j \in \mathbb{S}^+ \tag{14.36}$$

This means that their algebraic and geometric multiplicities are 1 and the associated eigenspace is completely described by the eigenvector. It would correspond to a Jordan block of $dim = 1$ that allows no generalised eigenvector. We can therefore say that $\mathbb{S}^+$ is only formed by a set of linearly independent eigenvectors and no generalised eigenvectors. This is a useful information to have when constructing the stability indicator in the next chapter. We have also checked the number the number of unstable directions that characterise a qESS. The results is shown in Fig. 14.2. As we can see a qESS typically has less than five dangerous directions out of 256, we will have to monitor when applying the forecasting method, which computationally represents a huge difference.

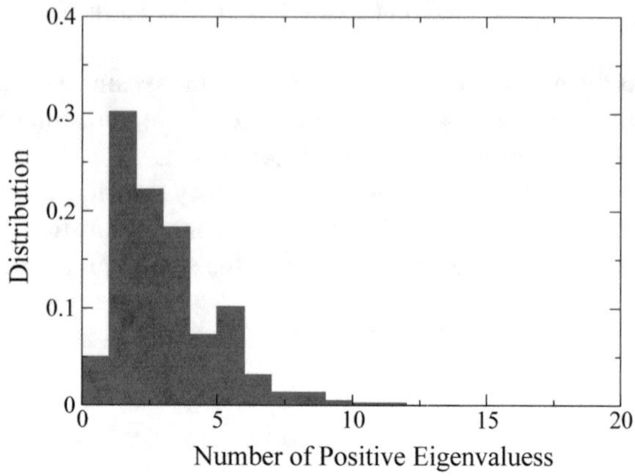

Fig. 14.2 We present the distribution of the number of unstable directions in the TNM, i.e. the number of λ with $\mathrm{Re}(\lambda) > 0$

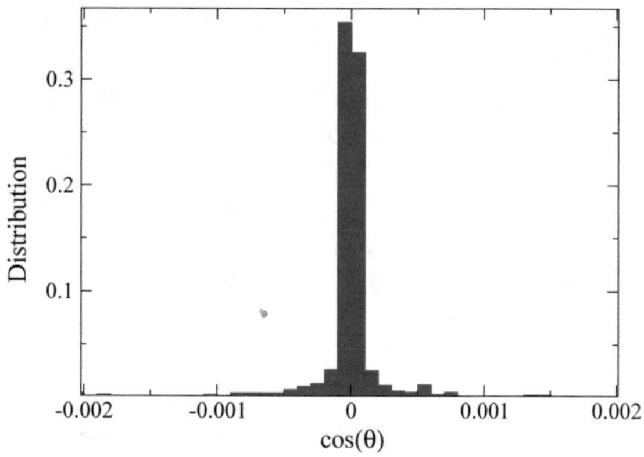

Fig. 14.3 In this figure we show the distribution of $\cos(\theta)$ between eigenvectors belonging the unstable subspace $\mathbb{S}^+$, for every fixed point. 82% of the couples have $\theta < 10^4$ which is technically zero and for $\simeq$90%

Moreover we have checked the distribution of the angles between the $\mathbf{e}^+$ belonging to the same Jabobian. Since $\mathbf{e} \in \mathbb{C}$ we have used the formula

$$\cos(\theta_{12}) = \frac{\mathrm{Re}(\mathbf{e}_1 \cdot \mathbf{e}_2)}{\|\mathbf{e}_1\| \|\mathbf{e}_2\|} \tag{14.37}$$

to calculate the $\cos(\theta)$ between the eigenvectors.

As we can see from Fig. 14.3 the eigenvectors are nearly orthogonal with 90% of the couples having $\theta < 10^3$ while 82% have $\theta < 10^4$.

14.4 SRM Mean Field Description

In Fig. 14.4 we show the typical behaviour of the quantity $\Delta(t)$. We can see how during the stable phases, which are clearly observable, the value fluctuates around

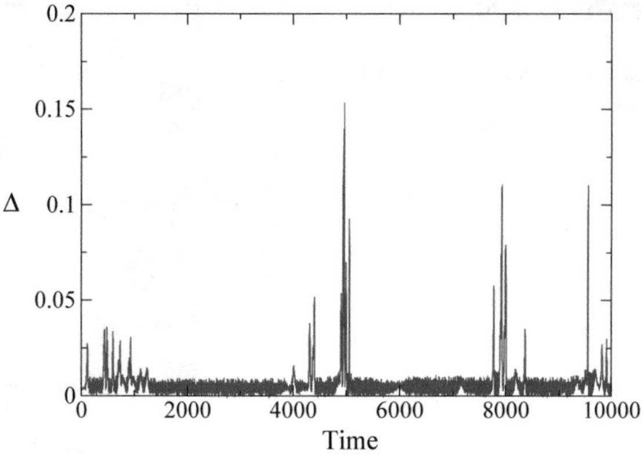

Fig. 14.4 The blue curve represents $\Delta(t) = \frac{\|n(t+1)-n(t)\|}{\|n(t)\|}$. From the figure one can clearly distinguish the stable phases form the transitions which correspond to the sharp peaks. During the stable phases $\Delta \sim 0.01$

$\sim$0.01. The slow rate of change in time allows us to treat the system in continuum time, just like we did for the TNM.

In this model the random mutations are the only source of stochasticity in the model's dynamics. To account for these stochastic events one has to consider the possibility that a strategy looses part of its frequency by mutating into other strategies and gaining frequency as a result of mutations happening elsewhere.

This implies that a given strategy may loose a fraction of players $\alpha^{mut}\mathbf{n}_i(t)$, which happens with probability p^{mut} or gain $\alpha^{mut} \cdot \mathbf{n}_j(t)$ which happens with probability $p^{mut} \cdot p_{i\to j}$ where

$$p_{j\to i} = \frac{e^{\frac{|i-j|}{\sigma}}}{\sqrt{2\pi\sigma}} \tag{14.38}$$

is the gaussian probability of j mutating into i. This second effect describes the probability of being populated by a mutation of some other strategy. The mean field equation therefore has the form:

$$n_i(t+1) \simeq n_i(t) + \left(\sum_j J_{ij}n_j(t) - \sum_{ij} J_{ij}n_i(t)n_j(t)\right) \cdot n_i(t)$$
$$- \alpha^{mut}p^{mut} \cdot n_i(t) + p^{mut}\sum_j \alpha^{mut}n_j(t)\, p_{j\to i} \tag{14.39}$$

One can see that the only difference with the stochastic update rule lies in the second part of the equation. We can express Eq. (14.39) in compact form as

$$\mathbf{n}(t+1) - \mathbf{n}(t) = \mathbb{T}(\mathbf{n}(t))\mathbf{n}(t) \tag{14.40}$$

where

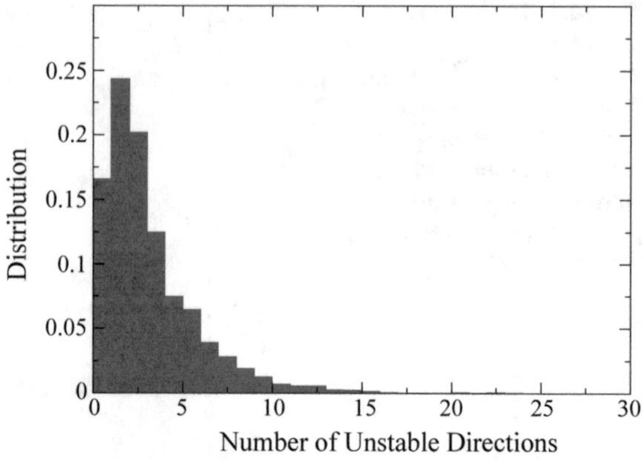

Fig. 14.5 Distribution of the number of unstable directions in the qESS of the SRM

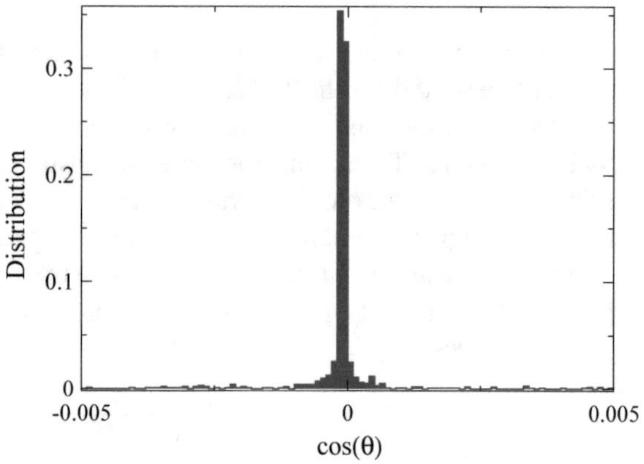

Fig. 14.6 We present the distribution of the $cos(\theta)$ between eigenvectors of the unstable susbace $\mathbf{e}^{+} \in \mathbb{S}^{+}$. We can see how they are almost orthogonal in most of the cases

$$\mathbb{T}_{ij} = \left(\sum_{j} J_{ij} n_j(t) - \sum_{ij} J_{ij} n_i(t) n_j(t) - \alpha^{\text{mut}} \right) \cdot \delta_{ij} \tag{14.41}$$

$$-\alpha^{\text{mut}} p^{\text{mut}} p_{i \to j} \cdot \left(1 - \delta_{ij}\right)$$

The stability matrix is obtained by substituting Eq. (14.41) in Eq. (14.10)

$$\mathbb{M}_{ij} = \mathbb{T}_{ij}(\mathbf{n}^{*}) + \left[J_{ij} - \sum_{k} (J_{ik} + J_{ki}) \mathbf{n}_k^{*} \right] \mathbf{n}_i^{*} \tag{14.42}$$

As one can see once again the $\mathbb{M}$ is not symmetric, $M_{ij} \neq M_{ji}$ so before limiting our analysis only the eigenspace we have to take a closer look at the structure of its spectrum.

The eigenspace of the jacobian of Eq. (14.42) is completely similar to the one found for the TNM's jacobian in Eq. (14.33). Once again

$$\lambda_i \neq \lambda_j \qquad \forall \lambda_i, \lambda_j \in \mathbb{S}^+ \tag{14.43}$$

so that the $\mathbf{e}_+ \in \mathbb{S}^+$ form a linear independent set of vectors which leaves no space for the generalised eigenvectors. This means that the unstable part of the jacobian is diagonalisable even in the SRM. In Fig. 14.5 we present the distribution of the number of unstable direction n_{λ_+} in each stable phase. We can see how $P(n_+)$ rapidly goes to zero, with most fixed points having under 5 unstable directions. Once again this means that through our method we can limit the monitoring of the dynamics from 256 directions to only 5, which is a strong improvement.

In Fig. 14.6 we show the distribution of the angles between eigenvectors belonging to the same subspace $\mathbb{S}^+$, using the formula once again in Eq. (14.37). We can see how $\mathbb{S}^+$ is very close to being an orthogonal space.

Chapter 15
Forecasting Procedure Based on Full Information

Duccio Piovani, Jelena Grujić and Henrik J. Jensen

In this chapter we will look at the results obtained when trying to forecast the arrival of transitions exploiting the theory introduced in the previous section. This procedure requires full knowledge on the system, and in order to apply it one needs to know both the full structure of the network and the weights of each link. Despite being unrealistic and of difficult application, this procedure was thought as a necessary first test of the general validity of forecasting method. We will start with a general outline of the method, and then show the results of its application to the two models.

15.1 Procedure

In the mean field approximation of the models the fixed point configurations are given as solutions to $\mathbb{T}(\mathbf{n}^*) \cdot \mathbf{n}^* = 0$. Because of the high dimensionality of the type of systems we have in mind, this equation will typically not be solvable analytically. In any case, the stochastic dynamics will not satisfy the fixed point conditions strictly.

D. Piovani
Head of Data Science, nam.R, 4 rue Foucault, 75116 Paris, France

J. Grujić
Department of Computer Science, Sciences and Bioengineering Sciences, Artificial Intelligence Laboratory, Vrije Universiteit Brussel, Pleinlaan 2, 1050 Brussels, Belgium

Computer Science Department, Faculty of Science, Machine Learning Group, Université Libre de Bruxelles, Boulevard du Triomphe, CP 212, 1050 Brussels, Belgium

H. J. Jensen (✉)
Department of Mathematics and Centre for Complexity Science, Imperial College London, South Kensington Campus, London SW7 2AZ, UK
e-mail: h.jensen@imperial.ac.uk

E. Altman et al. (eds.), *Multilevel Strategic Interaction Game Models for Complex Networks*, https://doi.org/10.1007/978-3-030-24455-2_15

Rather we will expect little time variation during a meta stable phase, i.e. $\mathbf{n}(t+1) \simeq \mathbf{n}(t) = \mathbf{n}^* + \delta\mathbf{n}(t)$.

In order to overcome this difficulty we approximate the fixed points of the mean field equation by local time averages over successive configurations in the quasi-stable phases of the full stochastic dynamics, namely:

$$\mathbf{n}^{\text{stoc}} = \frac{1}{T}\sum_{t=0}^{T}\mathbf{n}(t) \simeq \mathbf{n}^* \tag{15.1}$$

for which

$$\mathbb{T}(\mathbf{n}^{\text{stoc}})\mathbf{n}^{\text{stoc}} \simeq 0 \tag{15.2}$$

This will be the configuration around which we will study the fluctuations of the system. Indeed we are interpreting the qESS as a fixed point of the stochastic dynamics. Our goal is indeed to study the stability of the qESS, and to do that we map the stability in the neighbourhood of $\mathbf{n}^{\text{stoc}}$. This will allow us to predict the system's reaction to the stochastic perturbations, given that to the extent that the mean field matrix correctly describes the system, the transitions will happen along unstable directions in the configuration space.

In order to check if our method is correct two are the quantities that must be monitored: the instantaneous distance from the fixed point

$$\delta n(t) = \|\delta\mathbf{n}(t)\| = \|\mathbf{n}(t) - \mathbf{n}^{\text{stoc}}\| \tag{15.3}$$

and the maximum overlap between the perturbation and the eigenvectors $\{\mathbf{e}^+\}$ of the unstable subspace

$$Q(t) = \|\delta\mathbf{n}(t)\cdot\mathbf{e}_i\|_{\max} \qquad \forall i : \mathbf{e}_i \in \{\mathbf{e}^+\} \tag{15.4}$$

We have tried several ways to quantify the overlap of the perturbation with the unstable subspace and they all give extremely similar results because usually the system leaves the metastable state parallel to only one of the eigenvectors. The quantity in Eq. (15.3) will tell us the magnitude of the instantaneous perturbation while Eq. (15.4) will tell us its direction. Understanding how these two quantities relate to each other will tell us if our hypothesis are correct. It will tell us if by knowing the direction of the perturbation one can truly predict the future behaviour of its magnitude.

We build our procedure so that it can be applied to the systems in real time. To understand when the system enters a fixed point, we average the occupation boldtor $\mathbf{n}(t)$ over time windows of $\Delta T = 100$ time units (i.e. generations in the case of the TaNa) to obtain $\mathbf{n}^{\text{stoc}}$. We then check if the system is stationary, i.e. $T[\mathbf{n}^{\text{stoc}}]\mathbf{n}^{\text{stoc}} \simeq 0$, repeating the process until the condition is satisfied. When that happens we linearize about the configuration $\mathbf{n}^{\text{stoc}}$, and obtain the specific form of the stability matrix $\mathbb{M}(\mathbf{n}^{\text{stoc}})$ and therefore its eigenspace. At this point we are able to compute both $Q(t)$

and the instantaneous deviation from $\mathbf{n}^{\text{stoc}}$: $\|\delta\mathbf{n}(t)\| = \|\mathbf{n}(t) - \mathbf{n}^{\text{stoc}}\|$. A transition in this description is pictured as an unbounded sudden growth of $\|\delta\mathbf{n}(t)\|$. Once a transition out of the current qESS has occurred, we average again $\mathbf{n}(t)$ to establish the new quasi stable configuration $\mathbf{n}^{\text{stoc}}$.

15.2 Results

We will now show the results of the method we have just presented when applied to the two models. In both cases the results have been very good. The perturbation's orientation in $\mathbb{M}(\mathbf{n}^{\text{stoc}})$'s eigenspace proved to be a powerful early warning for the transitions. The forecasting success rate has been 85/90%. We have seen that the 15/10% of *missed* transitions, i.e. not forecasted, are the those happening along directions considered stable in our framework.

As we will see in the following paragraphs the picture that comes out from our analysis is one in line with a dynamical system embedded in a heterogeneous complex energetic landscape. Imagine a stable phase as a local minimum in a such high dimensional space. Once could interpret the *slopes* as the unstable eigendirections while the *barriers* as the stable ones. Therefore if there is indeed a much higher probability of leaving the local minimum through a slope, in our method there is still a non vanishing possibility of jumping over a barrier. This effect that takes into account of a stochastic perturbation being large enough to push the system out of the fixed point in a *wrong* direction, and indeed, as we will see, increases for increasing noise.

15.2.1 Tangled Nature

All the results presented in this paragraph are obtained with the paramount set $L = 8$, $p^{\text{mut}} = 0.2$, $p^{\text{kill}} = 0.4$, $K = 40$ and $\mu = 0.07$, unless otherwise stated. Given the big computational advantage of avoiding the calculation of the generalised eigenspace, and the particular form of $\mathbb{S}^+$ we have decided to construct the stability indicator only considering the eigenvectors $\mathbf{e}^+ \in \mathbb{S}^+$.

The first step is to take a look at the two quantities in Eqs. (15.3) and (15.4) in one single transition. In Fig. 15.1 we show Q (blue curve) and δn (red curve) as a function of the microscopic time steps. One microscopic time step is equal to one killing and one reproduction attempt, and as said one generation is 1 gen $= N(t)/p^{\text{kill}}$. From the figure it is possible to see that this roughly equals to 1 generation being roughly 10^4 single time steps. It is important to bear this in mind when interpreting the forecasting results that will be showed in this section.

Is is possible to clearly distinguish between the stable phase and the arrival of the transition. We observe that $\delta n(t)$ fluctuates during the qESS around a constant value. It seems that the fluctuations have a characteristic magnitude through out the qESS.

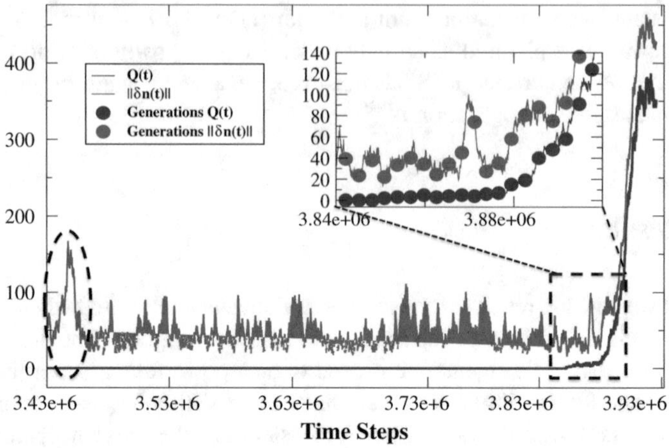

Fig. 15.1 Typical behavior of $Q(t)$ and $\|\delta\mathbf{n}(t)\|$ in a single run of the TaNa in time steps. Clearly $Q(t) \simeq 0$ even for more rare strong fluctuations (dashed circle) inside the qESSs, whereas it starts to increase rapidly before the actual transition. In the inset, we zoom on the transition and indicate with markers the points observed at the coarse-grained level of generations. Notice that between two generations many time steps (events) are present

The dashed circle indicates a perturbation that exceeds the normal values. On the other hand Q has no reaction to the fluctuations, and even for the stronger perturbation its value stays zero. This means that the perturbations $\delta\mathbf{n}(t)$ are happening in $\mathbb{M}$'s stable subspace $\mathbb{S}^-$. Q only grows when a transition is about to occur. Typically Q starts to increase several generations prior to the transition corresponding, in this particular case, to thousands of single update events.

As we can see from the inset, when Q starts peaking the values of $\delta n(t)$ are well within its characteristic values. An observer sitting at time step $t = 3.84 \cdot 10^6$ would have no clue of the arrival of a transition by only observing the $\delta n(t)$ time series. But the knowledge of the bump in Q would be interpreted as a warning sign for the arrival of an extreme rearrangement.

A more systematic analysis is showed in Fig. 15.2. We denote t^* the time at which the transition begins, which is set by the $\delta n(t)$ crossing a reasonably chosen threshold T_δ and staying consistently above this threshold $T_\delta = 150$. Given the sharp increase of $\delta n(t)$ when approaching the transition, t^* doesn't depend strongly on the precise choice of the threshold as long as its is chosen larger than the characteristic fluctuations of $\delta n(t)$ during the metastable configurations. To qualitatively understand the relation between $\delta n(t)$ and Q we studied the joint probability density $P(\delta n(t^* - \tau), Q(t^* - \tau))$ for τ generations before the t^*.

From the way the region of largest support move in the $Q - \delta n$ plane as the transition is approached we qualitatively see to what extent monitoring Q allows one to predict the transition. The bins in the Q axis are set to $b_Q = 10$ while in the $b_{\delta n} = 15$. Note that a significant support for values of Q starts to develop from around $\tau = 5$. At these times the deviation δn is still most often below the inherent qESS

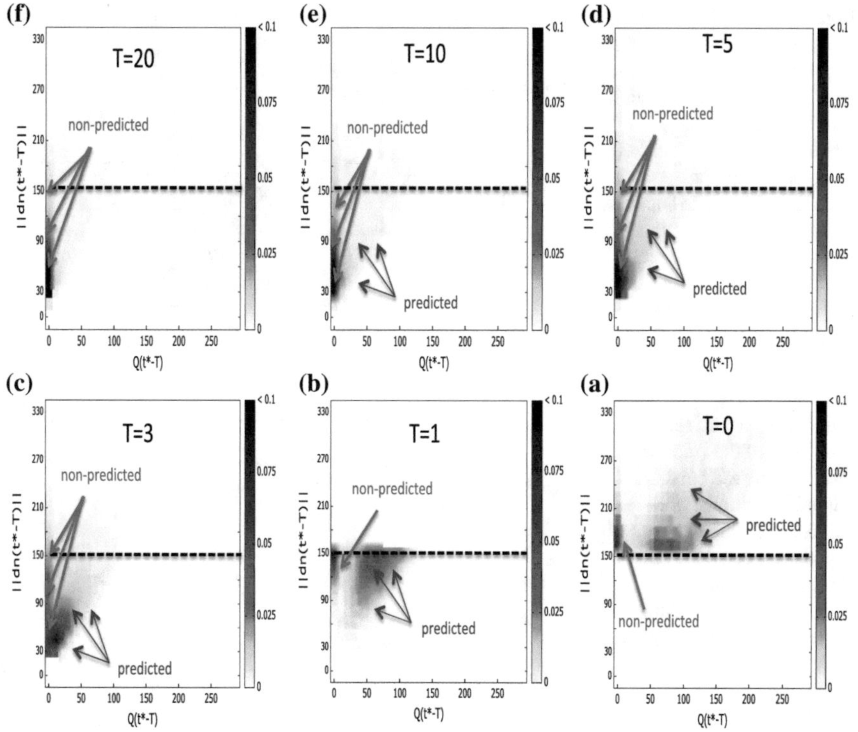

Fig. 15.2 2D distribution $P(\delta n(t^* - \tau), Q(t^* - \tau))$ averaged over 13000 transitions for different values of τ. The predictive power of Q is evident: typical fluctuations inside the qESSs are not signaled by Q (panels (**e**–**f**)), whereas dangerous perturbations leading to a transition are recognized by the increasing of Q away from zero (panels (**a**–**d**)). This is already seen for $\tau = 5$, which is still remarkably far from the transition. Examples of predicted/non predicted transitions are then shown with arrows in panels (**d**–**a**). The other plots can be interpreted in a similar way

fluctuation level of T_δ. We may encounter situations where Q gives a false signal, by increasing significantly in correspondence to small amplitude perturbations of $\delta\mathbf{n}(t)$. Such events will be analysed further down in the chapter.

In this first analysis we consider predicted transitions those for which the $Q(t)$ has moved at least to the second bin before $\tau = 0$. There is quite a neat separation happening around $\tau = 3$ and $\tau = 1$, between the transitions that have been predicted and those who have not. Despite a good success rate, approximately 85–87%, non-predicted transitions do occur and are related to the system leaving the qESS following a direction that which is weakly stable (negative eigenvalues close to zero). This is shown in Fig. 15.3, where the distribution of the real parts of the eigenvalues responsible for the transitions is plotted.

By looking at the distribution in Fig. 15.3 we can see show not only the sign of the real part but even the norm $\|\mathrm{Re}(\lambda)\|$ gives us information on the particular eigendirection of the fixed point. Indeed for negative values the larger the norm

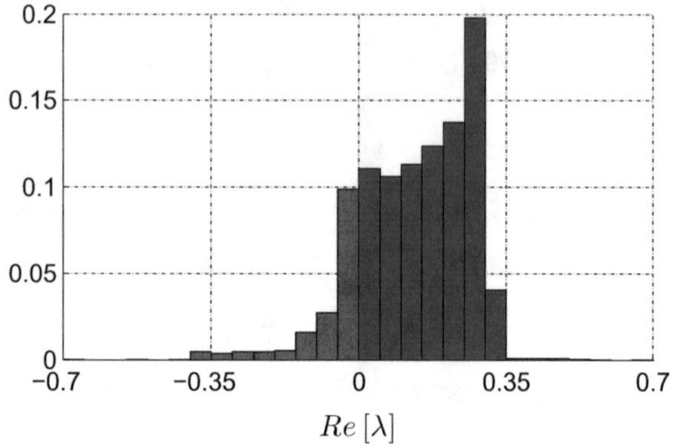

Fig. 15.3 Distribution of the real part (red/blue boxes for negative/positive one) of the eigenvalues correspondent to eigendirections with maximum overlap with $\delta\mathbf{n}(t)$ at the beginning of a transition. The distribution is clearly dominated by the unstable eigenspace, but a significant probability (≈17%) of weak stable eigenvalues is found

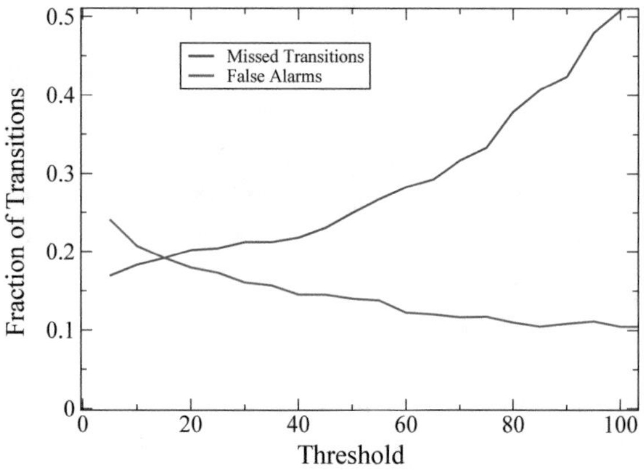

Fig. 15.4 We can see the behaviour of the fraction of false alarms, red curve, and missed transitions, blue curve, for different values of A_Q. As expected the two curves have opposite behaviour for increasing values of A_Q. From this figure a reasonable choice seems $A_Q = 20$

the less likely it is for a transition to occur in such direction. This result in full agreement with our analogy of an a complex energetic landscape: the larger the norm the higher barrier for negative values, and the steeper the slope for positive values. This purely stochastic phenomenon explains why we find with non vanishing probability transitions together with $Q \simeq 0$ (see Fig. 15.2, panel (a)).

In order to quantitatively study the problem we define an alarm signal. To do so we determine an appropriate threshold A_Q on $Q(t)$ and compare the number of false alarms with the number of missed transitions generated by different values of the chosen threshold A_Q. A false alarm is when the $Q(t)$ crosses A_Q but then goes back under its value before any transition occurs. On the other hand a missed transition corresponds to situations where $Q(t)$ remained below A_Q even though the given metastable configuration did become unstable and therefore a transition did occur.

In Fig. 15.4 we show these two quantities for different A_Q. The red curve is the fraction of missed transitions while the blu is the fraction of transitions that have produced false alarms. I have to spend a few lines to elaborate on this. If a given fixed point produces one or many false alarms it will be treated in the same way in our analysis, in that they are both considered as transition that have produced false

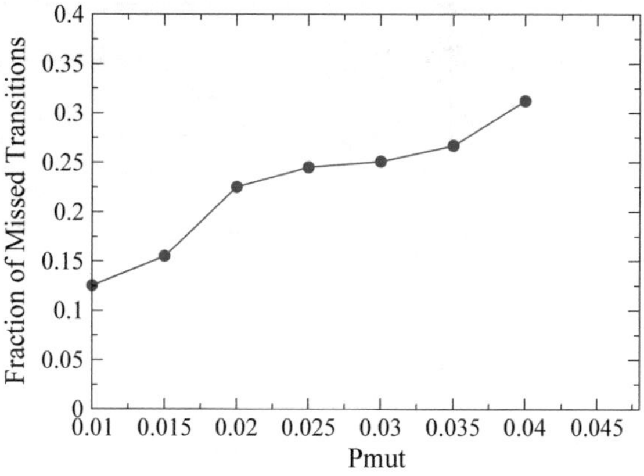

Fig. 15.5 We present in this figure for both models the fraction of missed transitions as a function of the noise in the system. We can see how for nosier systems its harder to forecast a transition

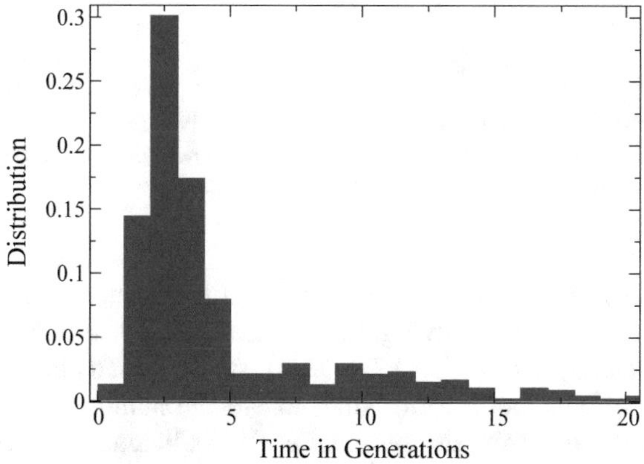

Fig. 15.6 Distribution of the respite of the alarms for a given threshold $A_Q = 20$

alarms. This is because we have found some fixed points whose analysis produced hundreds of false alarms but very few times the number was one or two. These events cannot absolutely be considered as independent because from our point of view this is due to the nature of the fixed point, or the mean field approximation more than due to the procedure. That said when increasing A_Q the fraction of false alarms decreases, as expected, while the fraction of missed transitions increases as one may expect.

As a further check on the complex landscape analogy we have checked the percentage of missed transitions for increasing levels of noise. To do this we have repeated the analysis for different values of the mutation probability p^{mut} for the same $A_Q = 20$. The parameter p^{mut} is our temperature like variable, so higher levels of p^{mut} imply larger stochastic fluctuations. As we can see for Fig. 15.5 the percentage of missed transitions increases indeed for increasing p^{mut}, meaning that the system for larger noise values jumps more often over the barriers represented by stable λ_-.

But what is the actual forecasting power of $Q(t)$? How many time steps before t^* does the Q indicator give the alarm? In order to measure this we have fixed $A_Q = 20$,

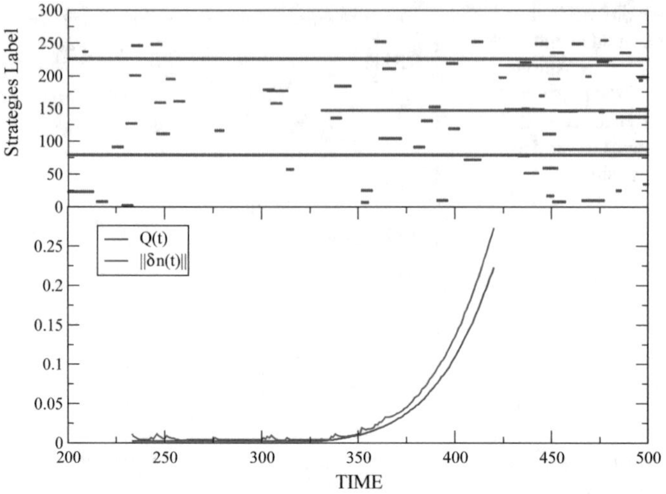

Fig. 15.7 We show the behaviour of our stability indicator $Q(t)$, in both the Replicator Model and the We compare Q's behaviour (blue curve) both to the displacement's $\delta n(t)$ (red curve) behaviour and the occupation plot. We can see how in both models the $Q(t)$ peaks only before the transition, while it doesn't feel the previous fluctuations

and then studied the distribution of the quantity $\Delta T = \|t^* - t_{cross}\|$, prior to $Q(t)$ goes above A_Q. In Fig. 15.6 we present the distribution of ΔT. We can see that more than 50% of cases $\Delta T \in [2, 5]$. As explained above when introducing the model, one generation corresponds to average number of time steps necessary to remove everyone from the system, i.e. $\frac{N(t)}{p^{\text{kill}}}$ individual updates. So even low values of ΔT can be considered to correspond to a strong forecasting power.

15.2.2 Stochastic Replicator Model

Proceeding in the same way as for the TNM we start by looking at the typical behaviour of the two quantities in Eqs. (15.3) and (15.4). In Fig. 15.7 we show how the two quantities behave before a transition. The parameter set in this model is $d = 256$, $\text{n}^{\text{ext}} = 0.001$, $\alpha^{\text{mut}} = 0.01$, $p^{\text{mut}} = 0.2$. We can see that $\delta n(t)$ red curve fluctuates around a constant value while $Q(t) \simeq 0$ during the stable phase. But when just before $t = 350$ a new player appears, and eventually drags the system out of the configuration, they both explode. This seems to mean that even for SRM the high values of $\delta n(t)$ are only possible in the unstable subspace of the eigenspace.

Once again a more systematic analysis is shown in Fig. 15.8 where we investigate the relation of the two quantities $Q(t)$ and $\delta n(t)$ just before the transition. The results are averaged over 10000 transitions. As we can see far from the transition (t = 20) most of the Q(t) are close to zero, but approaching the transition (t = 0) the average

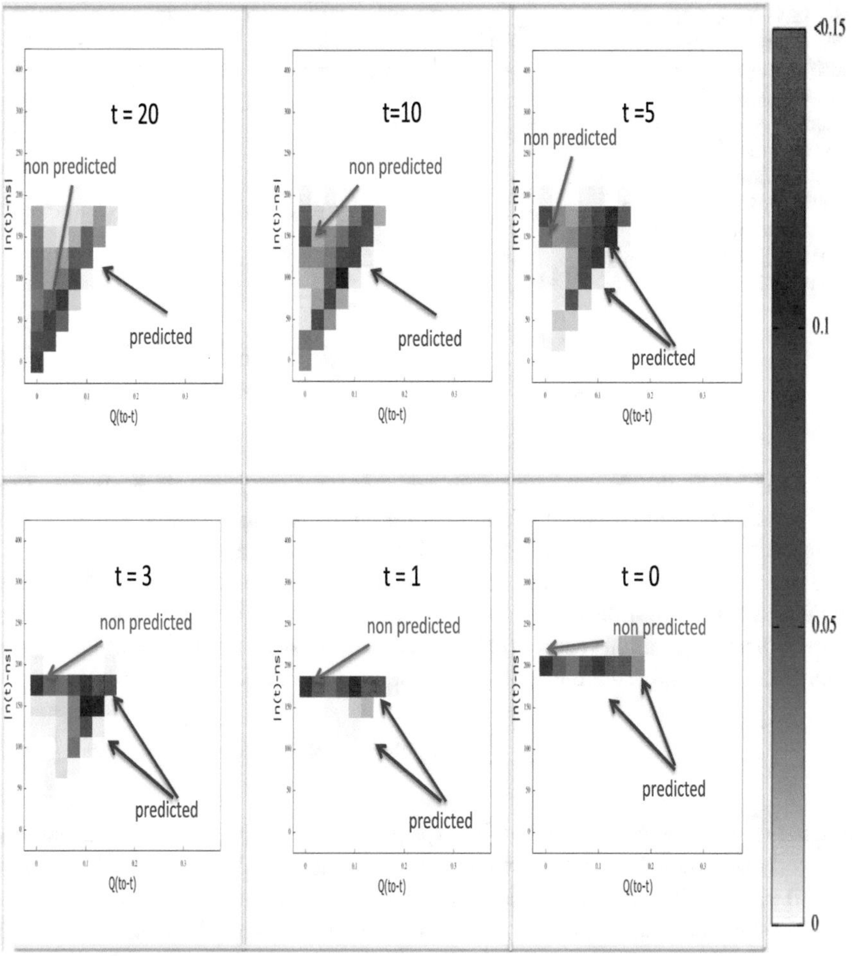

Fig. 15.8 The figures compare the indicator $Q(t^* - t)$ (x-axis) and $\delta n(t^* - t)$ (y-axis) where t^* is the time the transitions begins. As we can see for $t = 20$, meaning 20 steps before the transitions most of the transitions are not predicted ($Q(t^* - t) \simeq 0$), but as the system approaches the transition the vast majority of transitions (80/90%) are predicted by an increase of $Q(t)$

values of $Q(t)$ start increasing while the $\delta n(t)$ stays more or less constant. It is clear from Fig. 15.8 that even in this case a fraction of the transitions are not predicted. We interpret the *missed transitions*, exactly in the same way as for the TNM and will give proof of this further down.

To define an alarm we determine an appropriate threshold A_Q on $Q(t)$. To do so we compare the number of false alarms with the number of missed transitions generated by different values of the chosen threshold A_Q. In Fig. 15.9 we show these two quantities for different A_Q. The red curve is the fraction of missed transitions

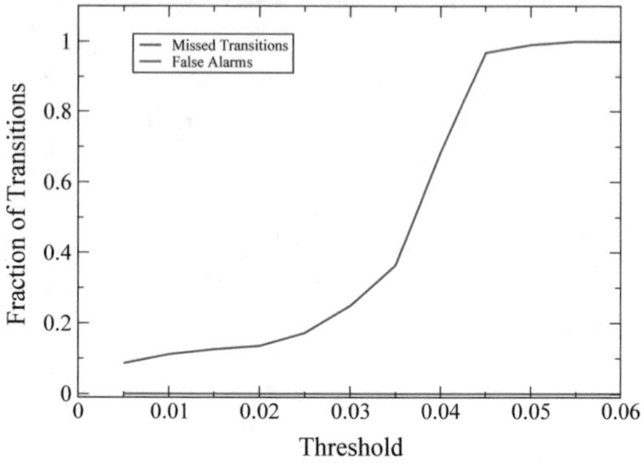

Fig. 15.9 We can see the behaviour of the fraction of false alarms and missed transitions for different values of A_Q. One can see how the procedure produced no false alarms in the Replicator Model which is consistent with what we expected given the Langevin nature of the model

while the blu is the fraction of transitions that have produced false alarms. The figure shows how the procedure, although for an increasing threshold is missing an increasing number of transitions, produces no false alarms at all.

The reason for this, we believe has to do with the Langevin nature of the dynamics in the SRM, i.e. deterministic dynamics + stochastic noise. Within this approach we expand the configuration vector $\mathbf{n}(t)$ in the $\mathbb{M}$' s eigenspace or generalised eigenspace plus noise. Remembering the form of the coefficients in Eq. (14.14) one gets

$$\mathbf{n}(t) = \sum_k (c_k(0)\exp(\lambda_k t) \cdot \mathbf{e}_k + \epsilon_k) \tag{15.5}$$

where $c_k(0)$ are the coefficients of the expansion and ϵ_k is the noise. This dynamics is clearly dominated by those components for which $\mathrm{Re}(\lambda_k) > 0$, but this is true only if $c_k(0) \neq 0$. When a node is populated by a mutation, in our framework this corresponds to setting $c_k(0) > 0$ for one or more k. If the coefficient is relative to an unstable direction from then on the term is suppressed if and only if the ϵ_k points in the opposite direction which given the high dimensions of the systems is highly unlikely. In other words in the framework of the SRM once the system is sliding down a slope it is incredibly unlikely for it to go back. In the TNM this picture does not hold because in that case there in no real deterministic part and the dynamics cannot be described in a Langevin style.

As a further check on the analogy with the complex landscape analogy we have checked the percentage of missed transitions for increasing levels of noise. To do this we have repeated the analysis for different values of the mutation probability p^{mut} for the same $A_Q = 0.01$. As we can see for Fig. 15.10 the percentage of missed transitions increases together with the noise, as one may expect.

In Fig. 15.11 we present the distribution of s, $\Delta T = \|t^* - t_{cross}\|$, for $A_Q = 0.01$, to check how many if time steps before the transitions is the alarm given. We can see that in the model the crossing times are tenths of time steps before the transition time.

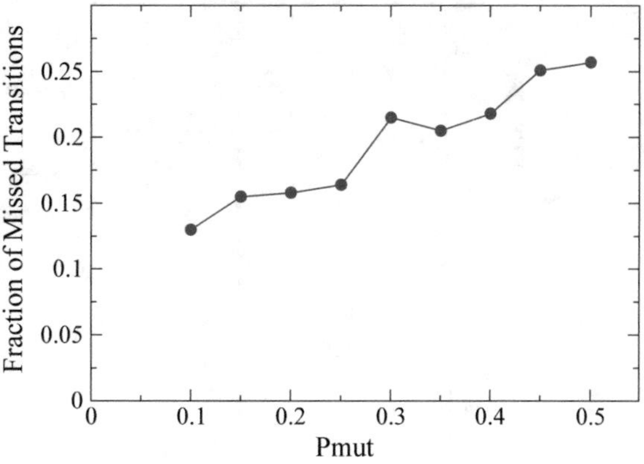

Fig. 15.10 We present in this figure for both models the fraction of missed transitions as a function of the noise in the system. In line with the complex landscape analogy the system jumps over the barriers more often for increasing values of p^{mut}

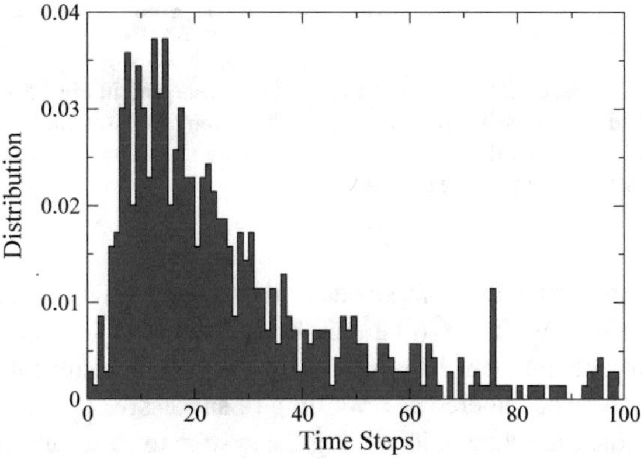

Fig. 15.11 Distribution of the respite of the alarms for a given threshold. The left panel refers to the Replicator model, for which $A_Q = 0.01$ a

Which means that in the SRM the system will go through many cycles of updates before the transition occurs.

15.3 Analysis of the Stability Indicator Q(t)

The stability indicator $Q(t)$, who's peaks we have used as early waring signals for forthcoming transitions is a measure of how much, at a given time, the occupation vector $\mathbf{n}(t)$ is embedded in the unstable subspace $\mathbb{S}^+$. In this section we want to understand the microscopic mechanisms that lead to this result, and how they translate in the model's details.

What $Q(t)$ is actually doing is measuring the activity of the occupancy on dangerous nodes, i.e. nodes that are *toxic* for a given stable configuration. Indeed every non zero component of the unstable eigenvectors $\mathbf{e}^+$ will tell us which nodes of the interaction network would bring instabilities in the system. Namely if $e_j^+ > 0$, where

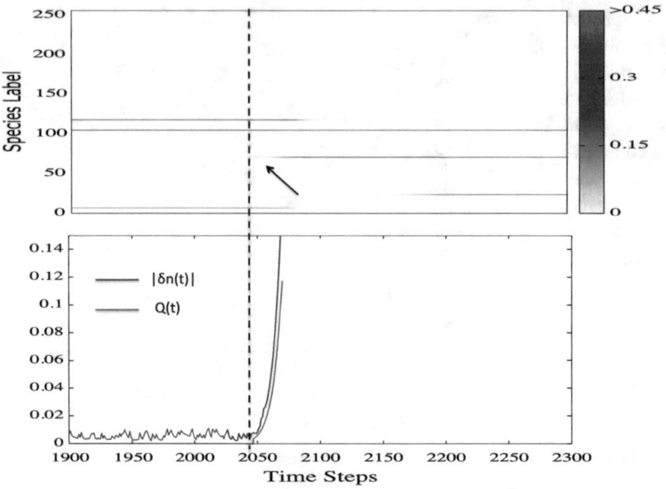

Fig. 15.12 This is the same type of figure showed in Fig. 15.13 for the SRM. Bottom panel $\delta n(t)$ and $Q(t)$, blue and red curve respectively, top panel weighted occupation plot. We can see how even in this model the transition is triggered but the arrival of a new fit mutant that my gaining weight disturbs the existing equilibrium

j indicates the component of the unstable eigenvector, this means the jth node is dangerous. The $Q(t)$ monitors the activity of such nodes. If one of these nodes were to become activa by mutations this would result into a rapid growth of $Q(t)$ and can be considered as a warning of successive transition. In other words the way the stochastic fluctuations bring the system towards unstable directions, is by activating the *toxic* components n_t of the occupation vector. And this is exactly what $Q(t)$ measures: the occupation of the toxic components.

Here we illustrate the temporal behaviour of $Q(t)$ and $\delta n(t)$ for both the TNM in Fig. 15.13 and the SRM in Fig. 15.12. In the top panels we present weighted occupation plots while the bottom figures show the behaviour the two quantities in $Q(t)$ and $\delta n(t)$. The arrow points at the new dangerous mutant that has entered the system, while the dashed bar indicates the moment it happens. Before the dashed line we can see how fluctuations in $\delta n(t)$ are bounded and $Q(t)$ essentially equals to zero. After the dashed line, when the new mutant has entered the system, we see an explosion of both quantities.

It is clear how the mutant once in the system quickly gains occupancy. One can see that by observing the shift from blue to red of the relative curve. As the curve becomes more red in the top panel, in the bottom panel the Q(t) increases its value and rapidly peaks. This is true for both models.

Mathematically the explanation is trivial. If one remembers that

$$Q(t) = \| \sum_j e^i_j n_j(t) \|_{\max} \quad \forall i : \mathbf{e}^i \in \mathbb{S}^+ \tag{15.6}$$

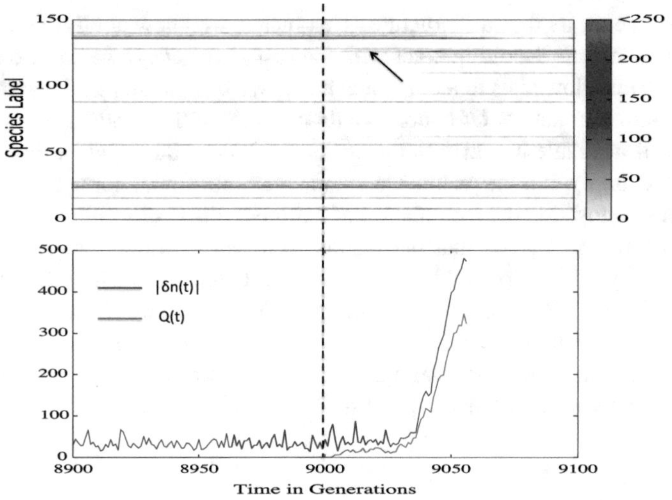

Fig. 15.13 In the bottom panel of this figure we show the behaviour of $\delta n(t)$ (blue curve) and $Q(t)$ (red curve) while approaching the transition in the Tangled Nature. In the top panel a weighted occupation plot is presented. We can see how the beginning of the transitions (dashed vertical black line) is triggered by a new mutant (black arrow) that quickly gains population. The arrival of the new dangerous mutant is singled by a peak in the Q(t)

then its clear that being the eigenvectors constant the only thing that changes are the components of the occupancy vector. Furthermore the only changes who will effect the indicator $Q(t)$ are those relative to components $j\,:e^{i}_{j} \neq 0$. Typically in the TNM there are $\simeq 5-10\;\,n_t$ for every stable configuration $\mathbf{n}^{\mathrm{stoc}}$, while the number for the SRM is slightly higher, oscillating between 15 and 30. Considering that the dimensions of the networks in our simulations is 256, monitoring 30 nodes at most is far more convenient computationally speaking.

So once again by use of the LSA mean field we are spotting the dangerous nodes, and then with $Q(t)$ we are monitoring the activity on these nodes. The moment by random mutations one of these nodes is activated we know there is a serious possibility that the system will be pushed away from its configuration.

15.4 Discussion

In this chapter we have presented the results of our forecasting method based on the LSA. We have tested it on a fully stochastic model like the TNM and a model with Langevin dynamics like the SRM. In both cases the results have been very good with the forecasting percentage reaching up to 90% of the transitions.

The procedure was based on full information on the system though. Not only we exploited the interaction relations of each node in the active network but we also used the knowledge on its the full structure. This means we knew at all time how every possible new mutant would interact with the rest of the active network once being populated. This allowed us to learn which nodes were to classify as dangerous while they were still non active. Of course we do realise that this amount of information will never be available in real systems, making this procedure quite *naive*. However this does not imply that the results we obtained are nor trivial nor useless. This was a required step towards the formulation of a more realistic procedure.

An encouraging sign is that the results were very similar for different levels of stochasticity. This is probably due to the fact that we are using the mean field approximation in the neighbourhood of a fixed point where the dynamics is stationary. It is tempting to say that for a stationary process a mean filed approximation is a good descriptive tool, and therefore claim that our results our exportable to other a broad range of systems with the same type of macroscopic behaviour.

Chapter 16
Procedures with Incomplete Information

Duccio Piovani, Jelena Grujić and Henrik J. Jensen

As mentioned at the beginning of the chapter an obvious short coming concerning application to real situations of the forecasting procedure as described so far is that we make use of complete knowledge of the entire space of agents and their interactions. To test the strength of the results against incomplete information a first attempt has been introducing an error in the interaction matrix used for the mean field treatment. This represents the situation in which an observer would have to measure the interactions between agents and does so with an error. This is possibly the biggest problem one would have to overcome when trying to describe real systems. As we will see the forecasting method has proven itself to be quite robust, yielding similar results in both models even in the presence of non negligible errors.

Furthermore we will discuss a new forecasting procedure, inspired by the one already presented, which doesn't need any knowledge about "in potentia" agents, which means that we don't need to know the complete structure of the underlying network. We only need to focus on the highly occupied nodes present in the system, and on their interactions. In other words in this new approach we only know what we see without making any use of the non active part of the interaction network, nor

D. Piovani
Head of Data Science, nam.R, 4 rue Foucault, 75116 Paris, France

J. Grujić
Department of Computer Science, Sciences and Bioengineering Sciences Artificial Intelligence Laboratory, Vrije Universiteit Brussel, Pleinlaan 2, 1050 Brussels, Belgium

Computer Science Department, Faculty of Science, Machine Learning Group, Université Libre de Bruxelles, Boulevard du Triomphe, CP 212, 1050 Brussels, Belgium

H. J. Jensen (✉)
Department of Mathematics and Centre for Complexity Science, Imperial College London, South Kensington Campus, London SW7 2AZ, UK
e-mail: h.jensen@imperial.ac.uk

E. Altman et al. (eds.), *Multilevel Strategic Interaction Game Models for Complex Networks*, https://doi.org/10.1007/978-3-030-24455-2_16

of the poorly occupied nodes. Once again the results have proven to be completely similar to the case of full information.

The good response to both these first attempts can be interpreted optimistically. Indeed it suggests that full knowledge of the structure of the interaction is not a necessary requirement to gain information on the stability of the fixed point's configuration.

16.1 Interactions with Error

We formalise what said in the introduction defining a new interaction matrix

$$J_{ij}^{e} = J_{ij}^{\mathrm{sim}} + \chi \tag{16.1}$$

where χ is $N(0, \sigma)$, i.e. a normally distributed random variable, of mean 0 and variance σ. We then repeat the exact same procedure outlined in the previous section but using $\mathbf{J}^{e}$ in the calculations. We will therefore study the stability around the fixed points using the an inaccurate Jacobian $M_{ij}(\mathbf{J}^{\mathbf{e}}) \equiv M_{ij}^{e}$.

In the limit that the mean field correctly describes the underlying dynamics, $\mathbb{M}$'s eigenspace is indeed a stability space, and its positive eigenvalues precisely spot where the instabilities and their direction. For this reason if we embed the model's dynamics in such space we are able to distinguish dangerous from non dangerous perturbations. If on the other hand the approximation fails to describe the dynamics, the eigenspace becomes just another vector space, and its eigenvalues loose their property. Embedding the dynamics in the eigenspace in this case will not give us any additional information on its stability. What we are doing by introducing an error in the interactions, is passing from a situation where we have showed that the mean field is a good approximation to one in which it will loose touch with the actual dynamics. In other words for sufficiently big errors eigenvectors $\mathbf{e}_{\lambda} : \mathrm{Re}(\lambda) > 0$ will not point to unstable directions. Here we want to check the robustness of the method to these errors.

In Fig. 16.1 we present the fractions of transitions we are not able to forecast and the fractions of false alarms we generate as function of the variance σ in the TNM. We are testing the performance of the method as function of how much the interaction matrix used for the stability analysis differs from the correct set of interactions. We can notice that for $\sigma < 0.2$ we are still able to forecast around 70% of the transitions and we generate less than 20% of false alarms. This is an encouraging result since a $\sigma = 0.2$ is clearly a significant error given that $J_{\mathrm{ij}} \in (-1, 1)$.

In Fig. 16.2 we present the same graph for the SRM. We can see how the results are very similar. In Fig. 15.9, in the previous chapter, we showed that in the SRM the method produced no false alarms, and gave an explanation of why this happened. Perhaps surprisingly this result holds even for large errors, and the fraction of false alarm stays $\sim$0.

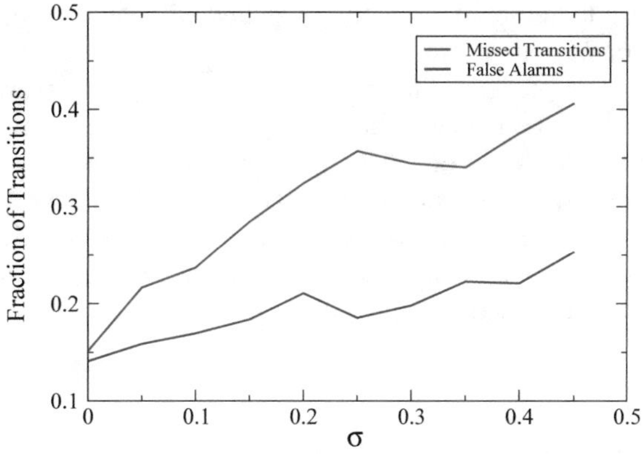

Fig. 16.1 In this figure we show the fraction of the transitions we are not able to forecast and the fractions of false positive, in function of the σ of the distribution of the random error in the interactions. Once again we have used $A_Q = 30$ for the TNM (right panel) and $A_Q = 0.01$ of the Replicator Model (left panel)

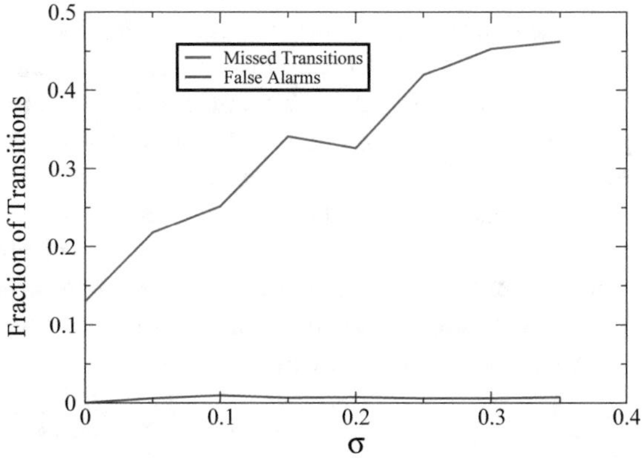

Fig. 16.2 In this figure we show the fraction of the transitions we are not able to forecast and the fractions of false positive, in function of the σ of the distribution of the random error in the interactions in the SRM. $A_Q = 0.01$

The fact that in both models, for errors that represent a considerable percentage of the actual interactions, we are still able to forecast the vast majority of transitions is definitely a good thing. It could imply that one is not forced to infer the exact interaction terms in the calculation of a Jacobian. Without this result an application to real systems would seemed not feasible.

16.2 New Procedure

We next consider a much simpler measure than the overlap function $Q(t)$. This new measure is inspired by the analysis presented above and leading to $Q(t)$ but avoids access to information about the *adjacent possible*, i.e. information about agents that are not extant in the system but could be populated through mutations. Our new measure only makes use of the time evolution of directly observable quantities and can therefore in principle be applied without the need of a dynamical model of the considered system.

For example if the method was to be used to analyse an economic environment it could be implemented just by observing the existing companies, without having to guess on the one that may appear. Translated into the model's jargon this means we will exploit only information coming from the active or occupied network. This implies that we reduce the dimensionality of the problem, which will vary depending on the specify configuration we need to analyse. The dimension will be given by the number of active nodes d_a:

$$\mathbf{n}^{\mathrm{stoc}} \longrightarrow \mathbf{n}^{\mathrm{a}} \qquad \mathbb{M}(\mathbf{n}^{\mathrm{stoc}}) \longrightarrow \mathbb{M}(\mathbf{n}^{\mathrm{a}}) \tag{16.2}$$

where $\mathbf{n}^a = (n_1^a, n_1^a, \ldots, n_{d_a}^a)$ is the active occupation vector where $n_i^a > 0$.

By applying the LSA to the active network $\mathbf{n}^a$ we can check that, during a stable phase, the configuration corresponds to a situation where the spectrum of the $\mathbb{M}(\mathbf{n}^a)$ consists of eigenvalues that all have negative real parts: the analysis of the Jacobian yields no unstable subspace. This means that evolving with the same dynamics but setting $p^{\mathrm{mut}} = 0$, this configuration would be stable for every and the system would never explore other areas of its phase space.

As the system evolves though, new mutants appear. As an indicator of approaching transitions we track the growths of the occupancy of these new agents, if their occupancy exceeds a certain threshold T_a we check the spectrum of the updated $\mathbb{M}$, in which the new agents are included. So every time a new mutant starts gaining occupancy we add it to the active network

$$\mathbf{n}^{\mathrm{new}} = \mathbf{n}^{\mathrm{stoc}} + \mathrm{mutant} \tag{16.3}$$

and compute the spectrum of $\mathbb{M}(\mathbf{n}^{\mathrm{new}})$. In case the spectrum now includes positive eigenvalue we take this as an indicator of, an approaching transition out of the present metastable configuration. This will be our new alarm.

16.3 Results

In the TNM we have implemented this procedure by computing the spectrum of the new Jacobian every time for new mutant $n_{\mathrm{mut}}(t) > T_a = 5$. In Fig. 16.6 we show the results of an application of this new procedure to the TNM. In both panels the red vertical lines indicate the times t^a of appearance of a species able to change the stability of the system, i.e. the alarm time. We can qualitatively see from the figure that just after the alarms the system actually undergoes a transition (Fig. 16.3).

In Fig. 16.4 we show the distribution of the quantity

$$\Delta t = t^* - t^a, \tag{16.4}$$

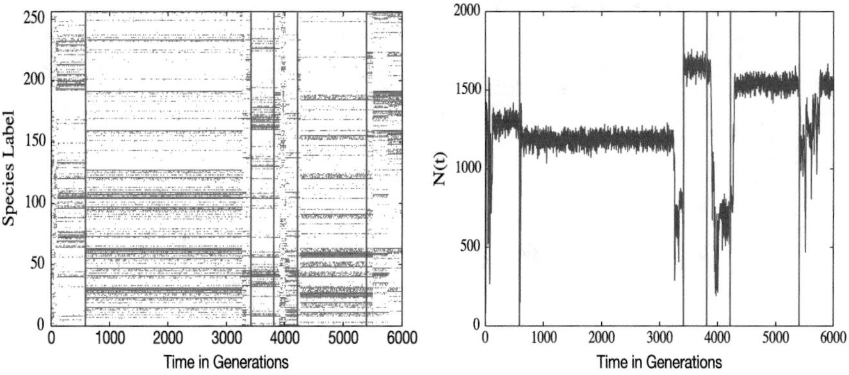

Fig. 16.3 Top left and bottom left respectively occupation plot and total numbers of individual $\sum_j n_j(t) = N(t)$ in the tangled nature model. The vertical red lines represent the alarm times. In the top and bottom right we compare the behaviour of the occupation plot and the frequencies of the most occupied strategies (blue curves) in the Replicator model with the alarms given by our new procedure. One can clearly see how after every alarm the system changes its configuration

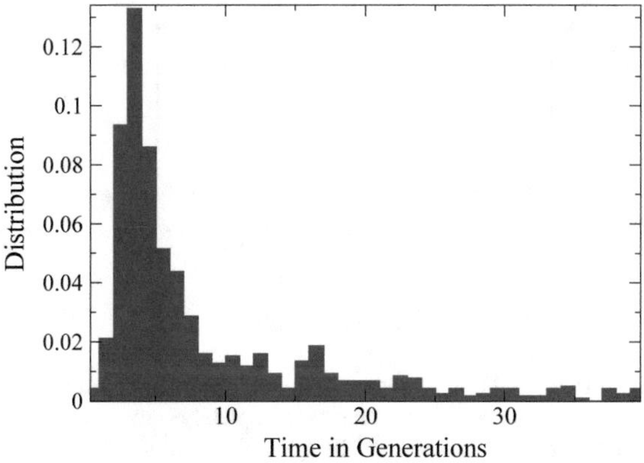

Fig. 16.4 Distribution of the time steps $\Delta t = t^* - t^a$ the alarm is given before the transitions

In the left panel we can see how the red lines appear right at the end of a stable phase and in the right panel we show the total number of individuals present in the system $N(t) = \sum_j n_j(t)$. A transition to a new metastable configuration is associated with a sudden change of this quantity. We notice that right after each alarm $N(t)$ exhibit a significant change. where t^* is the time at which the transition starts and t^a is the alarm time. If we compare this figure with Fig. 15.6 from the previous chapter we notice that with this new procedure one gains a considerable amount of forecasting power. With the old procedure most of the transitions were forecasted with less than 5 generations of advance, while the figure shows a big percentage of transitions forecasted with more than 5 generations before their arrival. This is a surprising result.

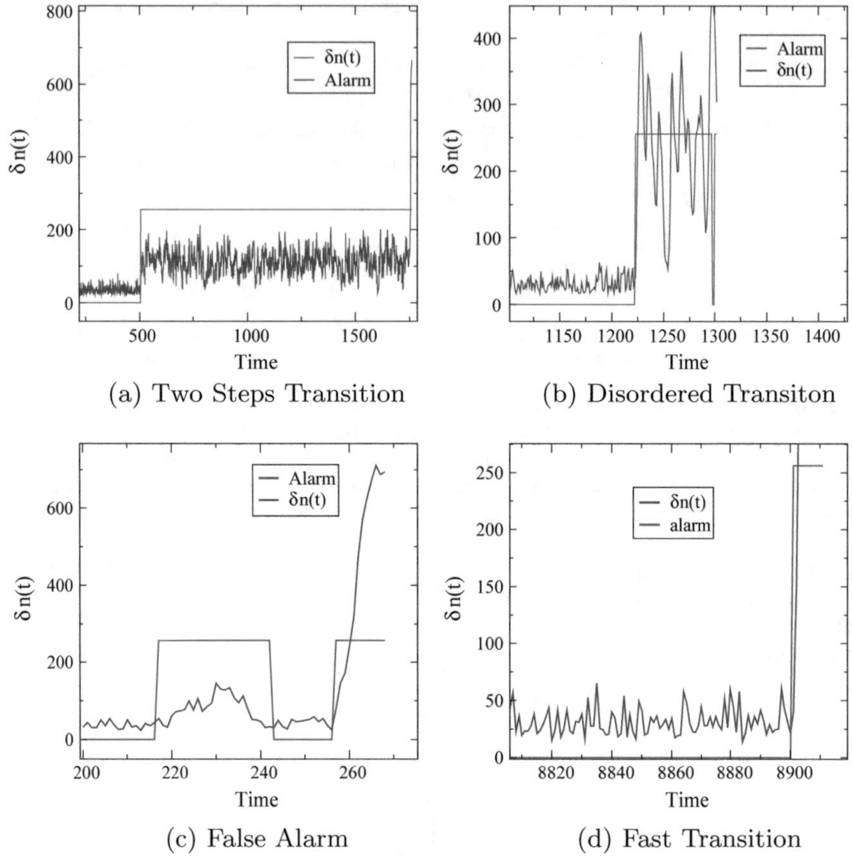

Fig. 16.5 In these figures we compare the quantity $\delta n(t) = \|\mathbf{n}^{\mathrm{stoc}} - \mathbf{n}(t)\|$ with the alarms in the red curve. The value of the alarm is completely conventional, and it was chosen to compare the alarm times with the behaviour of the system. When $a(t) > 0$ it means a dangerous mutant has entered the system. In **a**, **b** we see how once the mutant enters the system the is a first small transition, and only after many time steps the system is pushed further away, and the algorithm recognises the transition. In **c** we show a false alarm, while in **d** a very quick transition

In this case the definition of missed transitions and false alarms are slightly different. We define a false alarm when the occupation of the species responsible for the alarm instead of growing, eventually causing the transition, goes back beneath the threshold T_a. Indeed with the disappearance of the particular species the instabilities go away and therefore the transitions does not occur despite the alarm given. On the other hand a missed transition is simply a transition is not preceded by an alarm.

$$\rho_{\mathrm{false}} = \frac{n_{\mathrm{false}}}{n_{\mathrm{transitions}}} = 0.21 \qquad \rho_{\mathrm{missed}} = \frac{n_{\mathrm{missed}}}{n_{\mathrm{transitions}}} = 0.0618034 \tag{16.5}$$

done on 1000 transitions By comparing these results with the ones showed in Fig. 15.5 we see that with this new procedure the fraction of missed transitions decreases tangibly while the fraction of false alarms slightly increases. Is very unlikely to miss a transition but at the same time one every three alarms will be false.

In order to better understand the results we have just shower it is very instructive to take a close look at a few single transitions. In Fig. 16.5 we show 4 transitions that happened in very different ways. Transitions like that in (**a**) and (**b**) are those responsible for the very high values of Δt in the distribution. In order for a transition to be recognised as one, $\delta n(t)$ has to go over a threshold T_δ and stay over it. When that happens we stop monitoring the system an wait for it to stabilise again. But the t^* is exactly the that time. In (**c**) we show a false alarm while in (**d**) a fast transition which is forecasted with a small Δt.

In general we can say that we have gained a significant forecasting power and decreased of an order of magnitude the fraction of missed transitions. On the other hand we produce slightly more false alarms. On the whole this has been obtained by neglecting a chunk of information of the system and by making the procedure far more applicably to real systems. This is a fantastic improvement of the procedure.

16.3.1 Stochastic Replicator Model

We have implemented the same procedure in the SRM and in Fig. 16.6 we show the results of its application. Every time for a new mutant $n_{\text{mut}} > T_a = 0.01$, we check the spectrum of the active network setting an alarm if it presents at least one unstable direction.

Once again the red vertical lines indicate the alarm time t^a. In the left panel of the Fig. 16.6 the blue curves represent the frequencies of the most occupied strategies in the Replicator model. We can see how right after the red lines, the alarm times, a new strategy starts gaining frequency and eventually puts an end to the stable configuration.

The distribution of the Δt is shown in Fig. 16.7. By comparing the result with the distribution obtained using the old procedure in Fig. 15.4 we will see that the new procedure yields a higher forecasting power for the SRM as well. While by giving the same definition of false alarms and missed transitions as the one used for the TNM we obtain

$$\rho_{\text{missed}} = \frac{n_{\text{missed}}}{n_{\text{transitions}}} = 0.0406958 \qquad \rho_{false} = \frac{n_{\text{false}}}{n_{\text{transitions}}} = 0 \tag{16.6}$$

By looking at Fig. 15.9 we can appreciate how ρ_{missed} has decreased while $\rho_{\text{false}} = 0$ again. So even when applied to the SRM this new procedure performs better with less information.

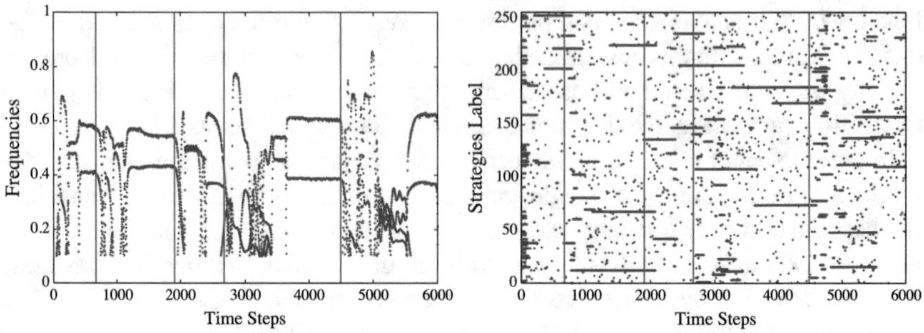

Fig. 16.6 We compare the behaviour of the occupation plot and the frequencies of the most occupied strategies (blue curves) in the SRM with the alarms given by our new procedure. One can clearly see how after every alarm the system changes its configuration

Fig. 16.7 Distribution of the Δt in the SRM

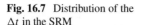

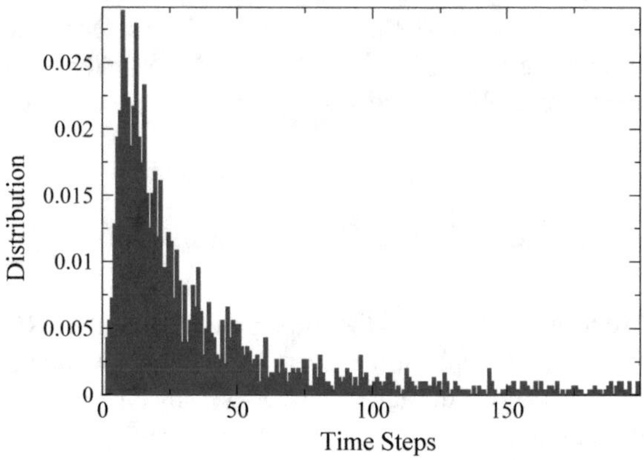

16.4 Discussion and General Conclusions for Part IV

The idea of this chapter was test the results we have obtained in the previous chapter against the lack of information on the system one wants to study. For this reason we have started by introducing an error in the interactions : observing a system one doesn't know precisely how to quantify the interaction between the components of the system and any estimation would be to some extent wrong. It was therefore important to check the robustness of the method to an error in their estimation. The fact that for relatively large errors the method's performance stayed good (80% of predicted transition with $\sigma = 0.2$), is indeed a promising result. This was a necessary condition (unfortunately not sufficient) for the method to be applicable to real systems.

Moreover when dealing with real systems, besides *wrong* interactions, one has to work with partial knowledge on the structure of the system. In the *naive* procedure of the previous Chapter we made use of complete knowledge on the possible components that could enter the system. This means we were able to guess the characteristics of the *toxic* components before they were even activated. These were the ones the $\mathbf{e}^{+}$ pointed at. Once again this is not a realistic situation. To overcome this problem we have changed the procedure making use of the same theoretical results. Indeed the new procedure is completely inspired by the previous one but only makes use of the information on the active network. Besides being realistic by analysing the results showed in this chapter and comparing them with the one of the previous one it is clear we have remarkably improved the performance of the stability indicator. For our future work we wish to test what presented here to a real data.

Of course the situation pictured in this chapter, even if closer to a real system, stays quite stylised and keeps making use of information that perhaps one would struggle to have in a real systems, like a descriptive mean field equation. But building a data driven model with the rapid development towards big-data sampling capacity in many areas of science is increasingly becoming a possibility.

Let us present some general conclusions for Part IV. A complex emergent property of large dynamical systems, formed by many interacting components, as we have seen appears to be the formation of rare extreme events, which lead the system to a transition from one configuration to another. These can be financial crashes, in socio-economical systems, punctuations in evolving biological ecosystems and of many other examples we have cited in this part. These transitions all share the role of primary importance they play in the evolution of such systems. It is now common knowledge that these are endogenously generated by the dynamic of the system themselves and not by external stochastic perturbations. Given the big changes theses bring and the often catastrophic rearrangements they cause, their study have gathered an ever growing interest in the past decades.

As we have seen, some physicists have seen the occurrence of a transition as a sign of a spontaneous organisation around a critical state where reorganisation of all sizes occur. In this optics transitions are often called *avalanches* and work as a release of tension, that is subject to a slow and spontaneous increase. Other describe them using analogies with physical systems going through a phase transitions. The comparison is indeed appealing and the great success statistical physics has had in describing these phenomena, bridging microscopic and macroscopic scales, made many people do, perhaps correctly, it was the right domain in which to study and analyse transitions in other contexts. Moreover others justify their occurrence with an internal change of the microscopic dynamics, and see in the super-exponential growth an early-warning signal of their arrival. Despite the countless efforts we are still far from a general theory capable of describing this particular intermittent evolution.

In this part we have proposed a new approach to solve this problem. Our claim is that in order to forecast such a complex and heterogenous process, one cannot simply rely, like very often people do, on the observation of a single macroscopic variable. Methods based on the analysis of a time series have constantly proven themselves either wrong or not right enough, and present often contradictory results. We believe that some information on the details of the system is necessary and must be possessed. Using Einstein's words "*make everything as simple as possible but not simpler*". We believe the approaches attempted in the past make things too simple. Our starting point has been exactly the opposite, and perhaps our method does suffer of the opposite problem.

We have described a new procedure for forecasting transitions in high dimensional systems with stochastic dynamics. Indeed our method is of relevance to systems where the macroscopic dynamics at the systemic level is not adequately captured by a well defined set of essentially deterministic collective variables, and by increasing the complexity we hope to make the description more realistic. We are dealing with situations that are not captured by the application of bifurcation theory such as those considered by Scheffer and collaborators. The authors of the papers themselves claim that in presence of stochastic noise their methods fail to adequately describe the evolution. Our hope is exactly that of describing high dimensional, highly connected, out of equilibrium complex systems, because these are the characteristics of the systems we want to study.

We have in mind complex systems in which the dynamics involves some evolutionary aspects, in particular situations where the dynamics generates new degrees of freedom. E.g. biological evolution, or economical and financial systems, where new agents (organisms, strategies or companies, say) are produced as an intrinsic part of the dynamics. We have demonstrated by use of two models of varying degree of stochasticity (the Tangled Nature model and the stochastic Replicator model) that a combination of analytic linear stability analysis and simulation allows one to construct a signal (overlap with unstable directions) which can be used to forecast a very high percentage of all transitions.

As already stated, the first procedure we have developed was quite naive, it required full information on the system. But nevertheless its results are of crucial importance. We have understood that a mean field description of these model is indeed descriptive of the underlying stochastic process. In the TNM for example the stochastic and deterministic dynamics are in fair agreement only during the quite phases. But this has proven to be enough. The majority of the times the system spontaneously chooses the unstable directions, indicated by the mean field approximation, to exit the stationary configuration. This is not a trivial result. Only after having learnt this, we were able to build a more realistic procedure which proved to be even more efficient despite making use of much less information.

Furthermore we have shown how the intermittent macroscopic behaviour can be obtained by use of the replicator equation, which people use in a broad range of different systems. Here the aim was to give a game theoretical interpretation of the TNM's results, given the growing presence game theory is gaining in the complexity environment. For this reason we have developed the SRM, which represents possibly the first attempt in this direction. Given the Langevin type of dynamics this model has it is not surprising that the results were even better than in the TNM. Indeed the dynamics in the SRM is partly deterministic with an added stochastic element.

The weakness of our procedure is that for real situations of interest (e.g. an ecosystem or a financial market) one may obviously not posses complete information. One will typically not have access to all the information about the interaction amongst the agents. This turns out to be less of a problem, since we have showed that even with a 10% inaccuracy in interaction strengths, we are still able to forecast a substantial percentage of transitions. Another short coming is that in real situations it can also be very difficult to know the nature of the new agents that may arrive as the system evolve. Our full mathematical procedure suggests a way to overcome this problem. Namely, the eigenvector analysis showed that transitions are often accompanied by the arrival of new agents, which exhibit a rapid growth in their relative systemic weight. We found that simply monitoring the rapidly growing new agents can enable prediction of major systemic upheavals. I.e. approaching transitions might not be apparent by focusing on the systemic heavyweights, but rather one should keep a keen eye on the tiny components to monitor whether they suddenly start to flourish. This can often be the signal of upcoming systemic changes.

A crucial test of course will be the application of the results and concepts presented in this part to raw data coming from the real world. We will find ourselves without the knowledge of the precise mechanisms that generate the stochastic process and of course the interactions between agents will not come from distribution known a priori. Furthermore a real world system is never completely isolated, and the interactions between the components could change due to the change in some external factor that one cannot control. This would indeed change the approach we have built, where the interactions where random and constant. Another problematic arises from the time scale of the observation. In both the TNM and the SRM we studied evolutionary time scales composed of many generations. Collecting data on such time scales is of course impossible, which implies that the system will be quite similar to itself, where few new *mutant* enter the system, and very few leave the system (are killed).

Our next step will be to test these findings on real data streams including high frequency financial time series. At the moment we are working, and plan to work, on an application of the method on financial time series. There is no other sector where the amount of data is so abundant, and it is therefore where we have planned to start our applications. In this type of system all the problems just cited are present. New laws, new legislations, new climates and new political alliances change the way products are correlated and the way the interact. Moreover the time scale of observation is much shorter in compare to the models we have dealt with. For this reason we have though to modify the procedure, introducing a time dependent interaction matrix that will be inferred directly from the data, and a constant set of assets. The variation of the interactions will lead to a variation of the jacobian and therefore of the eigenspace.

This is only the first of many applications that could be developed from what written here. Of course every application will require specific adjustments and consideration. Our hope is both of having created a new tool kit that could be useful to people in many different domains.

References

1. Accongiagioco, G., Altman, E., Gregori, E., Lenzini, L.: A game theoretical study of peering vs transit in the internet. In: NetSciCom 2014 INFOCOM Workshop (2014)
2. Accongiagioco, G., Altman, E., Gregori, E., Lenzini, L.: Peering vs transit: a game theoretical model for autonomous systems connectivity. In: IFIP Networking (2014)
3. Accongiagioco, G., Gregori, G., Lenzini, L.: Inferring Interdomain Internet Traffic Using DNS Data: Pros and Cons. Technical report (2014) [Online]. http://tinyurl.com/nujychk
4. Accongiagioco, G., Gregori, E., Lenzini, L.: Comput Netw. A structure-based internet topology gEnerator, S-BITE (2014)
5. Acemoglu, D., Ozdaglar, A.: Competition and efficiency in congested markets. Math. Oper. Res. **32**(1), 1–31 (2007)
6. Ager, B., Chatzis, N., Feldmann, A., Sarrar, N., Uhlig, S., Willinger, W.: Anatomy of a large european IXP. In: Proceedings of the 2012 ACM SIGCOMM Conference, pp. 163–174 (2012)
7. Aguilar, C.O., Gharesifard, B.: On almost equitable partitions and network controllability. In: American Control Conference (ACC), 2016, pp. 179–184. IEEE (2016)
8. Alboszta, J., Miekisz, J.: Stability of evolutionarily stable strategies in discrete replicator dynamics with time delay. J. Theor. Biol. **231**(2), 175179 (2004)
9. Alexa—actionable analytics for the web. http://www.alexa.com
10. Allen, L.J.S., Bolker, B.M., Lou, Y., Nevai, A.L.: Asymptotic profiles of the steady states for an SIS epidemic patch model. SIAM J Appl Math **67**(5), 1283–1309 (2007). SIAM
11. Altman, E., Altman, Z.: S-modular games and power control in wireless networks. IEEE Trans. Autom. Control **48**, (2003)
12. Altman, E., Hayel, Y., Kameda, H.: Revisiting collusion in routing games: a load balancing problem. In: Proceedings of Netgcoop, Paris (2011)
13. Altman, E., Mora, J.R., Wong, S., Hanawal, M.K., Xui, Y.: Net neutrality and quality of service. In: Proceedings of the Game Theory for Networks, GameNets (2011)
14. Altman, E., Basar, T., Jiménez, T., Shimkin, N.: Competitive routing in networks with polynomial costs. IEEE Trans. Automat. Contr. **47**(1), 92–96 (2002)
15. Altman, E., Boulogne, T., El Azouzi, R., Jimenez, T., Wynter, L.: A survey on networking games. Comput. Oper. Res. **33**(2), 286–311 (2006)
16. Altman, E., Hanawal, M.K., Sundaresan, R.: Regulation of off-network pricing in a nonneutral network. ACM Trans. Internet Technol. **14**(2–3), 21 (2014)
17. Andelman, N., Feldman, M., Mansour, Y.: Strong price of anarchy. Games Econ. Behav. **65**(2), 289–317 (2009)
18. Anderson, E.P., Jensen, H.J.: Network properties, species abundance and evolution in a model of evolutionary ecology. J. Theor. Biol. **232**, 551–558 (2005)
19. Anshelevich, E., Dasgupta, A., Kleinberg, J.M., Tardos, É., Wexler, T., Roughgarden, T.: The price of stability for network design with fair cost allocation. SIAM J. Comput. **38**(4), 1602–1623 (2008)

E. Altman et al. (eds.), *Multilevel Strategic Interaction Game Models for Complex Networks*, https://doi.org/10.1007/978-3-030-24455-2

20. Armstrong, M.: Competition in two-sided markets. RAND J. Econo. **37**(3), 668–691 (2006)
21. Auger, P.: Hawk-dove game and competition dynamics. Mathl. Comput. Model. **27**, 89–98 (1998)
22. Aumann, R.: Acceptable points in general cooperative n-person games. Lecture Notes Econ. Math, Syst (1979)
23. Aumann, R.J.: Subjectivity and correlation in randomized strategies. J. Math, Econ (1974)
24. Avrachenkov, K., Filar, J., Haviv, M.: Singular perturbations of Markov chains and decision processes. In: Handbook of Markov Decision Processes International Series in Operations Research and Management Science. Springer (2002)
25. Avrachenkov, K., Filar, J., Howlett, P.: Analytic perturbation theory and its applications. In: SIAM (2013)
26. Avrachenkov, K., Neglia, G., Singh, V.V.: Network formation games with teams. J. Dyn. Games **3**, 303–318 (2016)
27. Azad, A.P., Altman, E., El Azouzi, R.: Routing games : From egoism to altruism. In: Proceedings of WiOpt'10, pp. 528–537 (2010)
28. Babaioff, M., Kleinberg, R., Papadimitriou, C.H.: Congestion games with malicious players. Games Econ. Behav. **67**(1), 22–35 (2009)
29. Bachrach, Y., Kohli, P., Graepel, T.: Rip-off: Playing the cooperative negotiation game. In: AAMAS (2010)
30. Bailey, N.T.J.: The Mathematical Theory of Infectious Diseases and its Applications. Hafner Press, New York (1975)
31. Ball, F., Mollison, D., Scalia-Tomba, G.: Epidemics with two levels of mixing. Ann. Appl. Prob. **46–89**, (1997)
32. Ball, F., Neal, P.: A general model for stochastic sir epidemics with two levels of mixing. Math. Biosci. **180**(1), 73–102 (2002)
33. Ball, F., Neal, P.: Network epidemic models with two levels of mixing. Math. Biosci. **212**(1), 69–87 (2008)
34. Ball, F., Sirl, D., Trapman, P.: Analysis of a stochastic sir epidemic on a random network incorporating household structure. Math. Biosci. **224**(2), 53–73 (2010)
35. Ball, F., Britton, T., House, T., Isham, V., Mollison, D., Pellis, L., Tomba, G.S.: Seven challenges for metapopulation models of epidemics, including households models. Epidemics **10**, 63–67 (2015)
36. Banner, R., Orda, A.: Bottleneck routing games in communication networks. IEEE J. Sel. Areas Commun. **25**(6), 1173–1179 (2007)
37. Barrat, A., Barthelemy, M., Vespignani, A.: Dynamical Processes on Complex Networks. Cambridge University Press (2008)
38. Becker, N., Sibani, P.: Evolution and non-equilibrium physics. A study of the tangled nature model
39. Bélair, J., Campbell, S.A.: Stability and bifurcations of equilibria in a multiple-delayed differential equation. SIAM J. Appl. Math. **54**, 1402–1424 (1994)
40. Bellman, R., Cooke, K.L.: Differential Difference Equations. Academic Press, New York (1963)
41. Ben Khalifa, N., El-Azouzi, R., Hayel, Y., Sidi, H., Mabrouki, I.: Evolutionary stable strategies in interacting communities. In: Proceedings of Valuetools, pp. 214–222, Torino, Italy (2013)
42. Ben Khalifa, N., El-Azouzi, R., Hayel, Y.: Delayed evolutionary game dynamics with non-uniform interactions in two communities. In: Proceedings of IEEE CDC, pp. 3809–3814, Los Angeles, California, USA (2014)
43. Berezansky, L., Braverman, E.: On stability of some linear and nonlinear delay differential equations. J. Math. Anal. Appl. **314**, 391–411 (2006)
44. Berezansky, L., Braverman, E.: Explicit exponential stability conditions for linear differential equations with several delays. J. Math. Anal. Appl. **332**, 246–264 (2007)
45. Blocq, G., Orda, A.: Worst-case coalitions in routing games. CoRR abs/1310.3487 (2015)
46. Boccaletti, S., Latora, V., Moreno, Y., Chavez, M., Hwang, D.: Complex networks: structure and dynamics. Physi. Reports **424**(4–5), 175–308 (2006)

47. Bollobás, Béla: Random Graphs. Springer (1998)
48. Bonaccorsi, S., Ottaviano, S., Mugnolo, D., De Pellegrini, F.: Epidemic outbreaks in networks with equitable or almost-equitable partitions. CoRR abs/1412.6157 (2014)
49. Bonaccorsi, S., Ottaviano, S., De Pellegrini, F., Socievole, A., Van Mieghem, P.: Epidemic outbreaks in two-scale community networks. Phys. Rev. E **90**(1), 012810 (2014)
50. Bonaccorsi, S., Ottaviano, S., Mugnolo, D., De Pellegrini, F.: Epidemic outbreaks in networks with equitable or almost-equitable partitions. SIAM J. Appl. Math. **75**(6), 2421–2443 (2015)
51. Bourreau, M., Kourandi, F., Valletti, T.M.: Net neutrality with competing internet platforms. Technical report, CEIS (2014)
52. Boyd, R., Gintis, H., Bowles, S., Fehr, E.: Moral Sentiments and Material Interests. MIT Press (2005)
53. Brunetti, I., El-Azouzi, R., Altman, E.: Altruism in groups: an evolutionary games approach. In: The Proceeding of the Netgcoop Conference, Trento (2014)
54. Busch, C., Magdon-Ismail, M.: Atomic routing games on maximum congestion. Theor. Comput. Sci. **410**(36), 3337–3347 (2009)
55. Buslowicz, M.: Simple stability criterion for a class of delay differential systems. Inter. J. Syst. Sci. **18**(5), 993–995 (1987)
56. Camerer, C.: Behavioral Game Theory: Experiments in Strategic Interaction. University Press Princeton, NJ (2003)
57. Caragiannis, I., Kaklamanis, C., Kanellopoulos, P., Kyropoulou, M., Papaioannou, E.: The impact of altruism on the efficiency of atomic congestion games. In: TGC, pp. 172–188 (2010)
58. Caroli, A., Piovani, D., Jensen, H.J.: Forecasting transitions in systems with high dimensional stochastic complex dynamics: a linear stability analysis of the tangled nature model. Phys. Rev. Lett. **113**, 264102 (2014)
59. Castellano, C., Pastor-Satorras, R.: Thresholds for epidemic spreading in networks. Phys. Rev. Lett. **105**(21), 218701 (2010)
60. Cator, E., Van Mieghem, P.: Susceptible-infected-susceptible epidemics on the complete graph and the star graph: exact analysis. Phys. Rev. E **87**, 012811 (2013)
61. Cator, E., Van Mieghem, P.: Nodal infection in Markovian SIS and SIR epidemics on networks are non-negatively correlated. Phys. Rev. E **89**(5), 052802 (2014)
62. Chakrabarti, D., Wang, Y., Wang, C., Leskovec, C., Faloutsos, C.: Epidemic thresholds in real networks. ACM Trans. Inf. Syst. Secur. **10**(4), 1:1–1:26, Jan 2008
63. Chang, H., Arbor, A., Ave, P., Park, F.: To Peer or not to Peer : Modeling the Evolution of the Internet ' s AS-level Topology. In: INFOCOM 2006 25th IEEE International Conference on Computer Communications. Proceedings (2006)
64. Chang, H., Jamin, S., Willinger, W.: Internet connectivity at the AS-level: an optimization-driven modeling approach. In: Proceedings of the ACM SIGCOMM workshop, MoMeTools '03, pp. 33–46 (2003)
65. Chen, P.-A., de Keijzer, B., Kempe, D., Schäfer, G.: Altruism and its impact on the price of anarchy. ACM Trans. Econ. Comput. **2**(4), 17:1–17:45 (2014)
66. Chen, P.A., Kempe, D.: Altruism, selfishness, and spite in traffic routing. In: 9th ACM conference on Electronic commerce, p. 140149 (2008)
67. Chen, P.-A., Kempe, D.: Altruism, selfishness, and spite in traffic routing. In: Proceedings EC'08, pp. 140–149
68. Chen, J., Latchman, H.A.: Asymptotic stability independent of delays: simple necessary and sufficient conditions. In: Proceedings of The American Control Conference, pp. 1027–1031, Baltimore, Maryland, USA, June 1994
69. Chen, M.K., Nalebuff, B.J.: One-way essential complement. Yale Economic Applications and Policy Discussion Paper **22**, (2006)
70. Chen, P.-A., de Keijzer, B., Kempe, D., Schäfer, G.: The robust price of anarchy of altruistic games. Proc. WINE **2011**, 383–390 (2011)
71. Christensen, K., di Collobiano, S.A., Hall, M., Jensen, H.J.: Tangled nature model: a model of evolutionary ecology. J. Theor. Biol. **216**, 73–84 (2002)

72. Chung, F., Lu, L., Vu, V.: Eigenvalues of random power law graphs. Ann. Combinatorics (1), 21–33
73. Cole, R., Dodis, Y., Roughgarden, T.: Bottleneck links, variable demand, and the tragedy of the commons. Networks **60**(3), 194–203 (2012)
74. Cremer, J., Rey, P., Tirole, J.: Connectivity in the commercial internet. J. Ind. Econ. **48**, 433–72 (2000)
75. Cressman, R.: Evolutionary game theory with two groups of individuals. Games Econ. Behav. **11**(2), 237–253 (1995)
76. Daqing, L., Kosmidis, K., Bunde, A., Havlin, S.: Dimension of spatially embedded networks. Nat. Phys. **7**(6), 481–484, 06 2011
77. Darabi Sahneh, F., Scoglio, F., Chowdhury, F.: Effect of coupling on the epidemic threshold in interconnected complex networks: a spectral analysis. In: American Control Conference (ACC), pp. 2307–2312 (2013)
78. Demange, G., Wooders, M. (eds.): Group Formation in Economics: Networks, Clubs, and Coalitions. Cambridge University Press (2005)
79. Deri, L., Luconi Trombacchi, L., Martinelli, M., Vannozzi, D.: Towards a passive dns monitoring system. In: SAC, pp. 629–630. ACM (2012)
80. Dhamdhere, A., Dovrolis, C.: The Internet is flat: modeling the transition from a transit hierarchy to a peering mesh. In: Proceedings of the 6th International Conference, Co-NEXT. ACM (2010)
81. Dickison, M., Havlin, S., Stanley, H.E.: Epidemics on interconnected networks. Phys. Rev. E **85**, 066109 (2012)
82. Doebeli, M., Hauert, C., Holmes, M.: Evolutionary games and population dynamics: maintenance of cooperation in public goods games. Proc. Royal Soc. B **273**(1605), 3131–2 (2006)
83. Durrett, R., Levin, S.A.: Stochastic spatial models: a users guide to ecological applications. Philos. Trans. R. Soc. Lond. B, Biol. Sci., 343:329350 (1994)
84. Dutta, B., Jackson, M.O. (eds.): Networks and Groups: Models of Strategic Formation. Springer, Berlin, Heidelberg (2003)
85. Dutta, B., Mutuswami, S.: Stable networks. J. Econ. Theory **76**, 322–344 (1997)
86. Economides, N., Tag, J.: Net neutrality on the internet: a two-sided market analysis **24**, 91–104 (2012)
87. Economides, N.: The Economics of the internet Backbone. Handbook of Telecommunications Economics, Majumar, S., Vogelsang, I., Cave, M. (Eds.), pp. 374–412 (2006)
88. Emergence of scaling in random networks: Science **286**(5439), 509–512 (1999)
89. Emmerich, T., Bunde, A., Havlin, S.: Diffusion, annihilation, and chemical reactions in complex networks with spatial constraints. Phys. Rev. E **86**, 046103 (2012)
90. Emmerich, T., Bunde, A., Havlin, S., Li, G., Li, D.: Complex networks embedded in space: dimension and scaling relations between mass, topological distance, and euclidean distance. Phys. Rev. E **87**, 032802 (2013)
91. ERDdS, P., A R&WI. On random graphs i. Publ. Math. Debrecen **6**, 290–297 (1959)
92. Fabrikant, A., Koutsoupias, E., Papadimitriou, C.H.: Heuristically Optimized trade-offs: a new paradigm for power laws in the internet. In: Proceedings of the 29th ICALP '02. Springer (2002)
93. Faloutsos, M., Faloutsos, P., Faloutsos, C.: On power-law relationships of the internet topology. In: SIGCOMM Proceedings of the Conference on Applications, Technologies, Architectures, and Protocols for Computer Communication (1999)
94. Fehr, E., Schmidt, K.M.: A theory of fairness, competition and cooperation. Quar. J. Econ. **114**, 817868 (1999)
95. fen Cheng, S., Reeves, D.M., Vorobeychik, Y., Wellman, M.P.: Notes on equilibria in symmetric games. In: Proceedings of the 6th International Workshop GTDT, pp. 71–78 (2004)
96. Fisher, R.A.: The Genetic Theory of Natural Selection. Clarendon Press, Oxford (1930)
97. Foster, D., Young, H.P.: Stochastic evolutionary game dynamics. Theor. Popul. Biol. **38**(2), 219–232 (1990)

98. Frank, B., David, S., Pieter, T., et al.: Threshold behaviour and final outcome of an epidemic on a random network with household structure. Adv. Appl. Probab. **41**(3), 765–796 (2009)
99. Freedman, H.I., Kuang, Y.: Stability switches in linear scalar neutral delay equations. Funkcialaj Ekvacioj **34**(1), 187–209 (1991)
100. Friedman, E., Henderson, S.: Fairness and efficiency in processor sharing protocols to minimize sojourn times. In: Proceedings of ACM SIGMETRICS, pp. 229–337 (2003)
101. Fudenberg, D., Imhof, L.A.: Imitation processes with small mutations. J. Econ. Theory **131**, 251–262 (2006)
102. Fudenberg, D., Nowak, M.A., Taylor, C., Imhof, L.A.: Evolutionary game dynamics in finite populations with strong selection and weak mutation. Theor. Popul. Biol. **70**, 352–363 (2006)
103. Funk, S., Jansen, V.A.A.: Interacting epidemics on overlay networks. Phys. Rev. E **81**, 036118 (2010)
104. Gale, D., Shapley, L.S.: College admissions and the stability of marriage. Am. Math. Mon. **69**(1), 9–15 (1962)
105. Ganesh, A., Massoulié, L., Towsley, D.: The effect of network topology on the spread of epidemics. In: Proceedings IEEE INFOCOM 2005. 24th Annual Joint Conference of the IEEE Computer and Communications Societies, vol. 2, pp. 1455–1466. IEEE (2005)
106. Ganesh, A., Massoulie, L., Towsley, D.: The effect of network topology on the spread of epidemics. IEEE INFOCOM **2**, 1455–1466 (2005)
107. Gao, J., Buldyrev, S.V., Stanley, H.E., Havlin, S.: Networks formed from interdependent networks. Nat. Phys. **8**(1), 40–48, 01 2012
108. Golubitsky, M., Stewart, I., Török, A.: Patterns of synchrony in coupled cell networks with multiple arrows. SIAM J. Appl. Dyn. Syst. **4**(1), 78–100 (2005)
109. Gopalsamy, K.: Stability and Oscillations in Delay Differential Equations of Population Dynamics. Kluwer Academic Publisher, London (1992)
110. Gould, S., Eldredge, N.: Punctuated equilibria: the tempo and mode of evolution reconsidered. Paleobiology
111. Goyal, S.: Connections: An Introduction to the Economics of Networks. Princeton University Press (2007)
112. Goyal, S.: Connections: An Introduction to the Economics of Networks. Princeton University Press, July 2007
113. Gregori, E., Lenzini, L., Mainardi, S.: Parallel k -Clique community detection on large-scale networks. IEEE Trans. Parallel Distrib. Syst. **1–11**, (2012)
114. Gregori, E., Lenzini, L., Orsini, C.: k-clique communities in the internet AS-level topology graph. In: 2011 31st International Conference on Distributed Computing Systems Workshops, pp. 134–139, June 2011
115. Gregori, E., Improta, A., Lenzini, L., Orsini, C.: The impact of IXPs on the AS-level topology structure of the Internet. Comput. Commun. **34**(1), 68–82 (2011)
116. Gross, T., Dommar D'Lima, C.J., Blasius, B.: Epidemic dynamics on an adaptive network. Phys. Rev. Lett. **96**, 208701 (2006)
117. Guo, D., Trajanovski, S., Van Mieghem, P.: From epidemics to information propagation: striking differences in structurally similar adaptive network models. to appear (2015)
118. Guo, D., Trajanovski, S., van de Bovenkamp, R., Wang, H., Van Mieghem, P.: Epidemic threshold and topological structure of susceptible-infectious-susceptible epidemics in adaptive networks. Phys. Rev. E **88**, 042802 (2013)
119. Haddai, H., Rio, M., Iannaccone, G., Moore, A., Mortier, R.: Network topologies: inference modeling and generation. IEEE Commun. Surv. **10**(2), 48–69 (2008)
120. Hahn, R., Wallsten, S.: The economics of net neutrality. Berkeley Econ. Press Econo. **3**(6), 1–7 (2006)
121. Hale, J.K., Huang, W.Z.: Global geometry of the stable regions for two delay differential equations. J. Math. Anal. Appl. **178**, 344–362 (1993)
122. Hall, M., Christensen, K., di Collobiano, S.A., Jensen, H.J.: Time-dependent extinction rate and species abundance in a tangled-nature model of biological evolution. Phys. Rev. E **66**, (2002)

123. Hamilton, W.D.: The genetical evolution of social behviour. I and II, J. Theor. Biol. **7**, 1–52 (1964)
124. Hamilton, W.D.: The evolution of altruistic behavior. Am. Naturalist **97**, 354–356 (1963)
125. Hanski, I., Ovaskainen, O.: Metapopulation theory for fragmented landscapes. Theor. Popul. Biol. **64**(1), 119–127 (2003)
126. Hara, T., Sugie, J.: Stability region for systems of differential-difference equations. Funkcialaj Ekvacioj **39**(1), 69–86 (1996)
127. Hara, T., Sugie, J.: Stability region for systems of differential-difference equations. Funkcialaj Ekvacioj **39**(1), 69–86 (1996)
128. Harks, T.: Stackelberg strategies and collusion in network games with splittable flow. Theory Comput. Syst. **48**(4), 781–802 (2011)
129. Harks, T., Klimm, M., Möhring, R.H.: Strong equilibria in games with the lexicographical improvement property. Int. J. Game Theory **42**(2), 461–482 (2013)
130. Harsanyi, J.C., Selten, R.: A General Theory of Equilibrium Selection in Games. MIT Press (1988)
131. Hauert, C., Doebeli, M.: Spatial structure often inhibits the evolution of cooperation in the snow-drift game. Nature **428**, 643646 (2004)
132. Hayrapetyan, A., Tardos, E., Wexler, T.: The effect of collusion in congestion games. In: STOC, May 2006
133. Hill, A.L., Rand, D.G., Nowak, M.A., Christakis, N.A.: Emotions as infectious diseases in a large social network: the sisa model. Proc. Royal Soc. London B: Biol. Sci. **277**(1701), 3827–3835 (2010)
134. Hoefer, M., Skopalik, A.: Altruism in atomic congestion games. ACM Trans. Econ. Comput. **1**(4), 21 (2013)
135. Hofbauer, J., Sigmund, K.: Evolutionary Games and Population Dynamics. Cambridge University Press, Cambridge, UK (1998)
136. Hofbauer, J., Sigmund, K.: Evolutionary Games and Population Dynamics. Cambridge University Press, Cambridge (1998)
137. Hofbauer, J., Sigmund, K.: Evolutionary Games and Population Dynamics. Cambridge University Press, UK (1998)
138. Hofbauer, J., Sigmund, K.: Evolutionary game dynamics. Bull. Am. Math. Soc. **40**(4), 479–519 (2003)
139. Hofbauer, J., Sigmund, K.: Evolutionary game dynamics. Am. Math. Soc. **40**(4), 479–519 (2003)
140. Hofbauer, J., Schuster, P., Sigmund, K.: A note on evolutionarily stable strategies and game dynamics. J. Theor. Biol. **81**, 609–612 (1979)
141. Huang, X., Shao, S., Wang, S., Buldyrev, S.V., Stanley, H.E., Havlin, S.: The robustness of interdependent clustered networks. EPL (Europhys. Lett.) **101**(1), 18002 (2013)
142. Iijima, R.: On delayed discrete evolutionary dynamics. J. Theor. Biol. **300**, 1–6 (2012)
143. Iijima, R.: On delayed discrete evolutionary dynamics. J. Theor. Biol. **300**, 1–6 (2012)
144. Jackson, M.O.: Social and Economic Networks. Princeton University Press (2010)
145. Jackson, M.O.: Social and Economic Networks. Princeton University Press (2010)
146. Jackson, M.O., van den Nouweland, A.: Strongly stable networks. Games Econ. Behav. **51**, 420–444 (2005)
147. Jackson, M.O., Watts, A.: The evolution of social and economic networks. J. Econ. Theory **106**, 265–295 (2002)
148. Jackson, M.O., Wolinsky, A.: A strategic model of social and economic networks. J. Econ. Theory **71**, 44–74 (1996)
149. Jimenez, T., Altman, E., Pourtallier, O., Kameda, H.: Symmetric games with networking applications. In: Proceedings of Netgcoop, Paris (2011)
150. Jimenez, T., Hayel, Y., Altman, E.: Competition in access to content. In: Proceedings of IFIP Networking (2012)
151. Jiménez, T., Hayel, Y., Altman, E.: Competition in access to content. In: Proceedings of IFIP Networking, pp. 211–222 (2012)

152. José, R.: Correa and Nicolás E. Stier Moses. Wardrop equilibria, In Wiley Encyclopedia of Operations Research and Management Science (2010)
153. Judd, J.S., Kearns, M.: Behavioral experiments in networked trade. In: ACM Conference on Electronic Commerce, pp. 150159 (2008)
154. Kandori, M., Mailath, G.J., Rob, R.: Learning, mutation, and long run equilibria in games. Econometrica **61**(1), 29–56 (1993)
155. Kemeny, J.G., Snel, J.L.: Finite Markov Chains. Springer (1976)
156. Klaus, B., Klijn, F., Walzl, M.: Stochastic stability for roommate markets. J. Econ. Theory **145**(6), 2218–2240 (2010)
157. Kokotovic, P., Khalil, H., O'reilly, J.: Singular perturbation methods in control. In: SIAM (1986)
158. Korilis, Y.A., Lazar, A.A., Orda, A.: Achieving network optima using Stackelberg routing strategies. IEEE/ACM Trans. Netw. **5**(1), 161–173 (1997)
159. Koutsoupias, E., Papadimitriou, C.H.: Worst-case equilibria. Comput. Sci. Rev. **3**(2), 65–69 (2009)
160. Kramer, J., Wiewiorra, L.: Network neutrality and congestion sensitive content providers: Implications for content variety, broadband investment, and regulation. Inf. Syst. Res. **23**(4), 1303–1321 (2012)
161. La, R.J., Anantharam, V.: Optimal routing control: repeated game approach. IEEE Trans. Autom. Cont. **47**, 437–450 (2002)
162. Lagorio, C., Dickison, M., Vazquez, F., Braunstein, L.A., Macri, P.A., Migueles, M.V., Havlin, S., Stanley, H.E.: Quarantine-generated phase transition in epidemic spreading. Phys. Rev. E **83**, 026102 (2011)
163. Lajmanovich, A., Yorke, J.A.: A deterministic model for Gonorrhea in a non-homogeneous population. Math. Biosci. **28**(34), 221–236 (1976)
164. Lan, T., Kao, D., Chiang, M., Sabharwal, A.: An axiomatic theory of fairness in network resource allocation. INFOCOM '10
165. Lawson, D., Jensen, H.J.: The species-area relationship and evolution. J. Theor. Biol. **241**, 590–600 (2006)
166. Ledyard, J.O.: Public Goods: A Survey of Experimental Research. Princeton University Press. Princeton University Press, p. 111194 (1997)
167. Leicht, E.A., D'Souza, R.M.: Percolation on Interacting Networks. arXiv, (0907.0894v1)
168. Levine, D.K.: Modeling altruism and spitefulness in experiment. Rev. Econ. Dyn. **1**(3), 593622
169. Li, D., Li, G., Kosmidis, K., Stanley, H.E., Bunde, A., Havlin, S.: Percolation of spatially constraint networks. EPL (Europhys. Lett.) **93**(6), 68004 (2011)
170. Li, D., Qin, P., Wang, H., Liu, C., Jiang, Y.: Epidemics on interconnected lattices. EPL (Europhys. Lett.) **105**(6), 68004 (2014)
171. Li, X., Ruan, S., Wei, J.: Stability and bifurcation in delay-differential equations with two delays. J. Math. Anal. Appl. **236**, 254–280 (1999)
172. Li, C., van de Bovenkamp, R., Van Mieghem, P.: Susceptible-infected-susceptible model: a comparison of n-intertwined and heterogeneous mean-field approximations. Phys. Rev. E **86**(2), 026116 (2012)
173. Li, W., Bashan, A., Buldyrev, S.V., Stanley, H.E., Havlin, S.: Cascading failures in interdependent lattice networks: the critical role of the length of dependency links. Phys. Rev. Lett. **108**, 228702 (2012)
174. Li, Q., Braunstein, L.A., Wang, H., Shao, J., Stanley, H.E., Havlin, S.: Non-consensus opinion models on complex networks. J. Stat. Phys. **151**(1–2), 92–112 (2013)
175. Liu, M., Li, D., Qin, P., Liu, C., Wang, H., Wang, F.: Epidemics in interconnected small-world networks. PLoS ONE **10**(3), e0120701 (2015)
176. Lodhi, A., Dhamdhere, A., Dovrolis, C.: GENESIS: an agent-based model of interdomain network formation, traffic flow and economics. In: 2012 Proceedings IEEE INFOCOM, pp. 1197–1205, March 2012
177. Marceau, V., Noël, P.-A., Hébert-Dufresne, L., Allard, A., Dubé, L.J.: Adaptive networks: coevolution of disease and topology. Phys. Rev. E **82**, 036116 (2010)

178. Masuda, N.: Effects of diffusion rates on epidemic spreads in metapopulation networks. New J. Phys. **12**(9), 093009 (2010)
179. Mathworks. Matlab—http://www.mathworks.it/products/matlab/
180. Meyer, C.D. (ed.): Matrix Analysis and Applied Linear Algebra. Society for Industrial and Applied Mathematics, Philadelphia, PA, USA (2000)
181. Minc, H.: Non-negative matrices. New York (1988)
182. MIX—milan internet exchange. http://www.mix-it.net
183. Motiwala, M., Dhamdhere, A., Feamster, N., Lakhina, A.: Towards a cost model for network traffic. ACM SIGCOMM Comput. Commun. Rev. **42**(1), 54–60 (2012)
184. Musacchio, J., Schwartz, G., Walrand, J.: A two-sided market analysis of provider investment incentives with an application to the net-neutrality issue. Rev. Netw. Econ. **8**, 22–39 (2009)
185. NaMeX—nautilus mediterranean exchange. http://www.namex.it
186. Net neutrality and investment incentives: RAND J. Econ. **41**(3), 446–471 (2010)
187. Newton, J.: Coalitional stochastic stability. Games Econ. Behav. **75**(2), 842–854 (2012)
188. Newton, J.: Recontracting and stochastic stability in cooperative games. J. Econ. Theory **147**, 364–381 (2012)
189. Newton, J., Angus, S.D.: Coalitions, tipping points and the speed of evolution. J. Econ. Theory **157**, 172–187 (2015)
190. Newton, J., Sawa, R.: A one-shot deviation principle for stability in matching problems. J. Econ. Theory **157**, 1–27 (2015)
191. Nisan, N., Roughgarden, T., Tardos, E., Vazirani, V.V.: Algorithmic Game Theory. Cambridge University Press, New York, NY, USA (2007)
192. Njoroge, P., Ozdaglar, A., Stier-Moses, N., Weintraub, G.: Investment in two-sided markets and the net-neutrality debate. Rev. Netw. Econ. **12**(4), 355–402 (2013)
193. Norton, W.B.: A Study of 28 Peering Policies. Dr. Peering White Paper, Technical report (2010)
194. Norton, W.B.: Internet Transit Prices-Historical and Projected. Dr. Peering White Paper, Technical report (2010)
195. Norton, W.B.: The Art of Peering-The IX Playbook. Dr. Peering White Paper, Technical report (2010)
196. Nowak, A.S., Raghavan, T.E.S.: A finite step algorithm via a bimatrix game to a single controller non-zero sum stochastic game. Math. Programm. **59**, 249–259 (1993)
197. Oaku, H.: Evolution with delay. Japanese. Econ. Rev. **53**, 114–133 (2002)
198. Ohtsuki, H., Nowak, M.: The replicator equation on graphs. J. Theor. Biol. **243**(1), 86–97 (2006)
199. Ohtsuki, H., Nowak, M.A.: Evolutionary stability on graphs. J. Theor. Biol. **251**, 698707 (2008)
200. Ohtsuki, H., Hauert, C., Lieberman, E., Nowak, M.A.: A simple rule for the evolution of cooperation on graphs and social networks. Nature **441**, 502–505 (2006)
201. On the evolution of random graphs: Erd6s, P., Rényi, A. Publ. Math. Inst. Hungar. Acad. Sci **5**, 17–61 (1960)
202. Orda, A., Rom, N., Shimkin, N.: Competitive routing in multi-user communication networks. IEEE/ACM Trans. Netw. **1**, 614–627 (1993)
203. Orda, A., Rom, R., Shimkin, N.: Competitive routing in multiuser communication networks. IEEE/ACM Trans. Netw. **1**(5), 510–521 (1993)
204. Ortiz, L.E., Pemantle, R., Suri, S., Kakade, S.M., Kearns, M.: Economic properties of social networks. Advances in Neural Information Processing Systems (2005)
205. Ottaviano, S., De Pellegrini, F., Bonaccorsi, S., Van Mieghem, P.: Optimal curing policy for epidemic spreading over a community network with heterogeneous population. J. Complex Netw. **6**(5), 800–829 (2017)
206. Parshani, R., Buldyrev, S.V., Havlin, S.: Interdependent networks: reducing the coupling strength leads to a change from a first to second order percolation transition. Phys. Rev. Lett. **105**, 048701 (2010)

207. Pastor-Satorras, R., Vespignani, A.: Epidemic dynamics and endemic states in complex networks. Phys. Rev. E **63**(066117) (2001)
208. Pastor-Satorras, R., Moreno, Y., Vespignani, A.: Epidemic outbreaks in complex heterogeneous networks. Eur. Phys. J. B **26**, 521–529 (2002)
209. Patriksson, M.: VSP (1991)
210. Paul, G., Stanley, H.E., Buldyrev, S.V., Parshani, R., Havlin, S.: Catastrophic cascade of failures in interdependent networks. Nature **464**, 1025–1028 (2010)
211. PeeringDB. http://www.peeringdb.com
212. Pellis, L., Ball, L., Trapman, P.: Reproduction numbers for epidemic models with households and other social structures. i. definition and calculation of r 0. Math. Biosci. **235**(1), 85–97 (2012)
213. Percolation on dense graph sequences: Ann. Probab. **38**(1), 150–183 (2010)
214. Prakash, B.A., Tong, H., Valler, N., Faloutsos, M., Faloutsos, C.: Virus propagation on time-varying networks: Theory and immunization algorithms. In: Balcázar, J.L., Bonchi, F., Gionis, A., Sebag, M. (Eds.), Machine Learning and Knowledge Discovery in Databases, vol. 6323 of Lecture Notes in Computer Science, pp. 99–114. Springer, Berlin, Heidelberg (2010)
215. Qu, B., Wang, H.: Sis epidemic spreading with heterogeneous infection rates. arXiv preprint arXiv:1506.07293 (2015)
216. Rahmani, A., Ji, M., Mesbahi, M., Egerstedt, M.: Controllability of multi-agent systems from a graph-theoretic perspective. SIAM J. Control Optim. **48**(1), 162–186 (2009)
217. Restrepo, J.G., Ott, E., Hunt, B.R.: Onset of synchronization in large networks of coupled oscillators. Phys. Rev. E **71**, 036151 (2005)
218. Rogers, T., Clifford-Brown, W., Mills, C., Galla, T.: Stochastic oscillations of adaptive networks: application to epidemic modelling. J. Stat. Mech.: Theory Exp. **2012**(08), P08018 (2012)
219. Rosen, J.B.: Existence and uniqueness of equilibrium points for concave n-person games. Econometrica **33**(3), 520–534 (1965)
220. Ross, J.V., House, T., Keeling, M.J.: Calculation of disease dynamics in a population of households. PLoS One **5**(3), e9666 (2010)
221. Roth, A., Sotomayor, M.: Two-Sided Matching: A Study in Game-Theoretic Modeling and Analysis. Cambridge University Press (1992)
222. Roth, A.: The price of malice in linear congestion games. In: Proceedings of WINE'2008, pp. 118–125 (2008)
223. Roughgarden, T.: Stackelberg scheduling strategies. SIAM J. Comput. **33**(2), 332–350 (2004)
224. Roughgarden, T., Tardos, É.: How bad is selfish routing? J. ACM **49**, 236–259 (2002)
225. Sandholm, W.H.: Population games and deterministic evolutionary dynamics. In: Young, H.P., Zamir, S. (Eds.) Forthcoming in the Handbook of Game Theory and Economic Applications, vol. 4, pp. 703–778, 12 2014
226. Sandholm, W.H.: Population Games and Evolutionary Dynamics. MIT Press (2010)
227. Sandholm, W.H.: Local stability under evolutionary game dynamics. Theor. Econ. **5**, 27–50 (2010)
228. Santos, F.C., Santos, M.D., Pacheco, J.M.: Social diversity promotes the emergence of cooperation in public goods games. Nature **454**, 213–216 (2008)
229. Saumell-Mendiola, A., Ángeles Serrano, M., Boguñá, M.: Epidemic spreading on interconnected networks. Phys. Rev. E **86**(026106), Aug 2012
230. Sawa, R.: Coalitional stochastic stability in games, networks and markets. Games Econ. Behav. **88**, 90–111 (2014)
231. Scheffer, M.: Critical Transitions in Nature and Society. Princeton University Press (2009)
232. Scheffer, M., Bascompte, J., Brock, W.A., Brovkin, V., Carpenter, S.R., Dakos, V., Held, H., Van Nes, E.H., Rietkerk, M., Sugihara, G.: Early-warning signals for critical transitions. Nature **461**(7260), 53–59 (2009)
233. Scheffer, M., Carpenter, S.R., Lenton, T.M., Bascompte, J., Brock, W., Dakos, V., van de Koppel, J., van de Leemput, I.A., Levin, S.A., van Nes, E.H., et al.: Anticipating critical transitions. Science **338**(6105), 344–348 (2012)

234. Schwenk, A.J.: Computing the characteristic polynomial of a graph. In: Graphs and Combinatorics, pp. 153–172. Springer (1974)
235. Selten, R.: A note on evolutionarily stable strategies in asymmetric animal conflicts. J. Theor. Biol. **84**, 93–101 (1980)
236. Shakkottai, S., Altman, E., Kumar, A.: The case for non-cooperative Multihoming of users to access points in IEEE 802.11 WLANs. In: IEEE Infocom, Barcelona, Spain (2006)
237. Shakkottai, A.K.S., Altman, E.: Evolutionary power control games in wireless networks. 14 J. Sel. Areas Commun., 1207–1215
238. Shakkottai, S., Srikant, R.: Economics of network pricing with multiple ISPs. IEEE/ACM Trans. Netw. **14**(6), December 2006
239. Sharma, Y., Williamson, D.P.: Stackelberg thresholds in network routing games or the value of altruism, games and economic behavior. Games Econ. Behav. **67**(1), 174–190 (2009)
240. Sibani, P., Jensen, H.J.: Stochastic Dynamics of Complex Systems. Imperial College Press (2013)
241. Sigmund, K., Hofbauer, J.: Evolutionary Games and Population Dynamics. Cambridge University Press (1998)
242. Sigurd, D., Opper, M.: Replicators with random interactions: a solvable model. Phys. Rev. A **39**, 4333 (1989)
243. Singh, S., Kearns, M., Littman, M.: Graphical models for game theory. In: Proceedings of UAI, 1 (2001)
244. Smith, J.M., Price, G.R.: The logic of animal conflict. Nature **24**(6), 15–18, 02 November 1973
245. Smith, J.M., Price, G.R.: The logic of animal conflict. Nature **246**(5427), 15–18
246. Smith, J.M.: Evolution and the Theory of Games. Cambridge University Press (1982)
247. Smith, M.: Game theory and the evolution of fighting. In: John Maynard Smith, on Evolution (Edinburgh University Press), pp. 8–28 (1972)
248. Smith, J.M.: Evolution and the Theory of Games. Cambridge University Press, UK (1982)
249. Sobel, M.J.: Noncooperative stochastic games. Ann. Math. Stat. **42**(6), 1930–1935 (1971)
250. Stewart, I., Golubitsky, M., Pivato, M.: Symmetry groupoids and patterns of synchrony in coupled cell networks. SIAM J. Appl. Dyn. Syst. **2**(4), 609–646 (2003)
251. Su, C.: Equilibrium problems with equilibrium constraints: stationnarities, algorithms, and applications. In Ph.D. thesis, Stanford University (2005)
252. Szabo, G., Fath, G.: Evolutionary games on graphs. Phys. Rep. **446**, 97–216 (2007)
253. Taylor, P.D.: Evolutionarily stable strategies with two types of player. J. Appl. Prob. **16**(1), 76–83 (1979)
254. Taylor, P.D., Jonker, L.B.: Evolutionarily stable strategies and game dynamics. Math. Biosci. **40**, 145–156 (1978)
255. Taylor, P.D., Jonker, L.B.: Evolutionary stable strategies and game dynamics. Mathe. Biosci. **40**, 145–156 (1978)
256. Tembine, H., Altman, E., El-Azouzi, R.: Asymmetric delay in evolutionary games. In: Proceedings of Valuetools, Nantes, France (2007)
257. Tembine, H., Altman, E., El-Azouzi, R., Hayel, Y.: Evolutionary games in wireless networks. IEEE Trans. Syst. Man Cybern. **40**, 634–646 (2010)
258. Tembine, H., Altman, E., El-Azouzi, R., Hayel, Y.: Bio-inspired delayed evolutionary game dynamics with networking applications. Telecommun. Syst. **47**, 137–152 (2011)
259. Top-IX—torino piemonte internet exchange. http://www.top-ix.org
260. Tunc, I., Shkarayev, M., Shaw, L.: Epidemics in adaptive social networks with temporary link deactivation. J. Stat. Phys. **151**(1–2), 355–366 (2013)
261. Valdez, L.D., Macri, P.A., Braunstein, L.A.: Intermittent social distancing strategy for epidemic control. Phys. Rev. E **85**, 036108 (2012)
262. van de Bovenkamp, R., Van Mieghem, P.: Survival time of the susceptible-infected-susceptible infection process on a graph. Phys. Rev. E **92**(3), 032806 (2015)
263. Van Mieghem, P., Omic, J.: In-homogeneous virus spread in networks. **1306**, 2588 (2013)

264. Van Mieghem, P., van de Bovenkamp, R.: Accuracy criterion for the mean-field approximation in sis epidemics on networks. Phys. Rev. E **91**, (2015)
265. Van Mieghem, P.: Graph Spectra for Complex Networks. Cambridge University Press (2010)
266. Van Mieghem, P.: Graph Spectra for Complex Networks. Cambridge University Press (2011)
267. Van Mieghem, P.: Graph Spectra for Complex Networks. Cambridge University Press, Cambridge, U.K. (2011)
268. Van Mieghem, P.: The n-intertwined sis epidemic network model. Computing **93**(2–4), 147–169 (2011)
269. Van Mieghem, P.: Approximate formula and bounds for the time-varying susceptible-infected-susceptible prevalence in networks. Phys. Rev. E **93**(5), 052312 (2016)
270. Van Mieghem, P.: Interconnectivity structure of a general interdependent network. Phys. Rev. E **93**(4), 042305 (2016)
271. Van Mieghem, P., Cator, E.: Epidemics in networks with nodal self-infection and the epidemic threshold. Phys. Rev. E **86**(1), 016116 (2012)
272. Van Mieghem, P., Cator, E.: Epidemics in networks with nodal self-infection and the epidemic threshold. Phys. Rev. E **86**, 016116 (2012)
273. Van Mieghem, P., Omic, J., Kooij, R.: Virus spread in networks. Netw. IEEE/ACM Tran. **17**(1), 1–14 (2009)
274. Van Mieghem, P., Omic, J., Kooij, R.: Virus spread in networks. Netw. IEEE/ACM Trans. **17**(1), 1–14 (2009)
275. Van Mieghem, P., Ge, X., Schumm, P., Trajanovski, S., Wang, H.: Spectral graph analysis of modularity and assortativity. Phys. Rev. E **82**, 056113 (2010)
276. Van Mieghem, P., Wang, H., Ge, X., Tang, S., Kuipers, F.A.: Influence of assortativity and degree-preserving rewiring on the spectra of networks. Eur. Phys. J. B **76**, 643–652 (2010)
277. Vazirani, V.V., Noam, N., Roughgarden, T., Tardos, E.: Cambridge University Press (2007)
278. Veltri, R., Kianercy, A., Pienta, K.J.: Critical transitions in a game theoretic model of tumour metabolism. Interface Focus **4**(4), 20140014 (2014)
279. Vincent, T.L., Brown, J.S.: Evolutionary Game Theory, Natural Selection, and Darwinian Dynamics. Cambridge University Press (2005)
280. Volz, E., Meyers, L.A.: Epidemic thresholds in dynamic contact networks. J. R. Soc. Interface **6**, 233–241 (2009)
281. Wan, C.: Coalitions in nonatomic network congestion games. Math. Oper. Res. **37**(4), 654–669 (2012)
282. Wang, X., Loguinov, D.: Understanding and modeling the internet topology: economics and evolution perspective. IEEE Trans. Netw. **18**(1), 257–270 (2010)
283. Wang, H., Li, Q., Agostino, G.D., Havlin, S., Stanley, H.E., Van Mieghem, P.: Effect of the interconnected network structure on the epidemic threshold. Phys. Rev. E **88**(2), 022801 (2013)
284. Wardrop, J.G.: Some theoretical aspects of road traffic research. In: Proceedings of the Institute of Civil Engineers, Pt. II, vol. 1, pp. 325–378 (1952)
285. Wardrop, J.G.: Some theoretical aspects of road traffic research. Proc. Inst. Civ. Eng. **2**, 325–378 (1952)
286. Watts, D.J., Strogatz, S.H.: Collective dynamics of /'small-world/' networks. Nature **393**(6684), 440–442, 06 1998
287. Watts, D.J., Strogatz, S.H.: Collective dynamics of 'small-world' networks. Nature **393**(6684), 440–2 (1998)
288. Weibull, J.W.: Evolutionary Game Theory. Cambridge University Press (1995)
289. Widder, A., Kuehn, C.: Heterogeneous population dynamics and scaling laws near epidemic outbreaks. arXiv preprint arXiv:1411.7323 (2014)
290. Wieland, S., Aquino, T., Nunes, A.: The structure of coevolving infection networks. EPL (Europhys. Lett.) **97**(1), 18003 (2012)
291. William, B.: Norton. The 2013 Internet Peering Playbook (2011)
292. Wright, C.C.: An explanation for some aspects of the behaviour of congested road traffic in terms of a simple model. Transp. Res. **9**(5), 267–273 (1975)

293. Yao, D.D.: S-modular games, with queueing applications. Queueing Syst. **21**(3–4), 449–475 (1995)
294. Yasutomi, K., Tokita, A.: Emergence of a complex and stable network in a model ecosystem with extinction and mutation. Theor. Popul. Biol. **63**, 131–146 (2003)
295. Yi, T., Zuwang, W.: Effect of time delay and evolutionarily stable strategy. J. Theor. Biol. **187**(1), 111–116 (1997)
296. Yook, S.H., Jeong, H., Barabási, A.L.: Modeling the Internet's large-scale topology. In: PNAS, pp. 1–15 (2002)
297. Young, H.P.: The evolution of conventions. Econometrica **61**(1), 57–84 (1993)
298. Youssef, M., Scoglio, C.: An individual-based approach to SIR epidemics in contact networks. J. Theor. Biol. **283**(1), 136–144 (2011)
299. Zanette, D.H., Risau-Gusmán, S.: Infection spreading in a population with evolving contacts. J. Biol. Phys. **34**, 135–148 (2008)
300. Zheng, Y., Feng, Z.: Evolutionary game and resources competition in the internet. In: Modern Communication Technologies, 2001. SIBCOM-2001. The IEEE-Siberian Workshop of Students and Young Researchers, pp. 51–54 (2001)
301. Zhou, D., Stanley, H.E., D'Agostino, G., Scala, A.: Assortativity decreases the robustness of interdependent networks. Phys. Rev. E **86**, 066103 (2012)

Printed in the United States
By Bookmasters